ATLAS of GREEN ENERGY TRANSITIONS

This edited collection of scholars and activists employs immersive first-person narrative descriptions and rich imagery to tell the oft-revealing stories of contestation, exploitation, and complication within the landscapes upon which the world's green energy transition depends: the unsanctioned cobalt mines of the Congo, the solar farms clearing vast tracts of the Mojave Desert, the scattered e-waste operations of Zimbabwe, among others.

Utilizing the global supply chain as an organizing structure—working backwards from consumption to extraction, and back again—each chapter is framed around an abiotic protagonist crucial for the widespread adoption of renewable energy technologies. Cast as our saviors in the face of climate change, cobalt, aluminum, and the many critical minerals needed for solar panels, wind turbines, and electric vehicles are explored through the biophysical environments and cultural contexts in which they are extracted, refined, or deployed. In challenge to these descriptive chapters are counternarratives that offer alternatives beyond silver bullet ecomodernism toward complicated futures built on just practices and reciprocity. In grounding all chapters amid host landscapes, the collection cultivates a heightened awareness of land relations and global interconnection. The project thus not only engages the green energy transition, a topic of accelerating importance and topical prevalence, but uniquely exhibits the skills of landscape architectural practice to communicate environmental challenges of global proportions, identify points of interdisciplinary intervention, and craft compelling futures to catalyze change.

The book targets an audience of students and practitioners of the built environment while seeking to inspire the inner designer in all readers.

Matthew Seibert's work aspires to encourage one to rethink their position and relation to the world as the first, fundamental step in a theory of change toward a just, more promising future.

As an Associate Professor of Landscape Architecture at the University of Virginia, his research and teaching challenge dominant modes of knowledge production with specific attention to land relations. This practice and pedagogy in support of a world where many worlds and worldviews are not only welcome, but desired, is built first by creatively interrogating conventional ways of knowing through strategic disorientation. In the design of novel tools and methods, presuppositions can be disassembled, cultural constructions confronted, and power structures dismanteld, enabling a radical rebuilding of self, community, and environment in new and powerful ways. One must look backward and inward to orient the march forward.

In engrossing first-person accounts of the far-flung locales where green capitalism collides with local ecosystems, landscapes, and communities—and interspersed with stunning data visualizations, creative cartography, and photographs of field sites—this atlas vividly depicts, critiques, and reimagines the extractive frontiers of green technologies and renewable energy. An essential resource for scholars, students, and organizers alike.

– Theo Riofrancos,
author of *Extraction: The Frontiers of Capitalism*

An informative and much needed challenge to modern design's complicity in capitalism and its conceptions of progress and development. Seibert's Atlas *offers compelling resources for rethinking our reliance on ecomodernist approaches to the green energy transition, and directs us towards a future more open to alternative ways of providing for human flourishing and prosperity.*

– Kate Soper,
Emeritus Professor of Philosophy, London Metropolitan University

Matthew Seibert's Atlas *draws a much-needed pathway out of green energy utopianism by taking readers into those landscapes and their attendant ecologies that have been conscripted into the green energy supply chain. It's an ambitious proposition backed by richly layered maps, embodied collages, charts, narratives, counternarratives, and a diverse suite of contributors. Both persistently disruptive and optimistic, the book crosses continents to illustrate the scope, scale, and urgency of what has all too easily been framed as simply a "transition." Still, the book dares to visualize a just, beautiful, and reciprocal green future with clear, tactical framings of our next steps. I dare readers to do the same.*

– Kristi Cheramie,
Professor and Head of Landscape Architecture, Ohio State University

This is important work. Atlas of Green Energy Transitions *will leave you with a deep grasp of the hidden costs behind the promises of tech solutionism. In the face of humanity's all-life-threatening problems, what was supposed to be a "green" transition ends up perpetuating 500 years of colonial extractivism. It's a powerful and beautifully illustrated book full of tools you can use to counter the narratives driving forward what was supposed to be progress.*

– Céline Keller,
artist activist

The growing "green economy" of solar panels, windmills, and lithium batteries implies new supplies of copper, nickel, cobalt, rare earths, and other materials. It also implies new forms of land grabbing and water grabbing. This empirical book goes fearlessly around the world looking at grassroots complaints against the abuses of the energy transition. The industrial economy is not circular, it is entropic. There is an enormous "circularity gap" or "metabolic rift", an "entropy hole" because less than ten percent of the materials entering the economy are recycled. Can this be amended? This book provides realistic answers.

– Joan Martinez-Alier,
ICTA Universitat Autónoma de Barcelona

Everybody is talking about an energy transition. But we rarely see where this transition is actually unfolding—or what it is leaving behind. This is precisely why Mathew Seibert's Atlas *is so profoundly necessary. With a creative and highly engaging format, the authors of this volume guide us through the landscapes of extraction and exploitation that fuel our anything-but-clean energy transition. A must-read for anyone seeking a radical alternative to the tired, recycled promises of "green" development.*

– Marco Armiero,
ICREA Research Professor at the UAB and
editor-in-chief of *Resistance: A Journal of Radical Environmental Humanities*

Power, Conflict, + Possibilities

ATLAS
of
GREEN ENERGY TRANSITIONS

Edited

by

Matthew Seibert

Designed cover image: Matthew Seibert. Sources: Esri, Maxar, TomTom, Garmin, FAO, NOAA, USGS, © OpenStreeMap contributors, and the Gis User Community.
Half title page image source: Maxar.

First published 2026
by Routledge
605 Third Avenue, New York, NY 10158

and by Routledge
4 Park Square, Milton Park, Abingdon, Oxon, OX14 4RN

Routledge is an imprint of the Taylor & Francis Group, an informa business

Library of Congress Cataloging-in-Publication Data
A catalog record for this title has been requested

ISBN: 9781032771687 (hbk)
ISBN: 9781032771670 (pbk)
ISBN: 9781003481614 (ebk)

DOI: 10.4324/9781003481614

Typeset in Helvetica
by Matthew Seibert

Publisher's Note
This book has been prepared from camera-ready copy provided by the editor.

To my son, Ozzy, and the next generation,

for they inherit the landscapes we steward and destroy.

A

DGEMENT

CKNOWLE

I am deeply grateful to the contributing authors whose diverse perspectives, grounded research, and critical imaginations shaped the contours of this book. Their work challenges dominant narratives and opens space for more just and nuanced understandings of the green energy transition. I am equally thankful to my graduate research assistants, whose intellectual and logistical support was invaluable—from organizing materials and sources to producing provocative imagery to engaging in rich, generative dialogue throughout the editorial process. Marissa Walrath, Joyce Fong, and Yujie Liu, thanks for putting up with me. Sophia Dreps and Catelyn Southwell, thanks for the last-minute editing. Julia MacNelly, I couldn't have done this without you. Thank you for pushing back when it was needed and lifting my head up when the finish line kept getting pulled further back. I'll forever be in your debt.

This book is also indebted to the many communities upon whose land, labor, and lived experience the green energy transition continues to unfold. Their resistance, resilience, and vision demand that we look beyond technical fixes and toward structural change. To engage these realities with care and accountability requires more than analysis. It requires relationship.

In the spirit of the many before me, this work affirms the necessity of coming into right relations with land—relations premised not on extraction or entitlement, but on reciprocity, humility, and responsibility. These principles guided this project and will continue to shape the work ahead.

S

Table

DEPLOYMENT

PROCESSING

DISPOSAL

of

Contents

EXTRACTION

INTRODUC

The Dark Side of Green

by Matthew Seibert

TION

Interstate 15 is a relatively stark stretch of highway connecting two of the nation's largest desert metropolises, Los Angeles and Las Vegas. "Stark" is doing this landscape a severe disservice, but at 80 mph, weaving around semis and moving aside for the periodic speed freak, the Mojave Desert is effectively blurred into a beige expanse, often only articulated by a mountainous horizon line or Walmart-sized gas station.

So, when a glowing volume of light hovers above a shimmering oasis amidst the austere backdrop, one takes notice.

Having just left Primm Valley Casino's hefty stable of Tesla Superchargers[1] at the border of California and Nevada, I rub my eyes, expecting the mirage to dissipate. It doesn't. It grows as I speed closer. I can now see this volume of light is not some alien spacecraft but is rather held aloft by a vertical structure, surrounded by an array of reflective surfaces. In fact, as I pull off the highway and roll deeper into drive-by country, there are two other towers and arrays. The tops of these towers, however, are dark. I now find myself immersed in some dystopian sci-fi thriller: a glowing tower on one side, its radiance outcompeting the sun, appearing to harness some massive source of energy, and deadened arrangements on the other, their mirrored surfaces in disarray, pointing this way and that, some even fallen to the desert floor, for as far as the eye can see.

This is the Ivanpah Solar Electric Generating System (ISEGS), a trio of three concentrated solar thermal plants using almost 200,000 heliostats, or moving mirrors, to concentrate the sun's power into a receiver. At Ivanpah these receivers are boilers nearly 500 feet in the air, converting the reflected sun's heat into steam and, in turn, driving turbines for a maximum capacity of 392 megawatts of energy. As I send up my drone for a closer look from above, there are times when my camera feed is blinded by the sun's reflection. With anxiety and a sweating brow, the video feed comes back online. I feel as if I am trying to take a closer look at the crater of a volcano, dodging spits of magma and fiery temperatures with fingers crossed. As it turns out, my fears are not unfounded. In 2016, federal biologists said "about 6,000 birds die from collisions or immolation annually while chasing flying insects around the facility's three 40-story towers."[2] Called "streamers" by the plant's staff, one can actually see little trails of smoke spiral to the ground from around the towers as insects and birds are burned alive by the solar flux that can reach 1,000 degrees Fahrenheit.[3] My drone and I are set to perform a modern day reenactment of Icarus it seems.

But the sky is not the only threat. In apparent complement to the intense radiation threatening incineration to airborne life, a much slower violence befalls those dwelling across the desert floor. The Ivanpah project covers 6 square miles of the Mojave Desert, the construction of which affected threatened populations of the desert tortoise. Though developers of the project hired around 100 biologists and spent some $22 million caring for the site's tortoises, many were believed to succumb to the stresses such an operation entails (see Chapter 1 of this collection for details).[4] As a surprising number still misbelieve, the desert is not a wasteland devoid of life, a vast expanse of carte blanche real estate just awaiting

1 Yes, I rented a Tesla Model 3. For research, I promise. This was also well before DOGE.
2 Los Angeles Times. "This Mojave Desert Solar Plant Kills 6,000 Birds a Year. Here's Why That Won't Change Any Time Soon," September 2, 2016. https://www.latimes.com/local/california/la-me-solar-bird-deaths-20160831-snap-story.html.
3 Sweet, Cassandra. "High-Tech Solar Projects Fail to Deliver." *Wall Street Journal*, June 12, 2015, sec. Business. http://www.wsj.com/articles/high-tech-solar-projects-fail-to-deliver-1434138485.
4 "Desert Tortoise Translocation." Accessed July 26, 2023. http://www.basinandrangewatch.org/TortoiseTranslocation.html. Hawaii Reporter. "Desert Tortoises Get Trumped by California's Solar Plants." *Hawaii Reporter*, November 1, 2010. https://www.hawaiireporter.com/desert-tortoises-get-trumped-by-california's-solar-plants/. CH2M Hill. "Biological Assessment for the Ivanpah Solar Electric Generating System (Ivanpah SEGS) Project." Bureau of Land Management, December 2009. chrome-extension://efaidnbmnnnibpcajpcglclefindmkaj/https://eplanning.blm.gov/public_projects/nepa/65894/79908/92801/ISEGS_Biological_Assessment_Dec09.pdf.

improvement. Deserts are rich ecosystems, abundant with diverse life. Even less known, the desert is an important sink for carbon. A 2000 study found that desert soils sequester about 800–1700 gigatons of carbon globally.[5]

Standing next to my zero-emission Tesla Model 3 rental, having just recharged its batteries a few miles away, perhaps fed from this very solar thermal plant, it is easy to think the answer to climate change is just a matter of innovation, cost, and consumer behavior—that technology can save us. But with a deeper look, and as will be comprehensively illustrated in this book, it is clear that a profound body of costs and implications lines up behind every green energy utopia. Don't get me wrong, renewable energy is without a doubt worlds better than continuing to burn fossil fuels—a global practice led by affluent nations resulting in the existential crisis of climate change threatening all of life on earth. But these clean energy technologies are in many ways repeating the same mistakes of fossil fuel capitalism: understanding land as merely a resource to be exploited, that short-term benefits trump long-term sustainability, and that quality of life is highest when luxuriously fueled by the energy dense remains of prehistoric organisms. If the reader takes anything away from this book, it should be that this does not have to be the case. The green energy transition does not have to be merely less bad than the world that led to increasing sea level rise, wildfires, floods, heat waves, ocean acidification, and biodiversity loss. The green energy transition can be just, ecological, and beautiful. It can be transformative. It is an immense opportunity. Perhaps our last.

Ivanpah was heralded by the Obama administration as a signature achievement in clean energy production. And though Ivanpah has been marketed to prevent "up to 500,000 metric tons of carbon each year"[6] and to have "created more than 2,600 jobs during construction, in addition to 61 operational jobs,"[7] not only are

---

5 "Carbon Sequestration in Our Desert Lands." Accessed July 25, 2023. https://desertreport.org/carbon-sequestration-in-our-desert-lands/.

6 Press Enterprise. "Ivanpah Solar Plant, Built to Limit Greenhouse Gases, Is Burning More Natural Gas," January 23, 2017. https://www.pressenterprise.com/articles/plant-823670-gas-ivanpah.html/.

7 "Ivanpah Solar Electric Generating System, California, USA." Accessed July 26, 2023. https://www.nsenergy-business.com/projects/ivanpah-solar-electric-generating-system/.

Ivanpah Solar Electric Generating System

the ecological costs palpable, but the facility's inefficiencies[8] have threatened its decommission, even discouraging future concentrated solar power plants as photovoltaics (PVs) or solar panels are deemed far superior.

And this is just the deployment of a green energy technology. If we trace the larger supply chain further back—to raw material origins—we reveal further costs and externalities.

Lithium, Evaporation Ponds, and the Atacama Desert

I follow a large black plastic pipe through a landscape perhaps best described as a mid-century popcorn ceiling flipped upside down, scaled up about 50-fold, and dusted with khaki-colored desert dirt. The jagged, visually violent surface of crystallized salt forms a formidable cap to the Salar de Atacama, Chile's top lithium-producing salt flat. The road I am traveling on—this time in a measly gas-powered Kia Rio—reveals the white of its salt as if a massive putty knife scraped its path smooth. At the lithium mine's gated entrance I can only see suggestions of enormous, white mounds on the horizon, the waste salts following evaporation of tremendous volumes of lithium-rich brine under the desert sun.

A lot more can be seen from 400 feet above (as my remote-controlled Icarus avatar enables). Hidden behind berms flattened within the horizon are hundreds of rectangular ponds the size of airport parking lots, brilliantly colored in marine green, cerulean blue, icy white, and every shade in between. The contrast against the taupes, tans, and beiges of the Atacama Desert are striking. Even beautiful.

In sum, the Salar de Atacama has likely produced more lithium than any other site in the world. Its annual production honors have now been usurped by Western Australia, but it is still a notable global player feeding the world's growing hunger

8 In addition to its burning of natural gas for operation, making Ivanpah a greenhouse gas emitter under California law and requiring it take part in the state's cap and trade protocols. Orange County Register. "It is Not Easy Being Green: Ivanpah Solar Plant near Nevada Burns a Lot of Natural Gas, Making It a Greenhouse Gas Emitter under State Law.," October 21, 2015. https://www.ocregister.com/2015/10/21/its-not-easy-being-green-ivanpah-solar-plant-near-nevada-burns-a-lot-of-natural-gas-making-it-a-greenhouse-gas-emitter-under-state-law/.

for lithium and all that it powers. Which is a lot. Most of our mobile electronics—our phones, laptops, tablets, and almost everything else you plug in to charge—contains lithium, the lightest metal on earth with a high electrochemical potential. Its rocketing demand is mostly fueled by projections of electric vehicles (EVs), however. The batteries needed to propel our obsessions with private transportation require a great deal of this silvery substance—especially the large trucks and SUVs Americans crave.

Back in the Mojave, under Ivanpah's Towers of Sauron,[9] my Tesla Model 3 certainly relies on lithium, with a good chance some of it came from Chile. The green energy transition is triggering an explosion in demand for lithium. Some forecasts predict that by 2035 we will need an almost 200% increase in the number of lithium mines.[10] Which is daunting since lithium mines take about ten years or more to finance, plan, and build.[11]

And this is just lithium. Green energy technologies require a bevy of what are called critical minerals.[12] Using copper as an example, it is said that in the next 20 years we will need to mine the same amount as we did in the entire history of humanity combined.[13] When one reflects on the storied results common in mining, this is sobering. Mining has been responsible for great environmental degradation and human exploitation throughout history.[14]

Of course, recent mining related to green energy technologies already exemplify these worries. The mining of lithium in the Salar de Atacama is argued by

9 Tolkien fans, that one was for you.
10 Benchmark Source. "More than 300 New Mines Required to Meet Battery Demand by 2035," September 6, 2022. https://source.benchmarkminerals.com/article/more-than-300-new-mines-required-to-meet-battery-demand-by-2035.
11 Clifford, Catherine. "The 'Land Grab' for Lithium Is Just Getting Started with GM Deal, Says EV Materials Expert." CNBC, January 31, 2023. https://www.cnbc.com/2023/01/31/land-grab-for-lithium-is-just-getting-started-with-gm-expert-says.html.
12 The term "critical minerals" is not neutral—it is deeply subjective and politically charged. What is deemed "critical" depends on who is asking, and for what purpose. They are defined by national security interests or economic growth agendas, not independent bodies or research. A fairer framing might be: critical to whom, and at what cost?
13 Neuman, Scott. "Forget about Rare Earth Minerals. We Need More Copper." World. *NPR*, March 16, 2025. https://www.npr.org/2025/03/16/nx-s1-5327095/copper-rare-earth-minerals-mining-electronics.
14 Marx, Karl. *Capital: A Critique of Political Economy*. Penguin Classics. London: Penguin books, 1990. McNeill, John Robert. *Something New under the Sun: An Environmental History of the Twentieth-Century World*. First published as a Norton paperback. New York: W. W. Norton, 2001. Moore, Jason W. *Anthropocene or Capitalocene? Nature, History, and the Crisis of Capitalism*. 1st ed. Kairos. Oakland: PM Press, 2016.

Lithium Mine in Salar de Atacama, Chile

residents and activists to be lowering neighboring lagoon and aquifer levels as brine is pumped up from below the salt flat's crust.[15] This has affected the brine shrimp who inhabit these peripheral lagoons and the endemic species of flamingos who feed on them. Farmers and pastoralists of the area also claim their wells have run dry following increased mining activities, threatening their way of life that existed well before European arrival in the 16th century. The mining of nickel in Indonesia has been rife with environmental harm and lacks safety standards: deforestation of rainforests, suffocation of coral reefs with sediment runoff, chemicals like sodium cyanide polluting waterways, and spates of miner deaths.[16] The list goes on.

The urgency with which these raw minerals are desired an attempt to wean the modern economy off fossil fuels will surely incentivize the cutting of corners, unfortunately. The fast-tracking of the Thacker Pass Lithium project, for example, slated to be the U.S.'s top producer of domestic lithium, has already bulldozed the local resistance of the People of Red Mountain, a group of Indigenous communities uniting under one banner,[17] as well as ranchers and activists.[18] It is difficult to be optimistic about what will unfold during the next 20 years if the same amount is to be mined as in all human history. If history is any indicator, it will not be pretty.

As calls grow for transitioning our global economy off fossil fuels amidst increasingly devastating climate catastrophes, the landscapes needed to support the green energy revolution—not just the critical mineral mines, but the refineries, the waste and recycling hubs, the solar farms, and more—are exploding in demand, and controversy. Though supporting cleaner technologies and sources of energy, the same power dynamics, environmental externalities, and cultural exploitation of fossil fuel capitalism are being repeated. If we are to cultivate not only a marginally inhabitable world under a 1.5 degrees Celsius temperature increase, but a wholly different one—one of justice, well-being, and ecological vigor—then we must understand how we got here and where we can turn. In response, this book seeks to 1) reveal this "dark side of green" through descriptive narratives based in rigorous fieldwork and analysis, teasing out the dynamics at play across sites of the global green energy supply chain, and 2) creatively chart a new course beyond the dominant modes of relating to the world—often based in Euro-American Western worldviews—by challenging conventional land relations and capitalist economies of living through a parallel series of illustrated counternarratives.

---

15 "Observatorio Plurinacional de Salares Andinos (Plurinational Observatory of Andean Salt Flats)," n.d. https://salares.org/.

16 BBC News. "The Rush for Nickel: 'They Are Destroying Our Future.'" July 10, 2023, sec. Asia. https://www.bbc.com/news/world-asia-66131451. Suroyo, Gayatri. "Exclusive: Indonesian President Pledges Nickel Mining Clean up amid EV-Led Boom." Reuters, March 30, 2023, sec. Sustainable Business. https://www.reuters.com/business/sustainable-business/indonesian-president-pledges-nickel-mining-clean-up-amid-ev-led-boom-2023-03-30/. Tan, Rebecca, Dera Menra Sijabat, and Joshua Irwandi. "To Meet EV Demand, Industry Turns to Technology Long Deemed Hazardous." Washington Post. Accessed July 27, 2023. https://www.washingtonpost.com/world/interactive/2023/ev-nickel-refinery-dangers/.

17 "People of Red Mountain – Protect Peehee Mu'huh." Accessed July 27, 2023. https://peopleofredmountain.com/.

18 "Protect Thacker Pass." Accessed July 27, 2023. https://www.protectthackerpass.org/.

Climate Crisis and the Green Energy Transition Paradox

As I first started writing this introduction in my home office (which is also my toddler's bedroom, his crib located within reach just behind me), it was July of 2023 and just declared the planet's hottest month on record, "by a significant margin." The hottest in 120,000 years, in fact.[19] "The era of global boiling has begun," said UN secretary general António Guterres.[20] Precedent-defying heat waves broiled regions around the world. Canadian wildfire smoke blanketed parts of North America for the better part of the summer, forcing many—including my young son—to cloister themselves indoors for fear of damage to their health. Devastating floods ravaged India, South Sudan, Italy, and others. The Gulf Stream was said to possibly collapse as early as 2025, bringing catastrophic climate impacts.[21] The list was quickly multiplying. It wasn't even August yet.

"Welcome to the new normal," as many have summed it.[22] Climate change is here. There is no denying it, even if many still try. Widespread climatic extremes will violently force our way of life to dramatically change in ways we can't even imagine. I often suffer from periodic guilt for bringing my young son into this world, a world where his environment will be a continually evolving threat to his very existence in increasingly brutal ways.[23]

Amidst all the smoke-ladened air and inundated lands, there were a few glimmers of hope, at least momentarily. In 2022 President Biden invoked the Defense Production Act and signed the Inflation Reduction Act (IRA) to accelerate the transition to renewable energy. Though certainly insufficient for the global crisis, this was a transformative win at the time,[24] and seemingly catalyzed much of the Global North into some form of action as the world's largest economy signaled material interest in confronting climate change. The IRA was scheduled to provide $370 billion in incentives, grants, and loan guarantees to clean energy, covering everything from heat pumps to hydrogen production, the majority benefiting red, conservative states.[25] The aim was to cut U.S. greenhouse gas emissions by 40% below 2005 levels by 2030. This triggered a massive shift in domestic and international markets. Within the U.S., more than 100,000 green jobs were created by 2023,[26] and "more than 185 clean energy and electric vehicle projects

19 "July 2023 Sees Multiple Global Temperature Records Broken | Copernicus." Accessed July 27, 2023. https://climate.copernicus.eu/july-2023-sees-multiple-global-temperature-records-broken#:~:text=DOWNLOAD%20IMAGE-,According%20to%20the%20ERA5%20dataset%2C%20the%20global%20mean%20surface%20air,and%20warmest%20month%20on%20record.

20 Niranjan, Ajit. Europe environment correspondent. "'Era of Global Boiling Has Arrived,' Says UN Chief as July Set to Be Hottest Month on Record." *The Guardian*, July 27, 2023, sec. Science. https://www.theguardian.com/science/2023/jul/27/scientists-july-world-hottest-month-record-climate-temperatures.

21 Ditlevsen, Peter, and Susanne Ditlevsen. Warning of a Forthcoming Collapse of the Atlantic Meridional Overturning Circulation. *Nat Commun* 14, 4254 (2023). https://doi.org/10.1038/s41467-023-39810-w.

22 Gelles, David. "Climate Disasters Daily? Welcome to the 'New Normal.'" *The New York Times*, July 10, 2023, sec. Climate. https://www.nytimes.com/2023/07/10/climate/climate-change-extreme-weather.html.

23 These are certainly privileged words, I am aware. I am a straight, white, man in America. I do not live under constant threat like that of most populations of color or other marginalized groups.

24 Aronoff, Kate, Timothy Noah, Liza Featherstone, et al. "The Bitter Triumph of the Inflation Reduction Act." *The New Republic*, August 8, 2022. https://newrepublic.com/article/167337/bitter-triumph-inflation-reduction-act.

25 Kirk, Karin. "Clean Energy Generates Major Economic Benefits, Especially in Red States" *Yale Climate Connections*, March 28, 2025. http://yaleclimateconnections.org/2025/03/clean-energy-generates-major-economic-benefits-especially-in-red-states/.

26 World Economic Forum. "How the Inflation Reduction Act Is Impacting Green Job Creation," March 14, 2023. https://www.weforum.org/agenda/2023/03/us-climate-bill-green-jobs/.

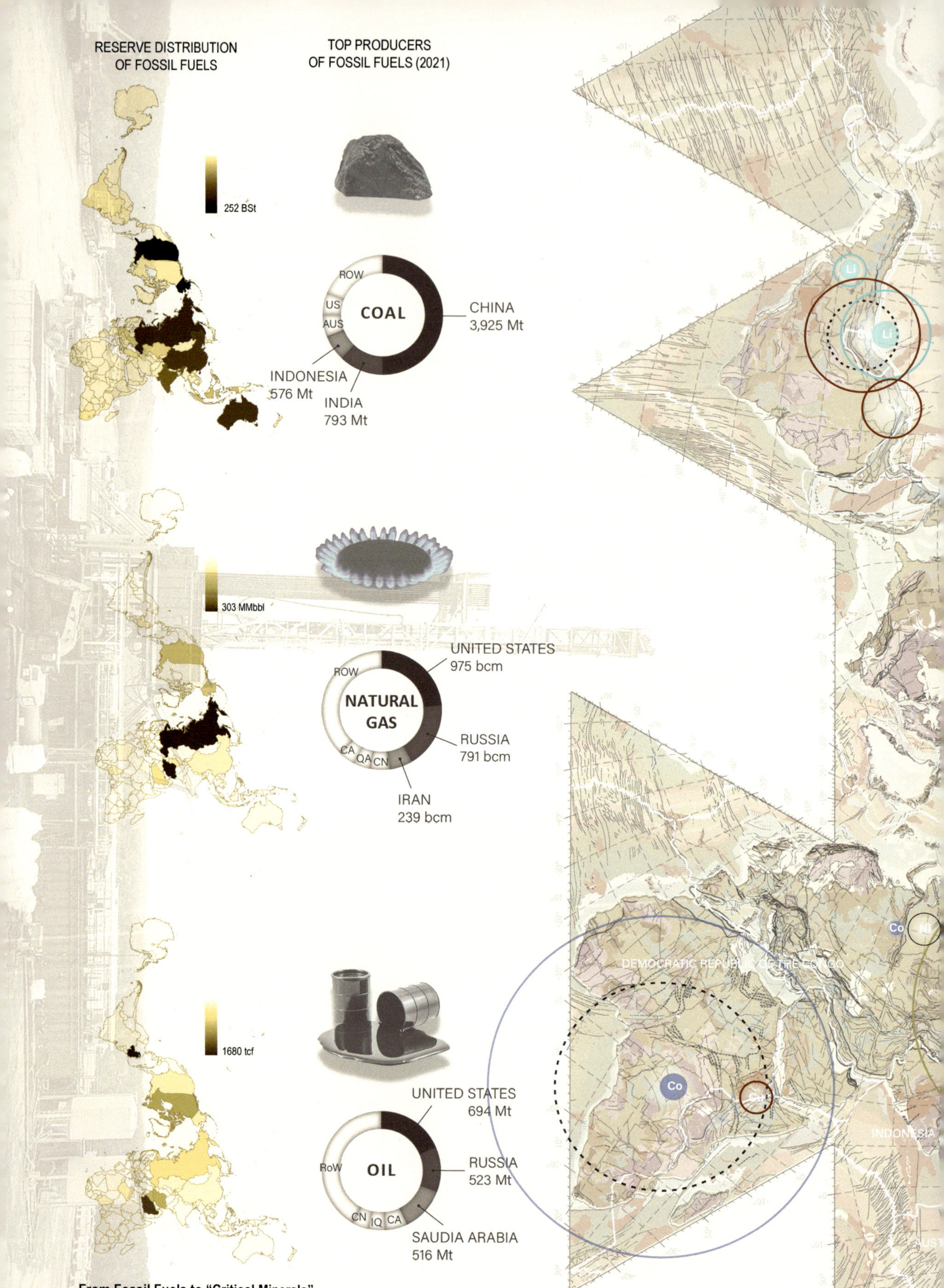

From Fossil Fuels to "Critical Minerals"

RESERVE DISTRIBUTION OF CRITICAL MINERALS

TOP PRODUCERS OF CRITICAL MINERALS (2021)

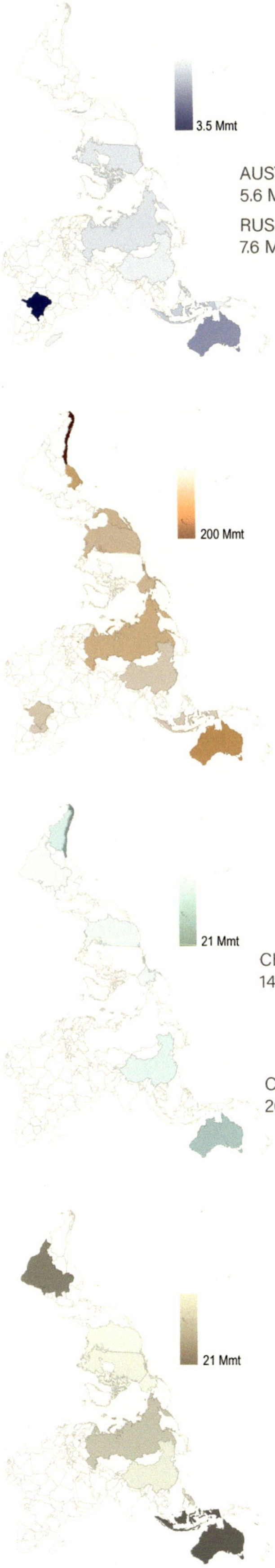

had been announced."[27] Moreover, as of May 2025, companies have announced plans to spend $843 billion on projects driven by these lucrative incentives of the IRA, with $321 billion of that already spent.[28] This includes solar farms, critical mineral mines, battery and EV manufacturing plants, charging stations, among others. Parallel developments were occurring internationally.[29]

Though, as signaled above, things are looking a lot less rosy now. Authoritarian leaders have made significant political gains across the globe recently, vowing to reverse course and expand the production and consumption of fossil fuels. In the U.S., the orange king and his followers even plan to undo much of the IRA's clean energy incentives, threatening the remaining $522 billion of the $843 billion in planned investments by private companies cited above, in addition to a general gutting of any kind of energy transition.[30]

Behind all these headlines and figures ripe for campaign trails is the fact that the U.S.'s—and much of the Global North's—answer to climate change is to either accelerate into the crisis in denial or adopt ecomodernism as savior. The former is suicidal and has been extensively proved as such.[31] I will save my breath for the latter. Ecomodernism is misguided, but at least deserving of engagement for the purpose of its uncrowning. This answer is a belief that technological innovation can beautifully temper the lethal certainty of climatic bedlam. Cities should be fed with large-scale solar farms, traversed in an army of EVs, linked with a smart grid containing batteries of mammoth proportions, and traveled between in hydrogen-fueled aircraft. But otherwise, the same: success means moving around in an SUV as the lone passenger, living in a big box house, well air-conditioned but largely devoid of life or community, food likely produced in ever-expanding monocultures. One of many things also left unsaid in such a scenario—this is only possible for a select few, generally at the costs of others, either unseen or deemed inconsequential. One might also call it a form of *green capitalism*[32] in the sense that it perpetuates capitalism's functionality and driving forces to maximize the accumulation of capital, or wealth.

The modern global economy's operation, centered around fossil fuels and unquenchable profit accumulation, appears to be already calcified into visions of the green energy transition, a transition marketed to save us from its

---

27 Natural Resources Defense Council. "Inflation Reduction Act to Drive Historic Clean Energy Transition," press release, June 29, 2023. https://www.nrdc.org/press-releases/inflation-reduction-act-drive-historic-clean-energy-transition.

28 Plumer, Brad and Harry Stevens. "A Clean Energy Boom Was Just Starting. Now, a Republican Bill Aims to End It," *New York Times*, May 21, 2025. https://www.nytimes.com/2025/05/13/climate/ira-repubican-tax-bill-clean-energy.html.

29 The International Renewable Energy Agency tracks such developments, including a wide range of publications on the topic. "Publications," May 16, 2025. https://www.irena.org/publications.

30 Plumer and Stevens, "A Clean Energy Boom Was Just Starting."

31 IPCC, 2023: Climate Change 2023: Synthesis Report. Contribution of Working Groups I, II and III to the Sixth Assessment Report of the Intergovernmental Panel on Climate Change [Core Writing Team, H. Lee and J. Romero (eds.)]. IPCC, Geneva, Switzerland, 184 pp., doi: 10.59327/IPCC/AR6-9789291691647.

32 Green capitalism is a concept that seeks to integrate environmental values in economic activities, but importantly, works within the existing capitalist system. Berghoff, Hartmut, and Adam Rome, eds. *Green Capitalism? Business and the Environment in the Twentieth Century*. 1st edition. Hagley Perspectives on Business and Culture. Philadelphia: University of Pennsylvania Press, 2017. Hawken, Paul, Amory B. Lovins, and L. Hunter Lovins. *Natural Capitalism: Creating the next Industrial Revolution*. 1st ed. Boston: Little, Brown and Co, 1999.

predecessor's very world order. As critical minerals and renewable energy landscapes are increasingly considered commodities and sites of national security across the world, the position of states and corporations has only been strengthened to push aside any obstacle in the name of maximum progress. This power dynamic, with its roots deeply established under colonialism, exists across all points of capitalist production: extraction, refining, manufacturing, consumption / deployment, and waste. As a result, sacrificial bodies have unofficially been declared and accepted in the path toward a clean economy: the indigenous ecologies and peoples entangled with the lithium mines of Chile's Salar de Atacama, the rivers of Iceland harnessed and obstructed to smelt aluminum, the forests outside Berlin cleared for a battery-producing Tesla Gigafactory, the electronic waste historically unloaded to Ghana's infamously named "Sodom and Gomorrah" landfill.

This is the dark side of green. There is a long list of costs—ecological, cultural, physiological; a substantial body of largely unseen externalities lurking behind every node in the green energy supply chain. Colonialism endures. The Global North perpetuates the exploitation of the Global South, the Indigenous, and the poor. In the U.S., the Shoshone and Paiute are dispossessed of their sacred lands for a material sacred to the green energy economy: lithium. In Greenland, communities still suffering the consequences of historical uranium mining are newly sought for their rare earth minerals. In the Democratic Republic of the Congo (DRC), the mushrooming demand for cobalt has constructed a profession of "artisanal miners," many of whom are children.[33]

As an owner of a small EV myself, I feel a small ounce of pride every trip I make to the grocery store or daycare, free of emissions, wondering why all don't make this transition. I am one of a growing number who wholeheartedly buy into the green energy utopia. I get it. But as this project demonstrates, there is much grotesquerie behind such a future suggestively veiled in green hues.

Moreover, the ecologically-prescient 19th-century English economist William Stanley Jevons would raise a red flag in warning. This demand for renewable energy, he would say, is only going to lead to an ever-growing hunger for it and all the material resources it requires. The Jevons paradox states that as efficiency in use of a resource increases, consumption of that resource will actually grow instead of decline. He observed this in steam engines and other coal-fueled applications. As steam engines improved in efficiency, paradoxically more and more coal was used to fire such engines as their deployment grew in ubiquity alongside dwindling costs. This efficiency can also accelerate economic growth as profits balloon, pushing demand higher still. The Jevons paradox is thus one such argument illustrating the fallacy of relying on technological innovation as eventual salvation. We cannot "out-tech" climate change as many might want to believe.

33 The United Nations Children's Fund (UNICEF) estimates that approximately 40,000 of the 255,000 Congolese mining for cobalt are children.

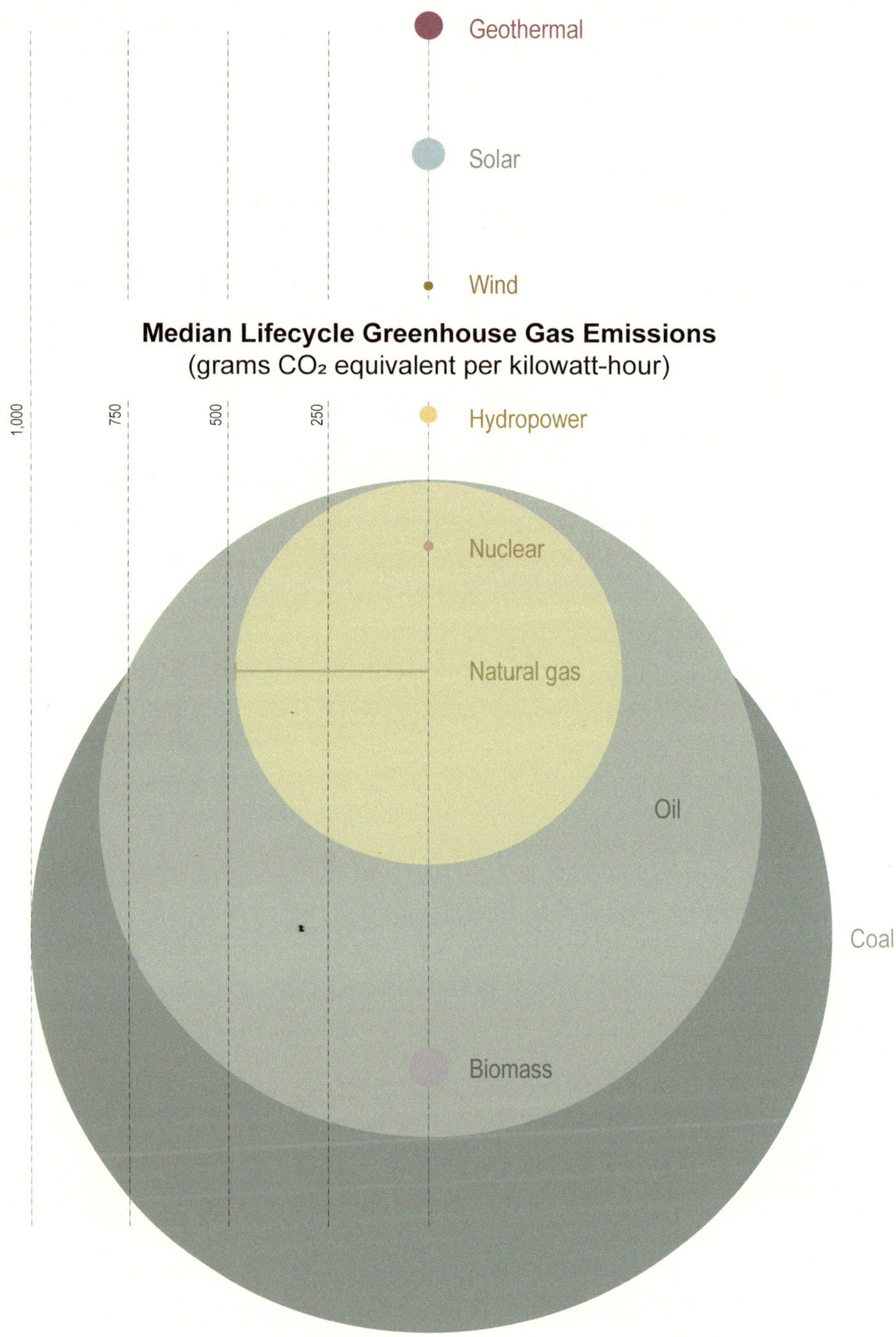

1930
With steam power and coal-fired power plants came an increase in coal usage

1859
The Drake Well, the first commercial oil well, was drilled near Titusville, Pennsylvania

BEFORE 1800
For the most part, biomass was used for heat and human / animal power for kinetic energy before the Industrial Revolution

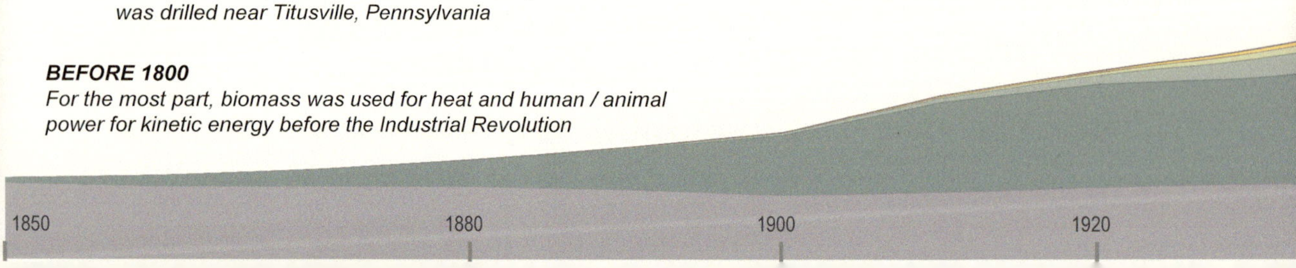

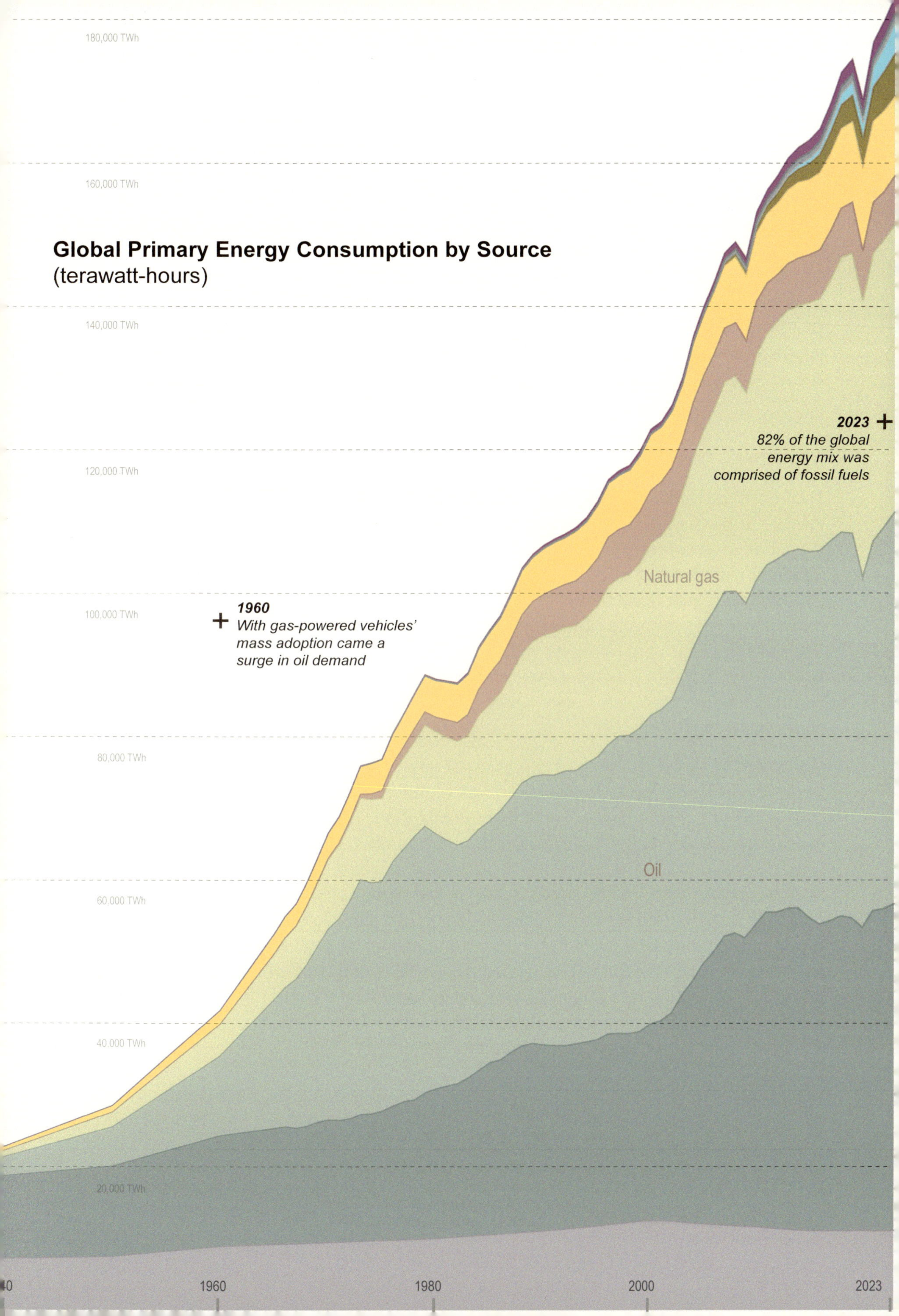
Global Primary Energy Consumption by Source
(terawatt-hours)
180,000 TWh
160,000 TWh
140,000 TWh
120,000 TWh
100,000 TWh
80,000 TWh
60,000 TWh
40,000 TWh
20,000 TWh
2023
82% of the global energy mix was comprised of fossil fuels
1960
With gas-powered vehicles' mass adoption came a surge in oil demand
Natural gas
Oil
1960
1980
2000
2023

The other side of this coin is the fact that renewable energy production, as much as it has dramatically grown in recent years,[34] has displaced next to nothing in our energy consumption diet, just added to it.[35] This is because our overall energy consumption keeps rising with population growth and economic development, particularly with the advent of artificial intelligence (AI) and its unquenchable hunger for more and more energy. It is also important to note that even renewable energy creates some greenhouse gas emissions in their technologies' material sourcing, manufacturing, and distribution. It takes energy to extract minerals out of the ground, assemble things like solar panels, and then transport them to where they are needed. And currently, much of this energy is from fossil fuels. Our ever-expanding desire for energy paired with the Jevons paradox above suggests unsustainable behaviors of consumption (and their resulting emissions) are almost baked into how we operate as a modern society.

Capitalism, Colonialism, and Chronoception

There are reasons why such behavior and patterns appear essentially inevitable. We live and operate amidst systems so common, so routinized, so a part of the fabric of everyday life they are veritably imperceptible. They are systems of relations. Relations of supply and demand, goods and services, production and consumption. Relations between lands and their cultures, between the superior and inferior, between the powerful and the wretched.[36] Relations with land itself. They are also our very relations with time and how they shade our relations with the world within which we live. These systems of relations might best be summarized as capitalism, colonialism, and (Western) chronoception. And they share a set of mutually reinforcing relations themselves.

Capitalism, as many know already, is an economic and political system where trade and industry are controlled by a select few private owners for the ultimate goal of capital accumulation, or profit maximization. The invisible hand of the market largely decides what goods and services cost, and where the profits and thus power are directed. One of capitalism's greatest strengths is its self-perpetuation. Investors feel greater faith in the higher power of the rational market over political leaders' human caprice, seeking out stability and long-term growth. Additionally, those who gain disproportionate power through this system are likely to invest in its continuation. The system feeds itself until it appears inevitable and practically invisible: "It is easier to imagine an end to the world than an end to capitalism," "easier to imagine a total catastrophe which ends all life on earth than it is to imagine a real change in capitalist relations."[37]

34 World Economic Forum. "Renewable Energy Capacity Surged Around the World in 2024." Accessed April 14, 2025. https://www.weforum.org/stories/2025/04/renewable-energy-transition-wind-solar-power-2024/#:~:text=Global%20renewable%20energy%20capacity%20grew%20by%20a%20record%2Dbreaking%2015.1,the%202030%20deadline%20is%20imminent.%E2%80%9D.

35 Energy Institute – Statistical Review of World Energy (2024); Smil (2017) – with major processing by Our World in Data. "Primary Energy from Other Renewables" [dataset]. Energy Institute, "Statistical Review of World Energy"; Smil, "Energy Transitions: Global and National Perspectives" [original data].

36 Fanon, Frantz. *The Wretched of the Earth*. New York: Grove Press, 2002.

37 Attributed to both Fredric Jameson and Slavoj Žižek.

But what about China you may ask! (The country that has secured international dominance in practically all parts of the green energy economy.[38]) It is certainly not conventional capitalism. But it sure does share a lot in common. Officially labeled a socialist market economy, for the sake of time it might be best understood as a type of bizarro[39] capitalism characterized by privately owned businesses under "forceful state controls."[40] "[Americans] don't know how capitalist China is," says Weijian Shan, economist, author, and Executive Chairman of PAG, an investment firm, born in China and educated in the U.S.[41]

Though perhaps capitalism's defining feature is that any limits are delineated by the market, regardless of the life-sustaining biophysical conditions of the environment within which it operates. It is a religion of infinite economic growth, an acute contradiction within the constraints of a finite world:

> *[The] basic contradiction of the capitalist system of control is that it cannot separate "advance" from the destruction, no "progress" from waste — however catastrophic the results. The more it unlocks the powers of productivity, the more it must unleash the powers of destruction, and the more it extends the volume of production, the more it must bury everything under mountains of suffocating waste.*[42]

The modern world, birthed by capitalism's alluring enrichment, has vastly exceeded its limited resources. We have crossed five planetary boundaries.[43] There are only nine, a couple of which have yet to even be quantified.

And so even in the context of climactic apocalypse and a drive toward clean energy like solar, the push for "the appropriation of 'cheap' nature and labor, [set] the stage for renewed forms of exploitation over peripheral geographies, communities, and species."[44] Political ecologist Sofia Avila-Calero calls this *solar capitalism*, as it aims to sustain and amplify systems of profit generation and capital accumulation within the ecological and material constraints of the present. Though largely responsible for increasing life expectancy and raising millions from poverty, if we are to continue down this capitalist path—myopically pursuing profit with every step—there will be a horrific population correction.[45] And the force of such an unimaginable tragedy will disproportionately fall atop those who

---

38 International Energy Agency. "Clean Energy Supply Chains Vulnerabilities." Accessed May 26, 2025. https://www.iea.org/reports/energy-technology-perspectives-2023/clean-energy-supply-chains-vulnerabilities.
39 Any DC comic book fans out there? DC Database. "Bizarro (New Earth)." Accessed August 17, 2023. https://dc.fandom.com/wiki/Bizarro_(New_Earth).
40 "China's Stunning Economic Turn – Michigan Journal of Economics," January 2, 2023. https://sites.lsa.umich.edu/mje/2023/01/02/chinas-stunning-economic-turn/.
41 Ignatius, Adi. "'Americans Don't Know How Capitalist China Is.'" *Harvard Business Review*, May 1, 2021. https://hbr.org/2021/05/americans-dont-know-how-capitalist-china-is.
42 Mészáros, István. *The Necessity of Social Control*. New York: Monthly Review Press, 2015.
43 "The Nine Planetary Boundaries." Text, September 17, 2009. https://www.stockholmresilience.org/research/planetary-boundaries/the-nine-planetary-boundaries.html. Persson, Linn, Bethanie M. Carney Almroth, Christopher D. Collins, Sarah Cornell, Cynthia A. De Wit, Miriam L. Diamond, Peter Fantke, et al. "Outside the Safe Operating Space of the Planetary Boundary for Novel Entities." *Environmental Science & Technology* 56, no. 3 (February 1, 2022): 1510–1521. https://doi.org/10.1021/acs.est.1c04158.
44 Avila-Calero, Sofia. "Solar Capitalism: Accumulation Strategies and Socio-Ecological Futures." *Sustainability Science*, March 21, 2025. https://doi.org/10.1007/s11625-025-01662-2.
45 Rees, William E. "The Human Ecology of Overshoot: Why a Major 'Population Correction' Is Inevitable." *World* 4, no. 3 (August 11, 2023): 509–527. https://doi.org/10.3390/world4030032.

are historically not responsible for breaking such planetary boundaries in the first place.

This is an example of colonialism. Defined as "control by one power over a dependent area or people,"[46] colonialism is more widely demonstrated as a practice of violence (often genocidal), control, and exploitation, historically in the pursuit of "Gold, God, and Glory," a practice that continues to this day. As introduced in the opening of this section, colonialism is about relations—between the colonizer and colonized. Lands are "discovered," local cultures oppressed, assimilated, or destroyed, and resources exploited by the superior, usually white power. The two historical waves of colonialism—the 15th century's Age of Discovery and colonization of the Americas and the 19th century's Scramble for Africa, both conducted by European powers—can still be read across global patterns of the powerful and the wretched:

> *It may be easy to brush colonialism off as a relic of the past, but we are all living in a world shaped by these histories of brutal and violent conquest. The wealth and prosperity of what were once the most powerful colonial nations in the world can be attributed to the theft of land, resources, and people from former colonies.*[47]

What we take for granted as a kind of endemic, locally produced poverty can be directly correlated to the exploitation and resource robbery of European conquest.[48]

Moreover, European or Western colonialism is not singular. A strong argument can easily be made for other nations exhibiting neocolonial power relations in

46 "Definition of COLONIALISM." Accessed August 16, 2023. https://www.merriam-webster.com/dictionary/colonialism.

47 Osman, Jamila. "How Colonialism Destroyed Cultures and Shaped the World." *Teen Vogue*, November 22, 2017. https://www.teenvogue.com/story/colonialism-explained.

48 Rodney, Walter. *How Europe Underdeveloped Africa*. Washington, D.C: Howard University Press, 1981.

today's world. China is a prominent example of this. China scholar Jean-Marc F. Blanchard writes:

> *the general features of China's relations with many countries today bear close resemblance to the European colonial powers' relations with African and Middle Eastern countries in the 19th and 20th century. Among other things, we witness countries exchanging their primary products for Chinese manufactured ones; China dominating the local economy; countries becoming heavily indebted to the [People's Republic of China]; China exerting greater weight on local political, cultural, and security dynamics; and Chinese abroad living in their own "expat enclaves."* [49]

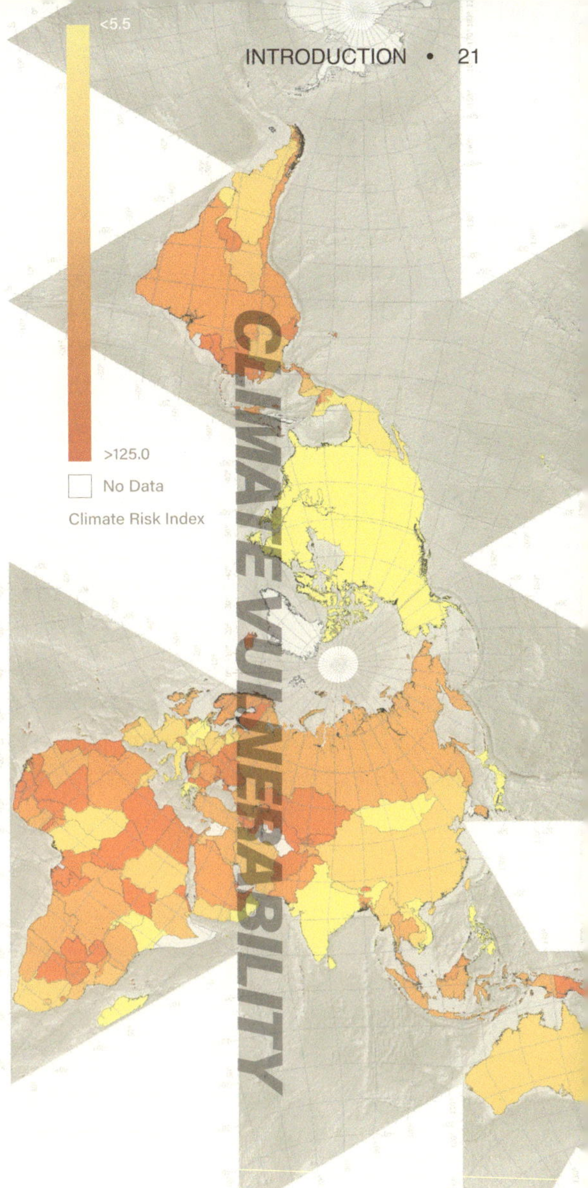

Africa has often been cast as ground zero for Chinese colonialism. Active in 39 countries within this continent, China is Africa's largest trade partner, with loans and investments in the tens of billions of dollars often severely lopsided to China's advantage.[50] This is not to say benefits don't exist for those countries who have developed strong ties with the Middle Kingdom. They do—as jobs, cash, and infrastructure. But China's dominance in the green energy supply chain is revealing of the scale at which they are tapped into African (and other non-Chinese) resources.[51] Due to smart planning and long-term strategizing formally beginning in 1990, China owns a disproportionate share of critical mineral reserves outside the U.S. Additionally, "processing of these minerals is also highly concentrated, with China being responsible for the refining of 90% of [rare earth elements] and 60–70% of lithium and cobalt," as well as the leading processor of copper and nickel.[52] Once gold,

---

49 Blanchard, Jean-Marc F., "Revisiting the Resurrected Debate About Chinese Neocolonialism." *The Diplomat*, February 8, 2018. https://thediplomat.com/2018/02/revisiting-the-resurrected-debate-about-chinese-neocolonia-lism/.

50 Etzioni, Amitai, "Is China a New Colonial Power?" *The Diplomat*, November 9, 2020. https://thediplomat.com/2020/11/is-china-a-new-colonial-power/.

51 Pitron, Guillaume. *The Rare Metals War: The Dark Side of Clean Energy and Digital Technologies*. Melbourne; London: Scribe, 2021.

52 IEA. "Clean Energy Supply Chains Vulnerabilities – Energy Technology Perspectives 2023 – Analysis." Accessed August 17, 2023. https://www.iea.org/reports/energy-technology-perspectives-2023/clean-energy-supply-chains-vulnerabilities.

then fossil fuels, the critical minerals required for green energy technologies are now increasingly the material plunder of modern day colonial conquest, perpetuating asymmetric power relationships.[53]

This exploitative practice continues, not only by foreign competitive powers, but within the very lands of nations who market themselves as humanitarian luminaries on the global stage. In the U.S., a country born amid the blood of 12 million Indigenous lives,[54] continues the dispossession of Native lands. Slated to be the biggest lithium producer in the country, the federal government has fast-tracked development of the Thacker Pass Lithium Mine in Nevada against considerable Indigenous resistance. The Shoshone and Paiute tribes claim this territory as sacred.[55] The examples of such resource grabs in the face of Indigenous groups are numerous, as will be exemplified in the following chapters.

Both capitalism and colonialism are indicative of how contemporary society has grown to relate to its host biosphere—something to be conquered, controlled, and exploited with little consideration of consequence. "Disregard market-driven planetary collapse and racial carnage; we are the chosen, as history has shown," seems to be the contention. This is due to many forces baked into the modern world, but one such force that is often overlooked, under-appreciated, and quietly running through many behavioral foundations is how the contemporary citizen relates to time.

Chronoception is the subjective experience of time. Actively researched by fields such as psychology and neuroscience, the way one experiences time directly informs one's approach to the world. For instance, studies have shown that time passes faster as you age due to a slowdown in image processing.[56] An individual in their later years might appreciate present moments more, long for yesteryears, or develop an increased interest in history. Chronoception is also interested in the existence of different modes of understanding time. The ancient Greeks, for instance, had two words for time: chronos and kairos, or what is often referred to as clock time and mind time, respectively. Chronos refers to linear, measurable, objective time. It is our standard understanding of time as something that perpetually marches forward irrespective of other subjects or forces. Kairos, on the other hand, is qualitative, referring to events or defined within the context of a narrative. It is time with meaning. If the mid-20th century is just 1,945 years into the Common Era, as it would be registered in chronos time, to kairos time 1945 is the year radionuclides were released into the atmosphere from atomic bomb detonations, for some marking the start date of the geological epoch known as

53 Llavero-Pasquina, Marcel, Grettel Navas, Roberto Cantoni, and Joan Martínez-Alier. "The Political Ecology of Oil and Gas Corporations: Total Energies and Post-Colonial Exploitation to Concentrate Energy in Industrial Economies." *Energy Research & Social Science* 109 (March 1, 2024): 103434. https://doi.org/10.1016/j.erss.2024.103434.

54 Smith, David Michael. "Counting the Dead: Estimating the Loss of Life in the Indigenous Holocaust, 1492–Present," Native American Symposium Proceedings, November 3, 2017. chrome-extension://efaidnbmnnnibpcajpcglclefindmkaj/https://www.se.edu/native-american/wp-content/uploads/sites/49/2019/09/A-NAS-2017-Proceedings-Smith.pdf

55 "People of Red Mountain – Protect Peehee Mu'huh." Accessed August 17, 2023. https://peopleofredmountain.com/.

56 Bejan, Adrian. "Why the Days Seem Shorter as We Get Older." *European Review* 27, no. 2 (2019): 187–194. doi:10.1017/S1062798718000741.

the Anthropocene, or the Age of Man named for humankind's outsized impacts on Earth's natural systems.[57] There are many ways to understand and employ the differences between chronos and kairos but what is important to take away is that there is a perception and interpretation of time beyond the measurements of a clock or calendar at the scale of a human life.

Geologist Marcia Bjornerud, author of *Timefulness: How Thinking Like a Geologist Can Save the World*, writes:

> *Most humans... have no sense of temporal proportion—the durations of the great chapters in Earth's history, the rates of change during previous intervals of environmental instability, the intrinsic timescales of "natural capital" like groundwater systems. As a species, we have a childlike disinterest and partial disbelief in the time before our appearance on Earth. With no appetite for stories lacking human protagonists, many people simply can't be bothered with natural history. We are thus both intemperate and intemporate—time illiterate. Like inexperienced but overconfident drivers, we accelerate into landscapes and ecosystems with no sense of their long-established traffic patterns, and then react with surprise and indignation when we face the penalties for ignoring natural laws.*[58]

Modern politics and capitalism give further examples of such disconnects and relational interconnections. In our economic system, productivity and efficiency often reign supreme, with professions that can't be computer-assisted or automated often undervalued—critical industries like education, healthcare, and childcare. Fiscal years and political terms reward short-term wins, discouraging long-term, generational work from gaining traction in a world of budget cycles and profit maximization. Our societal systems have exacerbated a modern trend to relate to only that which can be perceived or intimately fathomed. Divided from the necessity of food production, we no longer relate to seasonal change beyond fashion trends. Cleaved from natural systems, we rarely witness environmental repercussions. Our technology contracts space and time. We increasingly live isolated lives focused around smaller and smaller time scales. Even entertainment has shrunk to seconds-long Instagram reels and TikTok videos.

This wasn't always the case. Nor does it have to remain that way. Other cultures, both former and current, can point to new and old ways of relating to the world

57 This date was proposed by the twenty-nine members of the Anthropocene Working Group in May, 2019 and is currently favored over others. However, dates have been proposed as early as the beginning of the Agricultural Revolution, 12 thousand–15 thousand years ago. Additionally, a convincing group of professors of earth system science, global change science, anthropology, geography, and Indigenous theory argue the start of the Anthropocene should be dated to the Orbis Spike, a dip in carbon dioxide levels associated with the arrival of Europeans in the Americas, likely as Native American farmlands were abandoned following a sharp population decline. Some put forward the name "the Kleptocene" as more appropriate for this epoch to call "attention to colonialism's ongoing theft of land, lives (both human and nonhuman), and materials," naming "the problem of colonialism as responsible for [the] contemporary environmental crisis." Keeler, Kyle. "Colonial Theft and Indigenous Resistance in the Kleptocene." *Edge Effects*, September 8, 2020. https://edgeeffects.net/kleptocene/. Davis, Heather, and Zoe Todd. "On the Importance of a Date, Or, Decolonizing the Anthropocene". *ACME: An International Journal for Critical Geographies* 16, no. 4 (2017): 761–80. https://acme-journal.org/index.php/acme/article/view/1539. Subramanian, Meera. "Anthropocene Now: Influential Panel Votes to Recognize Earth's New Epoch." *Nature*, May 21, 2019. https://doi.org/10.1038/d41586-019-01641-5.

58 Bjornerud, Marcia. *Timefulness: How Thinking like a Geologist Can Help Save the World*. Princeton: Princeton University Press, 2018.

through alternative temporal literacies.[59] Additionally, in a more scientific vein, "[j]ust as the microscope and telescope extended our vision into spatial realms once too minuscule or too immense for us to see," Bjornerud contends "geology provides a lens through which we can witness time in a way that transcends the limits of our human experiences."[60] In other words, there are ways of viewing the world—disciplinarily, culturally, methodologically—that can begin to attune one to time scales beyond humans; at the very least, beyond capitalism's focus on short-term profit. These larger time scales are ecological, climatic, geological. With greater attunement might follow greater reciprocity between humans and their biophysical host environment.

Define and Design

The mission of this book is thus two-fold: 1) reveal the many paradoxes of the green energy transition through first-person narratives based in rigorous fieldwork and analytical descriptions, teasing out the oft-exploitative and harmful dynamics at play across sites around the world, and 2) chart a new course beyond the dominant modes of relating to the world—based in Euro-American Western worldviews—by challenging conventional land relations and perceptions of time through counternarrative explorations. The first will be achieved by calling on scholars, artists, and activists to tell stories about contemporary landscapes they have walked within and researched around. The second, projective objective is achieved by extending that timeline forward into alternative futures organized around ecology, justice, and reciprocity. As a whole, the risks of green energy are revealed and roadmaps for new more desirable futures are propagated.

Collectively, these two types of chapters—nonfictional descriptions and speculative vignettes—are conceptualized in parallel but navigated as a loop, as the table of contents illustrates. The nodes of capitalist production are thus traced, working backwards from consumption (and deployment) to raw material extraction and then back again. This not only illustrates the tensions, synergies, and dependencies across the global supply chain but facilitates a conversation of expanding time scales as the reader moves from present day to geological epochs. A chronoceptive recalibration is to occur. Additionally, imagery is employed throughout as more than mere supplement, but rather is instrumental in navigating the reader through the many entanglements of the dark side of green. The descriptive chapters utilize a collage of site photography and cartographic visualization to bring the reader deeper into a site's environment and community of actors. Unusual map projections are employed to challenge habitual conventions of how we look at the world.[61] By disrupting familiar orientations, these projections invite readers to reconsider power relations and

59 Deloria, Vine. *God Is Red: A Native View of Religion*. 3rd ed. Golden, Colo: Fulcrum Pub, 2003. Kimmerer, Robin Wall. *Braiding Sweetgrass: Indigenous Wisdom, Scientific Knowledge and the Teachings of Plants*. New York: Milkweed Editions, 2013. Rifkin, Mark. *Beyond Settler Time: Temporal Sovereignty and Indigenous Self-Determination*. Durham, NC: Duke University Press, 2017.

60 Bjornerud, "Timefulness."

61 A map projection is a method used to translate the 3D surface of the Earth onto a 2D map. The maps in this atlas use unconventional projections, mostly Fuller and Peirce Quincuncial.

hierarchies embedded in traditional cartography, particularly those reinforcing divisions between the Global North and Global South. Though contrasting in their aesthetics, the projective counternarratives cultivate an intentionally ambiguous and impressionistic attitude in their depictions of a different future. The reader thus has room to incorporate further details from their own imagination and positional specifics within the global community.

The book is intended for a general audience, as the climate crisis and current energy transition do not obey demographic silos or national borders. But, as the editor and designer of the project with a professional background in design, the book will be of particular interest to those in the design fields, including students of the built environment generally. And, with any luck, it will trigger design sensibilities in those outside the field. Though many of the authors are not designers in the traditional sense, their selection, positioning within the text, and contextual graphics contribute to a spatial intelligence and temporal literacy that the book as a whole seeks to cultivate. This intelligence is one of spatial relationships, between urban core and hinterland, between site and surroundings, between ostensibly disconnected production nodes across the globe. It is also one of temporal relationships, between generations, between geology, biology, and toxicology, between waste and material decomposition. This intelligence involves first understanding such relationships to subsequently leverage them with specific intention. Design theoretician Keller Easterling might call this intention *medium design*, or “the art of working on the world,” where design is less about objects or space but about intervening in systems.[62] As such it is “protocols of interplay—not things, but parameters for how things interact with each other.”[63] The descriptive chapters identify such systems and parameters. The speculative counternarrative then illustrates an example of rewriting their protocols of interplay. In effect, the green energy transition’s current perpetuation of the fossil fuel capitalist model—widespread, unjust exploitation of people and place for maximum profit by a few—is critically challenged.

It should be stated clearly that this book does not exist in isolation. It stands on and aside the shoulders of many. Perhaps most notably, in solidarity with and indebted to the work of Spanish ecological economist Joan Martinez-Alier and his collaborative project, the *Global Atlas of Environmental Justice*.[64] As a vast and dynamic online resource documenting 4,320 cases of environmental conflict (as of May 2025), it is collaboratively run by a network of researchers and activists. Built through the contributions of hundreds of people worldwide, it brings together firsthand accounts of resistance and observations from those bearing witness to environmental injustice. Moreover, the entries on conflicts around green energy technologies and the raw resources they require[65] grow in alarming numbers. The collection you hold in your hands is a highly artful and intentionally curated selection amongst a sadly expanding constellation of conflict.

62 Easterling, Keller. *Medium Design: Knowing How to Work on the World*. London: Verso, 2021.
63 Easterling, *Medium Design*.
64 “Ej Atlas.” Accessed May 25, 2025. https://ejatlas.org/.
65 Walter, Mariana, et al. “Mapping the Impacts and Conflicts of Rare-Earth Elements.” Debt Observatory in Globalization, 2023. Accessed May 25, 2025. chrome-extension://efaidnbmnnnibpcajpcglclefindmkaj/https://odg.cat/wp-content/uploads/2023/11/Mapping-Impacts-Conflicts-Rare-Earth-Elements.pdf.

Tracing the Green Energy Supply Chain

The first two chapters of the collection begin where the green energy transition often culminates: fields of renewable energy production. With sun and wind leading the race in production-to-cost efficiency, photovoltaic and wind power stations (aka solar and wind farms) are multiplying in number and scale. In Chapter 1, Co-Director of the Mojave Desert Land Trust, Kelly Herbinson, walks us through the desert ecosystem's extraordinary biodiversity and its equally extraordinary threat: solar farms. With close to a decade working as a field biologist managing large-scale translocations of threatened Mojave Desert tortoises to make way for solar panels, the author walks us through the complexities of renewable energy development in the California desert and outlines the difficult and necessary shifts we as a species need to make to successfully face climate change. Our guides in Chapter 2 then take us south to the Biobío region of Chile where wind farms continue historic injustices between Global North and South. Under a forest of giant wind turbines, the Latin American Observatory of Environmental Conflicts' (OLCA) Javier Arroyo Olea and María Paz López Ponce ask, "Is this the transition we need?"

Chapter 3 has us trace back a key material of solar panels and wind turbines to the fjords of Iceland where aluminum is refined. Theodore Teichman explores the complex relationship between aluminum smelting and hydroelectric energy in this island nation. In recent decades, Iceland's aluminum industry has been hailed as a symbol of progress and economic prosperity. The country's abundance of renewable energy sources, particularly hydropower, has attracted multinational corporations seeking to capitalize on the green energy revolution. However, beneath the shiny surface, the resources powering the "green" aluminum economy in Iceland are directly the result of climate change itself. The aluminum industry in Iceland is a story not just of factories and ports, but of dams and power lines and glaciers—it is a story of how the search for energy solutions can have dramatic consequences to the landscape at the territorial scale.

Production and deployment obscure the latter lifecycles of such technology. In Chapter 4, circular economy researchers Assia Boukhatmi and Roger Nyffenegger explore the current linear value chain in the photovoltaics industry, characterized by extractive production, premature disposal, and limited end-of-life (EoL) options. They debunk common myths about the industry's perceived sustainability, its dependence on global supply chains, and the underappreciated waste challenges ahead. By interrogating these myths and offering a systems-level view of the solar industry's lifecycle, this chapter invites a more nuanced discussion about what it will truly take to make solar energy the cornerstone of a sustainable future.

Expanding the conversation to e-waste, or electronic waste, Zimbabwean Professor Vusumuzi Maphosa tackles the technological refuse of our age that resists decomposition in Chapter 5. As devices become quickly obsolete, critical mineral demand surges, and e-waste flows from wealthier to vulnerable regions, perpetuating neo-colonial patterns masked as green solutions. In emerging economies, informal e-waste workers prioritize survival over safety, leaving behind a toxic legacy. Formalizing operations, closing policy gaps, and embracing restorative systems are essential steps toward aligning technological progress with environmental justice.

With Chapter 6, we begin a series of journeys to the material sources of the green energy transition. Douglas Robb locates us within Canada's Peace River Valley, a growing appendage for the production of green hydrogen. Indigenous peoples have lived along the Peace since time immemorial; for them, the river is a source of significant cultural and spiritual meaning.[66] Following a history of agricultural and oil and natural gas extraction—Canada now seeking to decarbonize its economy—a new commodity is poised to once again rewrite the landscapes of the Peace River: hydrogen.[67] By situating hydrogen within relational geographies shaped by centuries of resource extraction and settler colonial violence, Robb argues that hydrogen fuel may not necessarily, or simplistically, contribute to a so-called "just" transition.

Billy Fleming keeps us in the northern latitudes for Chapter 8. Home to the planet's second-densest collection of rare earth elements and critical minerals, South Greenland is a central node within a global supply matrix of mineral deposits, waste sites, processing and shipping facilities, manufacturing plants, and industrial-scale technological deployments (for example, wind and solar farms) at the core of the energy transition. Arctic ecosystems, cultural practices, and financialized ore come together in a mine near Narsaq known as Kvanefjeld. This chapter explores the web of relationships that form South Greenland through a set of fieldwork and related research conducted in the fall of 2022 through a design studio at Yale University.

From the Arctic to the tropics, Fitsum Fareguy charts an investigation into the emerging lithium mines of Brazil's Jequitinhonha Valley, revealing meaningful ties to the country's historical and highly problematic gold mining. Drawing on letters of protest, ancestral cosmologies, and community rituals like the Congado, the chapter traces a legacy of resistance rooted in quilombo traditions and sustained through contemporary grassroots alliances. The struggle for ancestral territories is not only about resource rights, but also the possibility of what Tiffany Lethabo King, Jenell Navarro, and Andrea Smith have called "otherwise worlds": spaces where Black and Indigenous life can persist beyond the timelines of techno-colonialism.

Lastly, art historian Ash Duhrkoop traces cobalt in Chapter 9, from its origins to its extraction and use in green technologies, while privileging the indigenous and local knowledge of the element. Today, *creuseurs*—or artisanal miners, many of whom are children—source a quarter of the cobalt extracted from the Democratic

---

66 Behn, Caleb, and Karen Bakker. "Rendering Technical, Rendering Sacred: The Politics of Hydroelectric Development on British Columbia's Saaghii Naachii/Peace River." Global Environmental Politics 19, no. 3 (2019): 98–119.

67 Natural Resources Canada. "Hydrogen Strategy for Canada: Seizing the Opportunities for Hydrogen." Ottawa: Natural Resources Canada, 2020.

Republic of the Congo (DRC), from where 70% of the world's supply is sourced.[68] Cast against the history of colonial-era mining, this chapter reveals the cycles of exploitation that are being resuscitated in the search for raw materials to fuel the green-energy revolution and ways local artists are confronting and challenging such realities.

We end our nonfictional travels across the DRC's Eastern border in Uganda, at a once prosperous copper-mining town, now struggling and longing for its boom years. Through residents and local specialists, the promise and pains of a mining economy are laid bare. Such truths inform what and how design can play a role in forming a desirable future in a context of global power imbalances. A practice of being-in-the-world, whether professional designer or quotidian citizen, is then positioned toward shaping a fundamentally different relationship to our shared world: one rooted in the dynamics of deep time, reciprocity, and the generative possibilities to which these give material form. Legal frameworks such as Ecuador's constitution that begins to acknowledge the rights of nature and economic models like best-selling author Kohei Saito's degrowth communism are highlighted as promising trajectories. But ultimately, this concluding chapter acts as a bridge to the collection's counternarratives. Landscape can be much more than mere setting or resource to lay claim. It can be a commons, a pedagogy, a tool, a time machine. It can be a spatial and material expression of histories held and futures rehearsed. Landscape can be the foundation through which a radical rethinking of identity, power, and existence challenges the existing order. As Cameroonian historian and political theorist Achille Mbembe contends in much of his postcolonial scholarship, the work of imagination demands that we invent new forms of life and new modes of political cohabitation.[69]

Let us now venture into the green energy supply chain with an eye toward change.

68 United States Geological Survey and the Cobalt Institute.

69 Mbembe, Achille. *Necropolitics.* Duke University Press, 2020. https://doi.org/10.1515/9781478007227. Mbembe, Achille. *On the Postcolony.* 2. repr. Studies on the History of Society and Culture 41. Berkeley: University of California Press, 2004. Mbembe, Achille. *The Earthly Community: Reflections on the Last Utopia.* Translated by Steven Corcoran. Rotterdam: V2_Publishing, 2022.

BIBLIOGRAPHY

Avila-Calero, Sofia. "Solar Capitalism: Accumulation Strategies and Socio-Ecological Futures." *Sustainability Science*, March 21, 2025. https://doi.org/10.1007/s11625-025-01662-2.

BBC News. "The Rush for Nickel: 'They Are Destroying Our Future.'" July 10, 2023. https://www.bbc.com/news/world-asia-66131451.

Behn, Caleb, and Karen Bakker. "Rendering Technical, Rendering Sacred: The Politics of Hydroelectric Development on British Columbia's Saaghii Naachii/Peace River." *Global Environmental Politics* 19, no. 3 (August 2019): 98–119. https://doi.org/10.1162/glep_a_00518.

Bejan, Adrian. "Why the Days Seem Shorter as We Get Older." *European Review* 27, no. 02 (May 2019): 187–194. https://doi.org/10.1017/S1062798718000741.

Benchmark Source. "More than 300 New Mines Required to Meet Battery Demand by 2035," September 6, 2022. https://source.benchmarkminerals.com/article/more-than-300-new-mines-required-to-meet-battery-demand-by-2035.

Berghoff, Hartmut, and Adam Rome, eds. *Green Capitalism? Business and the Environment in the Twentieth Century*. 1st edition. Hagley Perspectives on Business and Culture. Philadelphia: University of Pennsylvania Press, 2017.

Bjornerud, Marcia, and Haley Hagerman. *Timefulness: How Thinking like a Geologist Can Help Save the World.* First paperback edition, with Discussion questions. Princeton, NJ Oxford: Princeton University Press, 2020.

Blanchard, Jean-Marc F. "Revisiting the Resurrected Debate About Chinese Neocolonialism." *The Diplomat*, February 8, 2018. https://thediplomat.com/2018/02/revisiting-the-resurrected-debate-about-chinese-neocolonialism/.

"Carbon Sequestration in Our Desert Lands." Accessed May 27, 2025. https://desertreport.org/carbon-sequestration-in-our-desert-lands/.

CH2M Hill. "Biological Assessment for the Solar Electric Generating System (Ivanpah SEGS) Project." Bureau of Land Management, December 2009. chrome-extension://efaidnbmnnnibpcajpcglclefindmkaj/https://eplanning.blm.gov/public_projects/nepa/65894/79908/92801/ISEGS_Biological_Assessment_Dec09.pdf.

Clifford, Catherine. "The 'Land Grab' for Lithium Is Just Getting Started with GM Deal, Says EV Materials Expert." CNBC, January 31, 2023. https://www.cnbc.com/2023/01/31/land-grab-for-lithium-is-just-getting-started-with-gm-expert-says.html.

Copernicus. "July 2023 Sees Multiple Global Temperature Records Broken." Accessed May 27, 2025. https://climate.copernicus.eu/july-2023-sees-multiple-global-temperature-records-broken#:~:text=DOWNLOAD%20IMAGE-,According%20to%20the%20ERA5%20dataset%2C%20the%20global%20mean%20surface%20air,and%20warmest%20month%20on%20record.

Davis, Heather, and Zoe Todd. "On the Importance of a Date, Or, Decolonizing the Anthropocene". *ACME: An International Journal for Critical Geographies* 16, no. 4 (2017): 761–80. https://acme-journal.org/index.php/acme/article/view/1539.

DC Database. "Bizarro (New Earth)." Accessed August 17, 2023. https://dc.fandom.com/wiki/Bizarro_(New_Earth).

"Definition of COLONIALISM." Accessed May 27, 2025. https://www.merriam-webster.com/dictionary/colonialism.

Deloria, Vine. *God Is Red: A Native View of Religion*. 3rd ed. Golden, Colo: Fulcrum Pub, 2003.

"Desert Tortoise Translocation." Accessed May 27, 2025. http://www.basinandrangewatch.org/TortoiseTranslocation.html.

Ditlevsen, Peter, and Susanne Ditlevsen. "Warning of a Forthcoming Collapse of the Atlantic Meridional Overturning Circulation." *Nature Communications* 14, no. 1 (July 25, 2023): 4254. https://doi.org/10.1038/s41467-023-39810-w.

Easterling, Keller. *Medium Design: Knowing How to Work on the World*. London: Verso, 2021.

"Ej Atlas." Accessed May 27, 2025. https://ejatlas.org/.

Energy Institute – Statistical Review of World Energy (2024); Smil (2017) – with major processing by Our World in Data. "Primary Energy from Other Renewables" [dataset]. Energy Institute, "Statistical Review of World Energy"; Smil, "Energy Transitions: Global and National Perspectives" [original data].

Etzioni, Amitai, "Is China a New Colonial Power?" *The Diplomat*, November 9, 2020. https://thediplomat.com/2020/11/is-china-a-new-colonial-power/.

Fanon, Frantz, and Constance Farrington. *The Wretched of the Earth*. New York: Grove Press, 2002.

Feffer, John. "Mapping the Impacts and Conflicts of Rare-Earth Elements – FPIF." Foreign Policy In Focus, November 28, 2023. https://fpif.org/mapping-the-impacts-and-conflicts-of-rare-earth-elements/.

Gelles, David. "Climate Disasters Daily? Welcome to the 'New Normal.'" *The New York Times*, July 10, 2023, sec. Climate. https://www.nytimes.com/2023/07/10/climate/climate-change-extreme-weather.html.

Harvard Business Review. "Americans Don't Know How Capitalist China Is." May 1, 2021. https://hbr.org/2021/05/americans-dont-know-how-capitalist-china-is.

Hawaii Reporter. "Desert Tortoises Get Trumped by California's Solar Plants." Hawaii Reporter, November 1, 2010. https://www.hawaiireporter.com/desert-tortoises-get-trumped-by-california's-solar-plants/.

Hawken, Paul, Amory Lovins, and L. Hunter Lovins. *Natural Capitalism: Creating the next Industrial Revolution*. 1. paperback ed., [Nachdr.]. A Back Bay Book. New York: Little, Brown and Co, 2000.

International Energy Agency. "Clean Supply Chains Vulnerabilities – Energy Technology Perspectives 2023 – Analysis." Accessed August 17, 2023. https://www.iea.org/reports/energy-technology-perspectives-2023/clean-energy-supply-chains-vulnerabilities.

International Energy Agency. "Clean Energy Supply Chains Vulnerabilities." Accessed May 26, 2025. https://www.iea.org/reports/energy-technology-perspectives-2023/clean-energy-supply-chains-vulnerabilities.

IPCC, 2023: Climate Change 2023: Synthesis Report. Contribution of Working Groups I, II and III to the Sixth Assessment Report of the Intergovernmental Panel on Climate Change [Core Writing Team, H. Lee and J. Romero (eds.)]. IPCC, Geneva, Switzerland, 184 pp., doi: 10.59327/IPCC/AR6-9789291691647.

Keeler, Kyle. "Colonial Theft and Indigenous Resistance in the Kleptocene." *Edge Effects*, September 8, 2020. https://edgeeffects.net/kleptocene/.

Kimmerer, Robin Wall. *Braiding Sweetgrass: Indigenous Wisdom, Scientific Knowledge and the Teachings of Plants*. New York: Milkweed Editions, 2013.

Kirk, Karin. "Clean Energy Generates Major Economic Benefits, Especially in Red States." *Yale Climate Connections*, March 28, 2025. http://yaleclimateconnections.org/2025/03/clean-energy-generates-major-economic-benefits-especially-in-red-states/.

Leshne, Max. "China's Stunning Economic Turn." Michigan Journal of Economics (January 2, 2023).

Llavero-Pasquina, Marcel, Grettel Navas, Roberto Cantoni, and Joan Martínez-Alier. "The Political Ecology of Oil and Gas Corporations: Total Energies and Post-Colonial Exploitation to Concentrate Energy in Industrial Economies." *Energy Research & Social Science* 109 (March 2024): 103434. https://doi.org/10.1016/j.erss.2024.103434.

Los Angeles Times. "This Mojave Desert Solar Plant Kills 6,000 Birds a Year. Here's Why That Won't Change Any Time Soon," September 2, 2016. https://www.latimes.com/local/california/la-me-solar-bird-deaths-20160831-snap-story.html.

Marx, Karl. *Capital: A Critique of Political Economy*. Penguin Classics. London: Penguin books, 1990.

Mbembe, Achille. *Necropolitics*. Duke University Press, 2020. https://doi.org/10.1515/9781478007227.

Mbembe, Achille. *On the Postcolony*. 2. repr. Studies on the History of Society and Culture 41. Berkeley: University of California Press, 2004.

Mbembe, Achille. *The Earthly Community: Reflections on the Last Utopia*. Translated by Steven Corcoran. Rotterdam: V2_Publishing, 2022.

McNeill, John Robert. *Something New under the Sun: An Environmental History of the Twentieth-Century World*. First published as a Norton paperback. New York: W. W. Norton, 2001.

Mészáros, István. *The Necessity of Social Control*. New York: Monthly Review Press, 2015.

Moore, Jason W., ed. *Anthropocene or Capitalocene? Nature, History, and the Crisis of Capitalism*. Kairos. Oakland, CA: PM Press, 2016.

Natural Resources Defense Council. "Inflation Reduction Act to Drive Historic Clean Energy Transition," media release, June 29, 2023. https://www.nrdc.org/press-releases/inflation-reduction-act-drive-historic-clean-energy-transition.

Neuman, Scott. "Forget about Rare Earth Minerals. We Need More Copper." World. *NPR*, March 16, 2025. https://www.npr.org/2025/03/16/nx-s1-5327095/copper-rare-earth-minerals-mining-electronics

"The Nine Planetary Boundaries." September 17, 2009. https://www.stockholmresilience.org/research/planetary-boundaries/the-nine-planetary-boundaries.html.

Niranjan, Ajit. Europe environment correspondent. "'Era of Global Boiling Has Arrived,' Says UN Chief as July Set to Be Hottest Month on Record." *The Guardian*, July 27, 2023, sec. Science. https://www.theguardian.com/science/2023/jul/27/scientists-july-world-hottest-month-record-climate-temperatures.

"Observatorio Plurinacional de Salares Andinos (Plurinational Observatory of Andean Salt Flats)," n.d. https://salares.org/.

Orange County Register. "It's Not Easy Being Green: Ivanpah Solar Plant near Nevada Burns a Lot of Natural Gas, Making It a Greenhouse Gas Emitter under State Law." *Orange County Register* (blog), October 21, 2015. https://www.ocregister.com/2015/10/21/its-not-easy-being-green-ivanpah-solar-plant-near-nevada-burns-a-lot-of-natural-gas-making-it-a-greenhouse-gas-emitter-under-state-law/.

Osman, Jamila. "How Colonialism Destroyed Cultures and Shaped the World." *Teen Vogue*, November 22, 2017. https://www.teenvogue.com/story/colonialism-explained.

Pass, Protect Thacker. "Protect Thacker Pass." Accessed May 27, 2025. https://www.protectthackerpass.org/.

"People of Red Mountain – Protect Peehee Mu'huh." Accessed May 27, 2025. https://peopleofredmountain.com/.

NEVADA
CALIFORNIA
Las Vegas
Primm
Ivanpah
Solar Farms (MW)
1
10
50
100
Mojave Basin and Range Ecoregion (III)
USA Census Urban Areas
Interstate
States Boundaries
Twentynine Palms

I wa

ded

through the indigo air of a pre-dawn Mojave Desert morning, navigating awkwardly up a cobbled hillside. In one hand I held a radio receiver tuned to a specific frequency, and in the other, an antenna. I crested the hill hoping for a better vantage point to pick up the radio waves emitting from a radio transmitter located somewhere nearby. The other side of the hill cascaded with creosote shrubs, their tiny yellow blooms shuddering in the breeze, yucca plants, unflapped of course, and pockets of limestone rock strewn as if a giant boulder had been crushed and scattered. Which it had; a million years ago. The rocks sunk in the soil just so, suspended in a million-year-old roll downhill. Like they've been holding up the weight of the world all this time. I pointed the antenna north, then east, then south, then west, assessing which direction produced the loudest beep. Beep. Beep. BEEP. I was just on this side of frantic as a searing crescent of sun appeared over the mountains to the east. A cactus wren puffed his breast and twittered from a cholla. I descended toward the beeps, which were now shrill and urgent BEEEP, BEEEP, BEEEEPS. And then I found him. A male desert tortoise loped out of his half-moon-shaped burrow. A radio transmitter affixed to the rear of his shell silhouetted against the sand like a short stack of quarters. I unhooked the plastic bin from my backpack, pulled on some latex gloves, and crept toward him. The fiery yellow ring in his eye met mine as I placed him in the bin along with a small shovel of soil. It was translocation day for all the desert tortoises living on what would soon become the Ivanpah solar project. We were all a bit nervous.

The Mojave Desert is a 30-million-acre ecosystem spanning parts of southern California, southern Nevada, western Arizona and a tiny corner of Utah. The Ivanpah Valley, where our story takes place, is in the heart of the Mojave, in California just west of the Nevada border. Or some folks may understand if I say it's right off Interstate 15 about 60 miles south of Las Vegas.

The desert's reputation of lifeless desolation is inaccurate, and a real problem. Just look at the definition of a desert, "a dry barren area of land without water or vegetation." But the Mojave Desert is nothing of the sort. It swells with life—thousands of kinds of plants, reptiles, insects, mammals, and birds call the desert home. The deserts are some of the most biodiverse regions in the United States.[1] But so much of life here is secret. It's tucked between granite crags or tunneled deep into the ground. It's laced in the soil crust or nesting in bundles of sticks lodged imperceptibly in the rocky rills of a mountainside. It's all adapted to extremes; it's hiding. And much of it exists nowhere else on the planet. But it's easy to look at the brown expanse on a map and assume it's just miles of sand waiting for someone to give it a purpose. It is not.

The Ivanpah Valley is not what one might expect to find in the Mojave Desert. It can be downright green and lush in the spring. The expansive dry basin at the bottom of the valley fills with sparkling blue water after a good rain. If the water lasts, it penetrates the cracked-mud floor, smoothing its craggy surface and awakening a secret desert microcosm. Moisture permeates thousands of tiny, hard-shelled eggs hidden just below the surface. Over time they flex and pop, releasing a swarm of, believe it or not, tiny shrimp. Fairy shrimp eggs can lie dormant in desert basins for years waiting for rain. Once they hatch, they look like tiny clams, others like miniature horseshoe crabs. They dance around in the shallow, warm, alkaline water simply to eat and mate and then deposit fertilized eggs back into the dirt just in time for it to dry up, eventually into the cracked mud plain you've probably noticed if you've driven through a desert. You would never know the life it contains by looking at it from far away. It's a perfect analogy for the desert. A place that can appear so desolate but contains unimaginable life.

Clark Mountain is the highest peak in the Mojave ranges at about 8,000 feet, and it cradles the Ivanpah Valley in its lap. The foothills undulate with cobble-crusted ridges, and jagged, sandy-bottomed washes. They're dense with waxy-leaved creosote bushes, low lying dusty bursage, and rusty red buckwheat. Cholla cactus rise amongst the shrubs with densely knit spines wrapping spindly arms. Brick-red barrel cacti protrude awkwardly and, if you're lucky enough to see them before you step on one, tiny pincushion cacti hide among the cobblestones. Joshua trees tilt toward the sun, their arms heavy with inflorescences of creamy velvet white flowers. In the springtime, sandy expanses winding between the vegetation fill with annual grasses and flowers. If it's been a wet winter, it'll super-bloom and boil over with desert dandelion and poppies, white flowering chicory, goldfields, orange mariposa lilies, and stalky purple lupine. If you clamor up

1 Basin and Range Watch. "A Nomination to the U.S. Bureau of Land Management for Area of Critical Environmental Concern (ACEC) Status, Ivanpah Valley, California and Nevada." Needles Field Office: Bureau of Land Management, October 23, 2011.

the rocky hillsides, you'll find agaves tucked in the rocky crevices with perfectly whorled succulent green leaves, blooming indigo bells and desert ferns hiding in the shade. And that's just the plants.

The golden eagles nest high up on the mountainside. They're probably the only ones who see the mountain lions. There are bobcats and coyotes and a couple species of fox too. You have your typical complement of rodents, jackrabbits, badgers, and mule deer. It's hard to state the exact biodiversity for the Ivanpah Valley alone, but the species certainly number in the thousands. There are more than twenty species of snakes in the Mojave Desert. Only a few of them are poisonous, like the speckled rattlesnake with dusty rose scales and banded constellations of charcoal and brown. You find those in the rocky boulders. The Mojave also holds more than 200 species of birds, including roadrunners, quail, and loggerhead shrikes. There are orange and white and tiny blue butterflies, harvester ants, honeypot ants and fire ants (to name a few), and lots and lots of beetles. In fact, invertebrate species number in the thousands. Many of these species are endemic to the area; they live nowhere else on earth. And then you have the desert tortoise, the Mojave's unofficial ambassador, keystone species—and a controversial character, if you ask around. And the reason I was there.

It's only been about 10,000 years since giant ground sloths strolled between the Mojave Desert's Joshua trees and creosote bushes, grazing on yucca fruits, cacti, and globe mallow. The ice age receded into the past, and the deserts slowly, slowly, very slowly over thousands of years, became hotter and drier, transforming into an arid but still thriving landscape. The sloths, along with mammoths, giant bears, and camels, some of which had occupied the region for millions of years, slowly died out, along with the vast network of lakes and rivers. The humans who arrived in North America around 20,000 years ago persisted, and so did a few other species, like desert tortoises, Joshua trees, and creosote shrubs. Change continued over the last 10,000 years of course: the re-routing of dry washes and lakebeds with fluctuations in rainfall; the growing sophistication of humans who could hunt, grow food, and manage the landscape. But the ecosystem, the composition of plants and animals and people, expanded, contracted, and reshuffled on the timescale of millennia. Until a couple hundred years ago. I don't need to tell you what happened next.

If you investigate the history of humans in the Ivanpah Valley, you'll find a lot of recent stories; gold and silver mining drew migrants there for decades in the 19th and early 20th centuries until it all but collapsed from fluctuating precious metal prices and a pretty good drought that left the region reeling in the 1920s. But indigenous people have existed as part of the land there for millennia. Thousands of them were forcibly removed or killed as land was taken over by the government, parceled and sold or even given away to mostly white business owners to promote economic development in the region through livestock grazing and precious metal mining. The Ivanpah Valley contains many culturally important sites for southwestern tribes, and they still don't always retain the rights to utilize or protect them. Although there's more recognition of the atrocities committed against indigenous people here, our systems of land acquisition and management are still structured to benefit wealthy, mostly white people and entities, and largely exclude native people from their ancestral lands.

The word Ivanpah is Chemehuevi for "clean water." A few springs percolate from the foothills of Clark Mountain. You can imagine how important springs are in a desert. Just about every drop of water is fundamental to life here, for the people and the wildlife, for the whole of the ecosystem. It's the foundation of being. The springs were mostly taken over by the bigger mining companies in the late 19th century. Some of them are still owned by mining claims, some are "protected" but have remnants of cement, pipes, and wooden structures. Almost all of them have some amount of pollution from prior use.

There is one "town" near Ivanpah. Primm, Nevada rises out of the Mojave Desert like a Mad Max pit stop. The first town you see as you cross from California into Nevada on Interstate 15 has everything and nothing you need. An old rollercoaster built on precarious stilts will whip your body around Buffalo Bill's Resort and Casino and send you screaming down a decent free fall, if the wind isn't too strong and you're in the mood. After your wild ride you can grab a drink inside the casino at a bar circling a giant plaster tree covered in dust and laced with twinkling lights. But you have options for lodging in Primm. You can also stay across the interstate at Whiskey Pete's, a less overtly western-themed hotel and casino. The main attraction is the Bonnie and Clyde car, riddled with bullet holes, parked on the casino floor. The fancier amongst us may want to opt for the third lodging option, Primm Valley Resort, which has a kind of Mary Poppins English cottage with a hint of western vibe, which I suppose signals luxury. A hulking outlet mall and event arena are attached; you can wind your way through the shops (slightly more than half of them are in operation at any given moment) and watch other Primm tourists behold the spectacle along with you, sipping giant boba teas or nibbling on soft pretzels wondering, collectively: what is happening here?

In May of 2011 I arrived in Primm in a sandstorm. I was starting work the next day as a field biologist on the upcoming Ivanpah solar project. Four days prior I had finished graduate school in Laramie, Wyoming, and I'd been driving for two days riding out the waves of elation, sadness, and relief. I stopped at a gas station in front of Whiskey Pete's, one of those massive truck stops catering to the needs of long-distance truck drivers. The aisles overflowed with chips, motor oil, vape cartridges, and Las Vegas tchotchkes. I picked through the food section securing some last-minute essentials: Fritos and bean dip, Hostess cupcakes, ice and water. I paid for a shower at the counter, and the cashier walked me back to stall #4. I stood with the hot water running over my body for a very long time still processing the life transition laid out in front of me. I ordered a burrito from the taqueria in the back of the truck stop and then walked back to my truck with a sweatshirt pulled up over my face to block the sand and wind. I had outfitted the back of my truck as best as I could with no money. I had a basic camper shell and I folded up an old mattress topper to sleep on half of the truck bed. On the other half I had a few plastic tote bins holding my old camp stove, a jet boil for hot water, coffee making supplies, food, clothes, and toiletries. I kept water and a cooler with food and ice in my truck cab.

After restocking and cleaning "house" a bit I headed to camp. I drove out around the truck stop and wound behind Whiskey Pete's, traveling along the chain link perimeter fence of the parking lot that serves as a boundary between the town of Primm and the open desert. I opened the latched gate and proceeded down a bumpy dirt road into the desert, leaving Primm behind me shrouded in dust. After a mile or so I pulled into a small inlet tucked up against some jutting rocky hills protected a tiny bit from the wind. The ground was covered with shotgun casings. Old sheets of metal and a giant wooden spool were propped up and riddled with bullet holes. I found a level spot to camp for the night. I opened the back of my camper shell and climbed into my bed so I could eat my burrito while watching the sun slowly descend to the west, casting the desert floor in orange then purple then black. The lights of Primm flickered on the horizon.

My alarm went off at 5:00 a.m. After getting dressed I clicked on my head lamp and pulled out my makeshift kitchen. I held my hands over the steam rising from the boiling water. A black throated sparrow's whistle cut through the sound of distant traffic. Coffee percolated into an old ceramic mug. I perched on my tailgate waiting for the sun to break over the New York Mountains to the east. The warm glow came slowly, spread down the mountain, across the dry lake, and over the expanse of creosote, yuccas, cactus and Joshua trees that make up Ivanpah Valley. It was good to be home.

Chukchansi
Nüümü Witü (Eastern Mono/Monache)
Nyyhmy (Western Mono/Monache)
PANAMINT RANGE
DEATH VALLEY
Tübatulabal
Kawaiisu
Bakersfield
Kashtik
Kitanemuk
Barstow
Yuhaaviatam/Maarenga
Los Angeles
Acjachemen (Juaneño)
Payómkawichum (Luiseño)

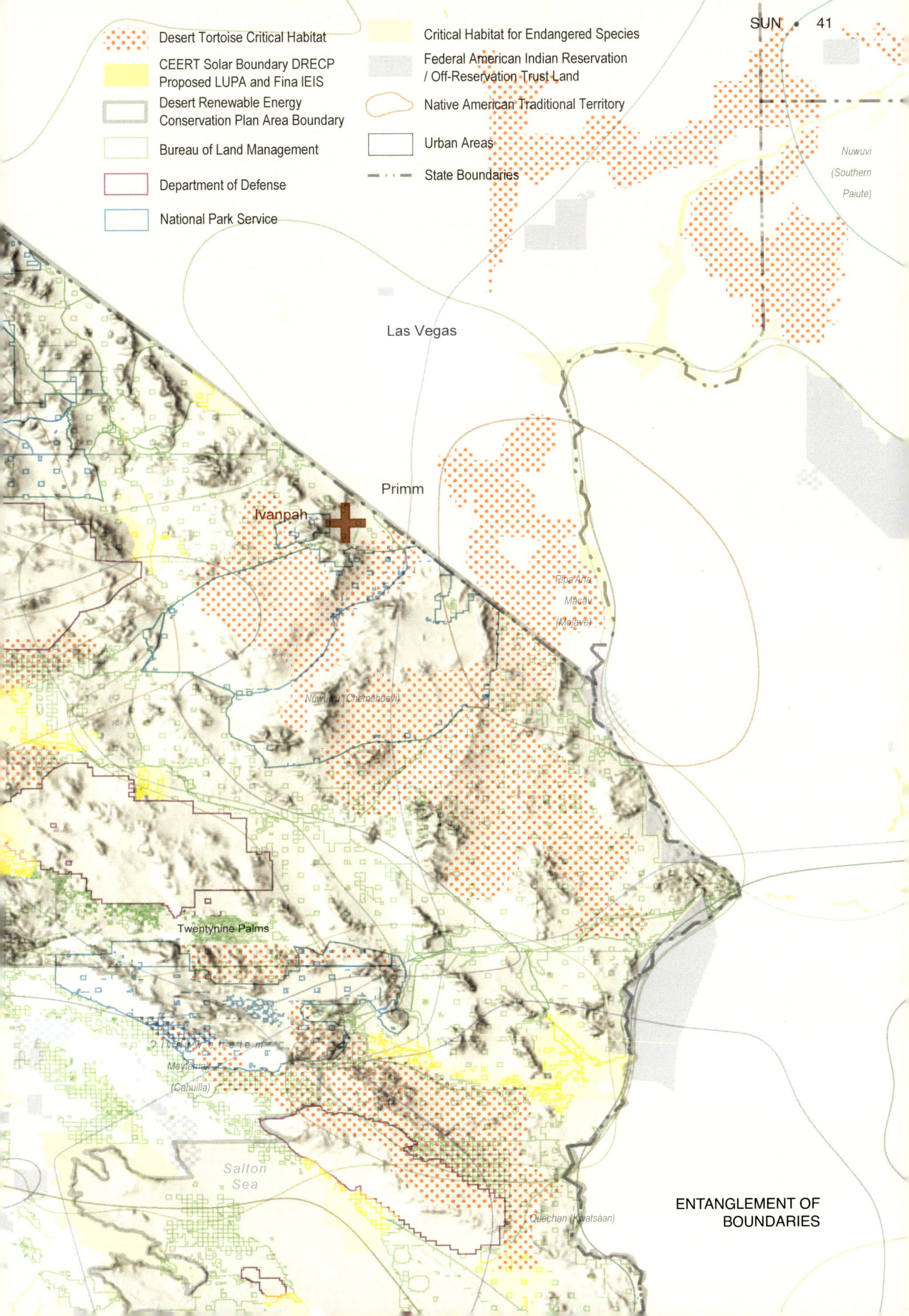

ENTANGLEMENT OF BOUNDARIES

The desert region of California is made up of three deserts, the Mojave, the Colorado, and a tiny fragment of the Great Basin, about 26 million acres. It's one quarter of the state. That's about the same size as New Hampshire, Vermont, and Connecticut combined. Most of that area is owned and controlled by the federal government. The three national parks make up about 5.8 million acres, the five major military installations make up about 3.2 million acres, and about 10.8 million acres are composed of public lands managed by the Bureau of Land Management (BLM). The Ivanpah Valley is situated in California, just south of the Nevada border and is mostly comprised of BLM land. BLM land is tricky to explain. Or perhaps it's not—it can be used for just about anything. There are many different types of designations allowing for different uses. One of them includes the lease of land to energy corporations for renewable energy development.

I first set foot in the Ivanpah Valley in the spring of 2003, a few years into my life as a desert tortoise field biologist. My field partner Pete and I wound our way through yucca plants, Joshua trees, and creosote shrubs searching for tortoises to count as part of an annual census. We summited a cobble-covered hill, and I remember seeing Interstate 15 carved like a thick channel through the center of Ivanpah Valley. I wondered how the tortoises crossed it, or if the sound made it hard for the birds to hear each other's songs. There were about twenty of us surveying for tortoises. After we finished our work for the day, we'd circle up our trucks on an old, abandoned airstrip near the Mojave National Preserve. We'd cook overly elaborate meals dreamed up while walking for hours subsisting on energy bars and warm, bruised apples left in the bottoms of our packs under the weight of tortoise measuring equipment—scales, calipers, and clipboards.

The Mojave Desert tortoise was listed as "Threatened" under the Endangered Species Act in 1990 because populations were declining rapidly from habitat loss, disease, predation by ravens, and a complicated web of other human-caused threats. As part of the recovery plan for the species, the U.S. Fish and Wildlife Service keeps tabs on the general abundance of tortoises throughout their range, which extends throughout the Mojave Desert region of California, Nevada, and into a small corner of Utah.

Tortoises have been in North America for many thousands of years; they knew the giant sloths, the camels, and the people. Like all desert life they've adapted to the dry and hot climate by developing some wild features. They spend about 90% of their time underground in burrows; and they have very minimal food and water needs satisfied by spending the spring, and if they're lucky enough to have summer rain, the fall, plodding around eating the tender green shoots of spring grasses and flowers, and opportunistically drinking from pools of rain water. Their shells help keep them warm and safe, and also make them impossible to discern from cobble on a typical desert slope.

Scientists began noting and raising concerns about the decline in tortoise numbers in the 1970s. By the 1990's some populations of tortoise had declined by as much as 90%. Their demise coincides with the last 100 years of human history in the desert, which has been characterized by habitat loss and destruction and increasingly, climate change. Despite the Mojave Desert being one of the more intact ecosystems in the lower 48 states, the desert has seen significant shifts in species composition with an influx of exotic plants and animals, expansive habitat loss from urban development, and more recently renewable energy, severe drought, and consequently more severe and numerous wildfires, which were historically infrequent. Unfortunately, desert tortoise numbers are still declining. In fact, a recent study showed tortoise numbers declined by 36% just in the past 20 years.[2] It's not surprising. Of the 352 species of turtle and tortoise on the planet, more than half are threatened with extinction. They are the second most imperiled group of vertebrates on the planet, outcompeted only by non-human primates. Our ecosystem's turtles and tortoises often function like canaries in the coalmine. Although desert tortoises adapted to the shifts in habitat that occurred over the last couple hundred thousand years, the abrupt changes occurring now are proving too much. After that first field season I fell in love with the desert and the tortoise and the life that came with being a desert field biologist. I spent the next five years living in my truck, working with tortoises and traversing the Mojave. My curiosity caught up with me and I ended up in graduate school for a while. As my program was wrapping up, I got a phone call from Pete. There was desert tortoise work in Ivanpah. A solar plant was going in and they needed experienced people to help manage the impacts to the tortoises. I didn't think twice. I bought a Tacoma with a camper shell and drove for two days from Wyoming to Primm.

It was 2011, and the Ivanpah Solar Electric Generating Station (ISEGS) had recently been permitted to eventually begin construction square in the middle of the Ivanpah Valley. This was a noteworthy moment for the state and the country because it represented a significant investment in shifting away from fossil fuels and toward renewable energy. ISEGS was a bold, audacious endeavor. When the project was first permitted in 2010, then California governor Arnold Schwarzenegger and Secretary of the Interior Ken Salazar gathered for a ribbon-cutting ceremony. Soon after, the long arduous environmental review process began. Battles ensued. This was a big deal, in part, because it would require the destruction of thousands of acres of mostly intact desert habitat including some of the best desert tortoise habitat around. The solar technology at ISEGS is different from most solar plants in the U.S. ISEGS uses a technology called concentrated solar power (CSP) to convert sunlight into electricity. The facility spans approximately 3,500 acres of federal public lands owned by the Bureau of Land Management near the California-Nevada border and the town of Primm. It's composed of three separate units, each roughly a 1,000-acre square equipped with a field of heliostats—large, flat mirrors—that track the sun's movement and reflect sunlight onto a central location at the top of a tower situated in the center of each square. As the concentrated sunlight strikes the tower, it heats a fluid to extremely high temperatures. This heated fluid generates steam, which drives

2 Zylstra, Erin R., Linda J. Allison, Roy C. Averill-Murray, Vincent Landau, Nathaniel S. Pope, and Robert J. Steidl. "A Spatially Explicit Model for Density That Accounts for Availability: A Case Study with Mojave Desert Tortoises." *Ecosphere* 14, no. 3 (2023): e4448. https://doi.org/10.1002/ecs2.4448.

turbines connected to generators, producing electricity. The generated electricity is fed into the power grid. The CSP technology used at Ivanpah is different from PV solar technology, which directly converts sunlight into electricity using solar panels. CSP was once a promising technology, but the focus has shifted more towards photovoltaic (PV) solar technology, which directly converts sunlight into electricity using solar panels. Photovoltaic solar has become the standard as it is now a more cost-effective and scalable choice.

Ivanpah was advertised as designed to produce 940,000 MWh of electricity per year. New thermal plants can take up to four years to reach full capacity. In its ninth year (2022), the annual production was 769,164 MWh. To begin operating each morning, the plant burns natural gas to the tune of about 1.2 million BTUs annually.

I was hired, along with about 70 other field biologists to manage ISEGS's impacts to desert tortoise. The state of California has some of the most stringent environmental regulations in existence, and developing habitat for a species listed under the Endangered Species Act is a big deal.

Although California has strict regulations, they do allow the loss of habitat for listed species, including the desert tortoise. The mitigation for this kind of a project includes paying to permanently protect similar habitat elsewhere, at a ratio of 2, 3, 4 or even 10 acres for every one acre destroyed.[3] The company also needs to pay to locate and move, or translocate, all tortoises from the project to a safe location. A large-scale translocation of tortoises is difficult and complex and can take years. The tortoises need to be located, affixed with radio transmitters, and monitored for about a year to learn about their behavior and ensure they are healthy, with the goal of a high chance of survival in a new location. Finding tortoises is difficult. They are cryptic, difficult to discern from the vegetation and rocky ground, and they spend most of their time in underground burrows. And when they hatch, they're not much bigger than a silver dollar.

We'd joke that these tortoises are made of gold because the cost of moving them was extraordinary. For ISEGS, there were three populations of tortoises that needed to be located, tracked with radio transmitters, and monitored for disease. The first were the project tortoises, the ones located within the boundaries of the proposed facility. They would eventually need to be moved to a recipient area. Because we moved them to an area that already had tortoises, we also needed to locate and monitor all the tortoises in that recipient area. Additionally, we needed to establish a control population separate from the translocated and recipient tortoises to be able to compare long-term outcomes so we could know definitively if this model was effective in at least giving the tortoises a chance at a similar lifespan compared to what they would've experienced under natural circumstances.

Why was this so complicated? Why couldn't we just put the tortoises from the project on the other side of the fence and walk away? We tend to grossly

3 U.S. Fish and Wildlife Service. "Biological Opinion for Activities in the California Desert Conservation Area." Memorandum. Carlsbad, California: United States Department of the Interior, September 1, 2017: 1–73.

oversimplify the inner lives of plants and animals. Tortoises are no exception. Desert tortoise males have an extraordinarily high level of testosterone. Like many animals, they're governed by a complex social hierarchy based on male dominance. An alpha male, usually the larger, older males, basically walk around their territory all day looking for other tortoises to mess with. If they find a male, they will beat them up in an epic combat mostly geared toward trying to flip one another over using the protruding part of their shell under their chins called a gular horn. They'll also try to mount them as if they were copulating to show dominance. This can go on for hours and it can be brutal. Tortoises are pretty good at righting themselves if they've been flipped over, but if a tortoise has been flipped in a sandy wash and it's a hot day, they can easily die from heat exposure. It's not uncommon to find smaller males dead in the wild on their backs. If the alpha male finds a female tortoise, they try to copulate with her by chasing her, knocking her about for a bit, and then mounting her from behind. Most females run the other way when they see a male approaching. Because tortoises are long-lived, they have well-established hierarchies governing each population. You can imagine the danger of dropping a couple dozen new tortoises into an area without giving it some thought first. We affixed radio transmitters to all the adult tortoises in the recipient area and tracked them for about a year, collecting data on which burrows they used, who they hung out with, and who appeared to be alpha. All that information was incorporated into a massive translocation plan that detailed which project tortoises go where to ensure their safety to the maximum extent possible.

We woke up early on translocation day. We dressed, boiled water for coffee, and brushed teeth in silence. It was dark and cool out. We pulled our trucks out around the corral and filed onto the dirt road, over the interstate and toward the project site. We met as a team to go over logistics, even though we'd gone over them several times before. Each of the project tortoises had a folder of information including their health status, ID number, and a map detailing their drop-off location. In the previous weeks we had dug burrows for each of the tortoises so they would have plenty of shelter sites to choose from. We knew they would be confused and restless, and we wanted to ensure they wouldn't struggle to find a safe place to retreat from the sun. Weeks earlier we drew blood samples and sent them out for verification that they had no antibodies indicating respiratory infection. We double checked to make sure temperatures were forecasted to stay cool for the requisite number of days. And then we began. We used radiotracking equipment to find a tortoise, placed it in a disinfected plastic tote bin, and drove it to its predetermined drop-off location. The tortoises were dropped off and were radio tracked daily and then bi-weekly, and then weekly for years to ensure their safety. Many of them stayed put, but some of them started walking back to where they came from. Some paced the fencing that kept them from coming back on site. But eventually, all of them settled down and established a new home.

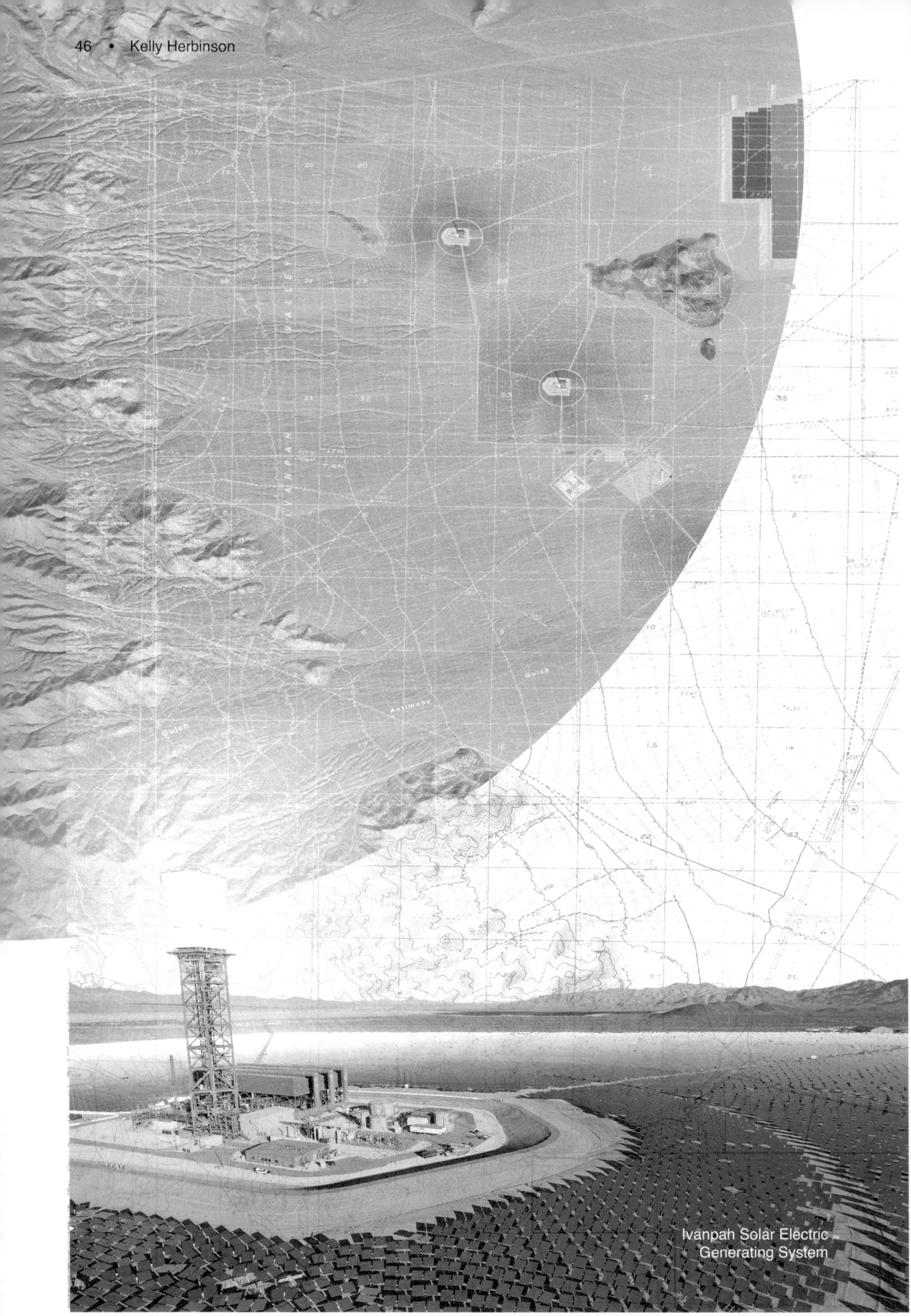

Ivanpah Solar Electric Generating System

And then construction began. Now that the tortoises were removed, the one listed species on site, the rest was to be destroyed. Rather than completely clear the vegetation, the developers agreed to mow it down to a height that would allow the installation of the mirrors, but not destroy the vegetation and the seed bank. This was a good thing. However, there were still miles of concentric circled roads to be graded, staging areas and flat expanses that would eventually hold offices. There was still a substantial amount of vegetation lost.

The months that followed saw the quick demolition of thousands of acres of desert in Ivanpah Valley. Many of the ancient Joshua trees, yuccas, and cholla cactus were mowed down. It's hard to know what happens to the snakes and the lizards, the ground squirrels. My hope is always that they were able to get out. The bigger animals that can move quickly, the jackrabbits, the kit fox, and the badgers certainly do. But it's hard to know the magnitude of what is lost. At the very least, and this is clear if you drive along Interstate 15 now, the ecological connectivity across the valley, already bisected by the interstate, is broken. The desert from Interstate 15 all the way west to the foothills of Clark Mountain is now a constellation of mirrors, thousands of them circling three looming towers reflecting the sun's powerful radiance up to a bright tip of each tower.

This system of mirrors reflecting the sun's rays toward a power tower is a technology not used often anymore for a few reasons. The main one being that photovoltaic panels produce more energy more efficiently. Another reason is that the sea of mirrors look like, well, a sea. They look like a large body of water if you are flying overhead, which is a problem if you are a bird looking for water. As the mirrors were installed, we started to notice a few things. Water birds, like American coots or eared grebes would land in the area, confused. An interesting fact about these types of water birds is they need water to take flight. They can't run on land and take flight, their anatomy is designed to only do this on water. So, a grebe landing in the desert is a real problem. We would have to catch it and drive it to the nearest body of water so it could take flight again. But that was only the beginning. The rays of light reflected from the mirrors to the towers were incredibly intense. So intense, that if a bird flew near the tower it would be incinerated immediately. Monitoring reports show that on average the ISEGS plant likely kills thousands of birds annually.[4] In 2018, a team of researchers resurveyed bird populations in the Mojave Desert, some near Ivanpah, to compare bird species richness and abundance with data collected historically by Josepf Grinnell 100 years ago. The results showed a 43% decline in the number of bird species present. The study cited climate change, and specifically a reduction in rainfall as the culprit.[5]

What does all this loss of our wildlife mean? Our ecosystem, the one humans have adapted to and evolved as part of for ten thousand years, creates the

---

4 Walston, Leroy J., Katherine E. Rollins, Kirk E. LaGory, Karen P. Smith, and Stephanie A. Meyers. "A Preliminary Assessment of Avian Mortality at Utility-Scale Solar Energy Facilities in the United States." *Renewable Energy* 92 (July 1, 2016): 405–414. https://doi.org/10.1016/j.renene.2016.02.041.

5 Iknayan, Kelly J., and Steven R. Beissinger. "Collapse of a Desert Bird Community over the Past Century Driven by Climate Change." *Proceedings of the National Academy of Sciences* 115, no. 34 (August 21, 2018): 8599. https://doi.org/10.1073/pnas.1805123115.

oxygen we breathe, provides the food we eat, the shelter we sleep in, and cycles nutrients, including carbon, in a cycle of life and death, waste, and reuse. All the millions of species living within our planet's systems contribute to the cycle. When we talk about protecting desert tortoises, or Joshua trees, or any species, we're talking about protecting one element of a whole inter-related system, our home and theirs. We're talking about protecting them, but we're also talking about protecting us. Our kids. Their kids.

In a relatively short amount of time, we managed to exploit a resource so spectacularly it disrupted the climate of an entire planet. And our answer, so far, has been to destroy a different resource, in this case land—plants, animals, insects, fungi, entire habitats—to build renewable energy infrastructure to offset the effects of the human-caused climate change. It seems we've missed one of the lessons from our mistake. Although we did learn to switch from utilizing a non-renewable resource, fossil fuels, to a renewable resource, sunlight or wind, there is an even more important lesson to which we seemed to have averted our eyes. We're just trading one planetary atrocity for another.

In 2023, a team of scientists determined the nine planetary boundaries that regulate the stability of planet Earth.[6] These nine "boundaries" represent the physical and biological processes and problems that weave together to inform the habitability of our planet. They include climate change, novel entities (pollution), land system change, freshwater change, biogeochemical flows, and biosphere integrity (functional and genetic biodiversity). Note that climate change, or fossil fuel use, is only one of them. Biosphere integrity, or biodiversity, was identified as one of the most critical to the stability of our planet. It was also one of the most severely transgressed. The nine boundaries demonstrate something relatively simple: we need clean air, clean water, and thriving plants and animals for humans to survive.

By commanding nature purely as a resource to use and profit from, we have created a false sense of separation between our survival and the health of the planet. In reality, it is not as simple as supplanting fossil fuels with renewable energy. If the development of renewable energy is adversely effecting another critical planetary function, we're just shifting pressure to another part of the planet, creating another problem to solve. And the worst part is, we're destroying all of this—our home, all these plants and animals that have been part of our shared history as humans for tens of thousands of years—without acknowledging the profound sacrifice of our living ecosystems, and our species' own future. It's collapsing because the ruling classes have spent centuries, one generation at a time, trying to control, extract, and profit from a planet like cowboys taking turns on a bucking bronco to win a cash prize. We've so thoroughly disrupted our planet's natural nutrient cycle that we've altered the climate. We've caused temperatures to rise, glaciers to melt, storms to grow stronger, droughts to

6 Richardson, Katherine, Will Steffen, Wolfgang Lucht, Jørgen Bendtsen, Sarah E. Cornell, Jonathan F. Donges, Markus Drüke, et al. "Earth beyond Six of Nine Planetary Boundaries." *Science Advances* 9, no. 37 (September 13, 2023): 1–16. https://doi.org/10.1126/sciadv.adh2458.

proliferate, and seas to acidify. The kind of planetary shifts we have ushered in over only a few centuries of recklessness are staggering.

Before ISEGS was finished being built, another 1,685-acre solar project went in directly next door, a small corridor separating their fences. Before that one was complete, another, this time 3,500 acres, went in just outside of Primm on the Nevada side. Solar projects began proliferating across the Mojave, in California and Nevada. I spent the next several years camping outside of Primm, managing desert tortoise translocations for those projects, and then for more solar projects further north in Nevada. It became a slog. A fight to do what was right for the tortoises despite this rampant destruction. We were exhausted and bewildered, broken down from the year-round, long field days watching the wildlife we cared so much for be taxed or even destroyed. I should note here the proliferation of solar in the Ivanpah Valley and in other parts of the desert alarmed enough people that federal and state agencies along with nongovernmental organisations (NGOs) rallied together to create the Desert Renewable Energy Conservation Plan (DRECP). The DRECP essentially took a landscape-scale view of renewable energy development in California, and identified development focal areas with minimal environmental impacts where permits would be fast tracked, and areas to be set aside for conservation. This was an enormous win for California and has been effective. No other solar developments have occurred in California with as dense tortoise habitat as Ivanpah. Nevada, on the other hand, is a different story. With no such policy in place, solar development proliferates, even in areas with a lot of tortoises.

Years of traversing beautiful parts of the desert and having the opportunity to see wild animals be wild kept me going for a long time. Taking breaks laying out on a stretch of desert pavement, pressing my hands on the warm brown rocks. We'd stop to identify tiny annual blooms, or exposed bird nests, or long-legged spiders stilting across sand. In the evenings we'd circle our trucks out by the shot-up wooden spool or tucked behind an old broken-down wooden corral. A dilapidated metal windmill creaked overhead as we cooked family style meals on camp stoves slid out of our truck beds. Some of us would set up horseshoes and fling them while talking about the things they found that day. Others would gather around a tailgate with old guitars and a fifth of whiskey, cackling about some dumb thing that one of us did. We'd build fires with old wooden pallets, eat voraciously, and watch the moon arc overhead. We all had been doing this awhile together, a decade or so. It was like coming home to family every time. There were no things to buy, no mirrors to consider your appearance, no electricity. Just us and the desert. I have thought about this for years now and I still don't know how to say it. Except that I know we stumbled upon something magical. For me at least, it felt like we unlocked a knowing. I felt overwhelmed with love and peace and connection to the desert and these people. Like this is how we are meant to live.

In 2018 I traded in my 17-year long life as a field biologist when my daughter was born. And in 2021 I took a job as the Executive Director of the Mojave Desert Land Trust, an organization dedicated to protecting desert habitat. I couldn't resist the opportunity to spend my days problem solving for this important place. It really boils down to one simple idea: we are nature. The desert wasn't just something we cared about or wanted to protect. It was woven into the fabric of who we are. I couldn't stand to watch its destruction because it became part of me. Funny how this all transpired under the shadow of those solar towers.

In January 2025, as I was wrapping up edits on this chapter, Pacific Gas and Electric announced they were ending their contracts with the ISEGS owners, closing two of the three power units in 2026. Southern California Edison, which buys the rest of the power, is also in discussions to terminate their agreement. These power purchase agreements were supposed to last through 2039 and potentially be renewed for another 30 years. But the solar technology at Ivanpah proved too expensive; the economics didn't work out. I sat reading the news story slack-jawed at the "well, we tried" sentiment of the whole thing. Like it could've just been one big shrug.

I puzzled for a moment over this. The permanent destruction of this land for the sake of solar had been a blow I've spent a decade and a half working to understand. I've approached this truth from every angle trying to make sense of it; trying to uncover and illuminate the true cost. But the permanent destruction of thousands of acres of land for what amounts to a temporary economic experiment—the thousands of birds killed, the hundreds of tortoises impacted, the acres of thriving, biodiverse desert that persisted for millennia—just...gone? And for what?

In February, a few weeks after this announcement, I drove past the ISEGS power towers on my way to the 50th Desert Tortoise Council Symposium in Las Vegas. The familiar glowing towers lit the way. A new gas station had emerged near our old campsite to the east. A new sprawling warehouse complex studded the highway north of Primm. At the symposium I listened to the familiar slate of speakers, scientists presenting data under instruction to not be too "doom and gloom." Federal workers presented on Zoom, having been grounded by a new presidential administration that only days before commanded a devastating set of lay-offs and budget threats. One speaker, shirking instruction, gave a talk essentially delineating the end of the world for tortoises, the desert, and of course, us. The bright side being that the planet, one day, perhaps a millennium from now, will be biodiverse and thriving once again. Absent from humans.

Renewable energy, he said, is perhaps a great con. A capitalist solution for a capitalist world, but so long as it perpetuates extraction it will never fix the problem.

I drove home, past the power towers again. I took the long way, through the Mojave National Preserve, a little-known National Park that encompasses part of the Ivanpah Valley to the south. I stopped in my favorite Joshua tree forest for a break. The looming, giant trees, some hundreds of years old, always seem to impart some kind of wisdom. Anything that just exists in one place for hundreds of years must know things. Scientists are finding that Joshua trees, like many other trees, communicate below ground through dense thatches of root systems lined with mycorrhizae. They can signal threat or even disburse resources to other trees in need. They remind me of humans, when we are at our best: capable of connection and reciprocity. It's easy to point outward; to complain and blame. It's hard to be still and to remember that we, too, are sustained by one another.

I got back into my truck and headed to work.

BIBLIOGRAPHY

Basin and Range Watch. "A Nomination to the U.S. Bureau of Land Management (Needles Field Office) for Area of Critical Environmental Concern (ACEC) Status, Ivanpah Valley, California and Nevada. Submitted by Basin and Range Watch on October 23, 2011 to: Mary Jo Rugwell District Manager Bureau of Land Management 4701 North Torrey Pines Drive Las Vegas, NV 89130." Needles Field Office: Bureau of Land Management, October 23, 2011.

Desert Tortoise Council. "Symposium Overview." Accessed April 5, 2025. https://deserttortoise.org/annual-symposium/overview/.

Iknayan, Kelly J., and Steven R. Beissinger. "Collapse of a Desert Bird Community over the Past Century Driven by Climate Change." Proceedings of the National Academy of Sciences 115, no. 34 (August 21, 2018): 8597–8602. https://doi.org/10.1073/pnas.1805123115.

Richardson, Katherine, Will Steffen, Wolfgang Lucht, Jørgen Bendtsen, Sarah E. Cornell, Jonathan F. Donges, Markus Drüke, et al. "Earth beyond Six of Nine Planetary Boundaries." *Science Advances* 9, no. 37 (September 13, 2023): 16. https://doi.org/10.1126/sciadv.adh2458.

Smith, Christopher Irwin, Lynn C. Sweet, Jeremy Yoder, Michael R. McKain, Karolina Heyduk, and Cameron Barrows. "Dust Storms Ahead: Climate Change, Green Energy Development and Endangered Species in the Mojave Desert." *Biological Conservation* 277 (January 1, 2023): 109819. https://doi.org/10.1016/j.biocon.2022.109819.

U.S. Fish and Wildlife Service. "Biological Opinion for Activities in the California Desert Conservation Area." Memorandum. Carlsbad, California: United States Department of the Interior, September 1, 2017.

Walston, Leroy J., Katherine E. Rollins, Kirk E. LaGory, Karen P. Smith, and Stephanie A. Meyers. "A Preliminary Assessment of Avian Mortality at Utility-Scale Solar Energy Facilities in the United States." *Renewable Energy* 92 (July 1, 2016): 405–414. https://doi.org/10.1016/j.renene.2016.02.041.

Zylstra, Erin R., Linda J. Allison, Roy C. Averill-Murray, Vincent Landau, Nathaniel S. Pope, and Robert J. Steidl. "A Spatially Explicit Model for Density That Accounts for Availability: A Case Study with Mojave Desert Tortoises." *Ecosphere* 14, no. 3 (2023): e4448. https://doi.org/10.1002/ecs2.4448.

IMAGE CITATIONS + CREDITS

Listed by Page Number

32 Photo: Matthew Seibert.

Map sources: "Power Plants." Accessed May 1, 2025. https://atlas.eia.gov/datasets/eia::power-plants/about; "State Highway Network Lines." Accessed May 1, 2025. https://gis.data.ca.gov/datasets/77f2d7ba94e040a78bfbe36feb6279da_0/about; "Mojave Desert Ecoregion." U.S. Geological Survey. Accessed May 1, 2025. https://catalog.data.gov/dataset/mojave-desert-ecoregion; Esri; U.S. Department of Commerce, Census Bureau; CGIAR; USGS.

34 Photo: Kelly Herbinson.

39 Photo: Kelly Herbinson.

40 Photo: Kelly Herbinson.

Map sources: "California Land Ownership." Accessed May 1, 2025. https://gis.data.ca.gov/datasets/CALFIRE-Forestry::california-land-ownership/about; "Desert Renewable Energy Conservation Planning Boundary [Ds501]." Accessed May 1, 2025. https://gis-california.opendata.arcgis.com/datasets/CDFW::desert-renewable-energy-conservation-planning-boundary-ds501/about; "ECOS: Species Profile." Accessed May 1, 2025. https://ecos.fws.gov/ecp/species/4481; "Mojave National Preserve Tract and Boundary Data." National Park Service. Accessed September 15, 2022. https://catalog.data.gov/dataset/mojave-national-preserve-tract-and-boundary-data; "Native-Land.ca I Our Home on Native Land." Accessed May 1, 2025. https://native-land.ca/; "Protected Areas Exclusion (Solar)." Accessed May 1, 2025. https://gis.data.cnra.ca.gov/datasets/CAEnergy::protected-areas-exclusion-solar-1/about; Esri; U.S. Department of Commerce, Census Bureau; CGIAR; USGS.

46 Photo: Matthew Seibert.

Map sources: Esri; TomTom; Garmin; FAO; NOAA; USGS; © OpenStreetMap contributors, and the GIS User Community; © 2013 National Geographic Society; i-cubed; Maxar.

50 Photo: Kelly Herbinson.

Wind Farms and Deepening Conflicts in Biobío, Chile's Imposed Transition

by Javier Arroyo Olea + María Paz López Ponce

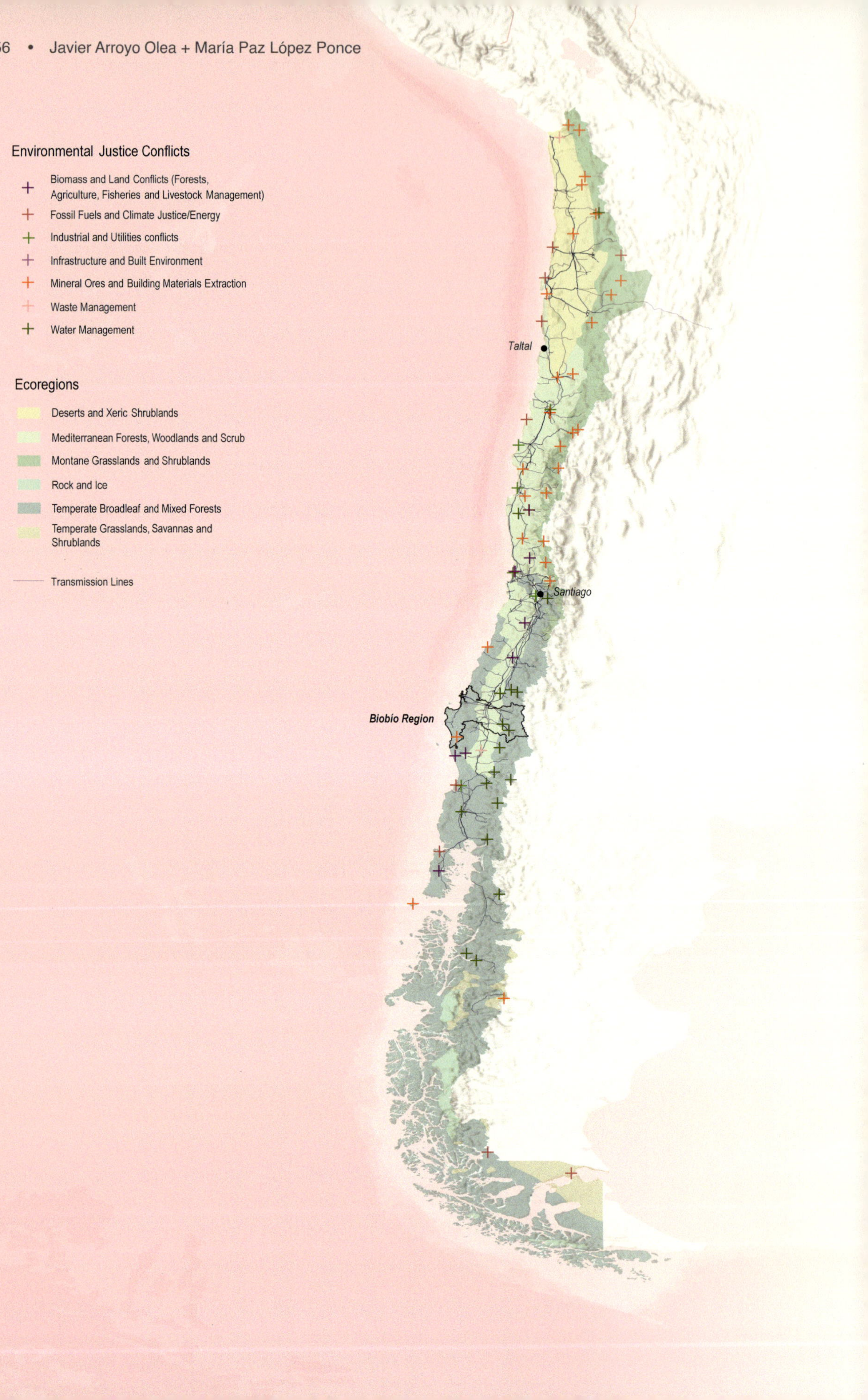
Environmental Justice Conflicts
Biomass and Land Conflicts (Forests, Agriculture, Fisheries and Livestock Management)
Fossil Fuels and Climate Justice/Energy
Industrial and Utilities conflicts
Infrastructure and Built Environment
Mineral Ores and Building Materials Extraction
Waste Management
Water Management
Ecoregions
Deserts and Xeric Shrublands
Mediterranean Forests, Woodlands and Scrub
Montane Grasslands and Shrublands
Rock and Ice
Temperate Broadleaf and Mixed Forests
Temperate Grasslands, Savannas and Shrublands
Transmission Lines
Taltal
Santiago
Biobío Region

Introduction

Chile is a long and narrow territory between the sea and the mountains; a country of forests, beaches, mountains, deserts, and great numbers of unique species. But it is also an exporting country and, therefore, a great part of the elements of Nature[1] are exploited to be sold as raw material—utterly changing the landscapes and lifeways of its communities. The vast deserts and their mountains wear their colors under a constantly blue sky; as we travel through them we find high Andean lagoons and salt flats that are oases of life for flamingos, vicuñas, and countless microorganisms. Unfortunately, it has been affected by mining and in recent years threats have increased due to lithium extraction projects.

To the south, the forests: mountains covered with vegetation are home to pumas, foxes, monitos del monte—one of the few marsupials in Chile, birds, reptiles, and insects, which together with plants and fungi form an impressively rich landscape. These forests are also home to indigenous and rural communities that find food and medicine there. However, industrial forestry has been felling native forests, planting in their place monocultures of pine and eucalyptus. These species, nonnative to Chile, transform the diverse and humid landscape into one that is homogeneous and dry, threatening both the biodiversity and the *lawen*.[2]

Throughout the country we can see the impacts of this economy that exploits and exports Nature, and we have become accustomed to bathing on beaches next to thermoelectric plants and calling the forests "forest plantations." Today, the multiple energy projects that threaten Chile's territories, justified by the climate crisis and the corporate energy transition, have expanded rapidly. The valleys and hills once dedicated mainly to agriculture are now plagued with solar plants and giant wind towers, intimidating to the eye: these are not small projects that bring energy to the surrounding communities, but megaprojects that cover the land, imposing themselves with force wherever you look. In the lands where we once saw hills, the horizon or the sea, where we once felt the breeze in our faces and moving our hair, where for an instant we were fully aware of the air, the wind and its infinite presence, now we see a population of wind towers, accompanied by noise, the constant flickering between the light of day and the darkness of the blades.

1 The word Nature is used with a capital letter in order to recognize it as a subject of rights.

2 *Lawen* is the ancestral medicine of the Mapuche people.

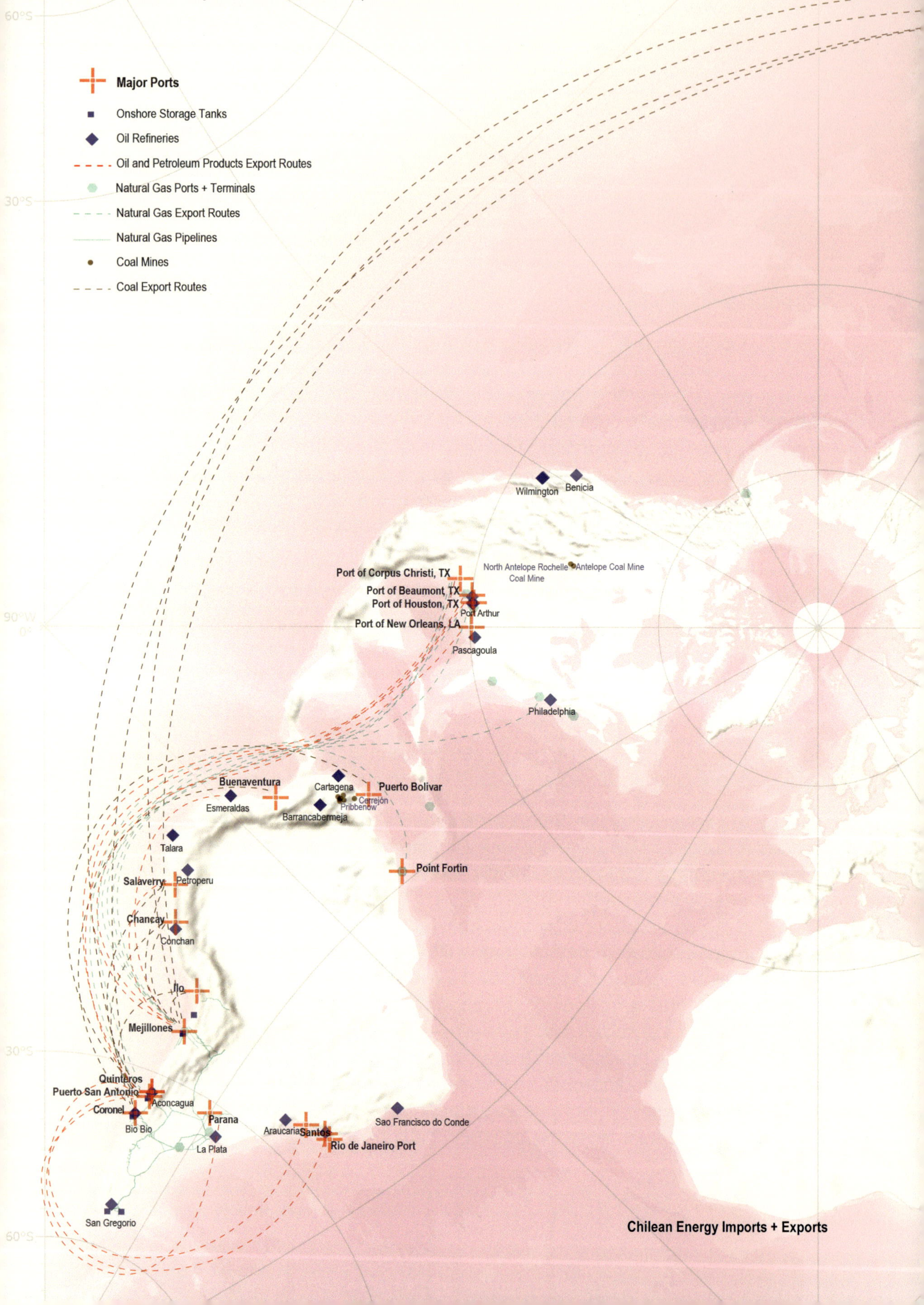

Chilean Energy Imports + Exports

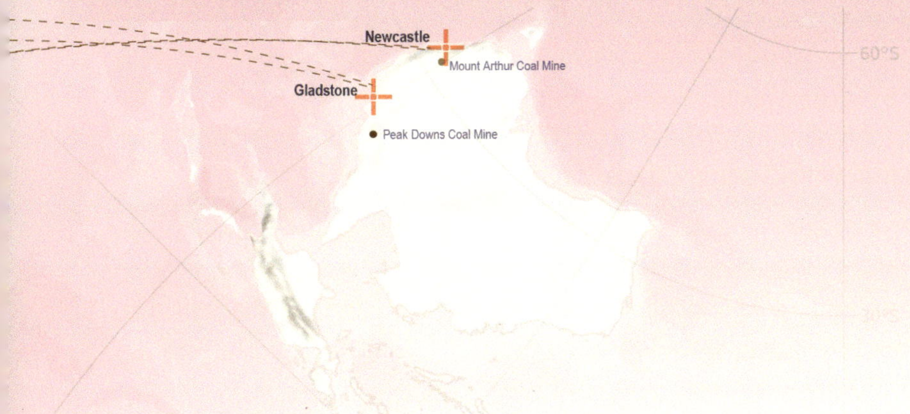

Wind tower projects have been a source of conflict during the last few years due to the effects experienced mainly by rural communities in Chile. The wind that blows between valleys and mountains today is harnessed, with promises of change, as the dream of a supposed energy transition. In reality, we are focused on expanding, rather than transitioning, energy production. The commodification of wind is an example of this: winds that could bring change if harnessed justly are neither transforming the extractivism at the core of the country's energy industry nor reducing the impacts of the climate crisis.

Thus, in this chapter we want to present some key elements that characterize the socio-environmental conflict that hampers the advance of wind energy projects, specifically in the Biobío region, a region located in south-central Chile. This work gathers stories of those who became neighbors of true wind giants and live day by day under the impacts of multiple wind turbines or suffer the threats of projects in development that could soon become their worst neighbors. In addition, we offer documentation of the effects and imposition of a new energy order: one that brings with it promises of a brighter future, but only for a few, where those within the so-called public-private alliance, consisting of Chilean state agencies and national and transnational businessmen, are the only beneficiaries.

The development of society since the industrial revolution has been strongly marked by hydrocarbons—gasoline, diesel, coal, and natural gas, all substances forming the backbone of global energy production. This, along with the values of unlimited growth and the accumulation of wealth, has led us to an ecological crisis at a scale never seen before. In Chile's case, most of the country's energy supply comes from the import of hydrocarbons, making it highly dependent on the international market.[3] Furthermore, 72% of the country's energy usage does not take place in our homes, but rather is attributed to the industrial, mining, and transportation sectors.[4] Electric energy is completely in private hands, considered a consumer good, rather than a social right necessary for heating, preparing food, and sustaining life.

---

3 Osorio-Aravena, Juan Carlos, Lorenzo Reyes-Chamorro, Jannik Haas, Jorge Aguilera, Rodrgo Palma-Behnke, and Christian Breyer. "Insights for Informing Energy Transition Policies – Are Decision Makers Listening to Science? The Case of Chile." *Energy Strategy Reviews* 58 (March 2025): 3. https://doi.org/10.1016/j.esr.2025.101644.

4 Ministerio de Energía, Gobierno de Chile. *Anuario Estadístico de Energía 2023*. (Santiago, Chile: Ministerio de Energía, 2023): 163.

Today, faced with the urgency of the climate crisis, the world is seeking to abandon hydrocarbons for renewable energies. In Chile, these renewable alternatives are being implemented and exported under the same privatized model that has brought humanity to this critical point.

The privatization of common goods, along with extractivism— the extraction of so-called "raw materials" in large volumes, destined mainly for export—has impacted ecosystems and local communities in our country and around the world for decades. In this context, it is precisely the energy sector that has the highest number of socio-environmental conflicts, where according to the National Institute of Human Rights (INDH), 37% of the conflicts in the country are linked to energy projects. This is mainly due to the installation of thermoelectric and hydroelectric power plants and energy transmission lines, which cause great social and environmental damage, violating the rights of communities and indigenous peoples through the contamination of soil, air, and water, increased chronic diseases, impacts on local economies, population displacement, and profound changes in culture and local ways of life.

For decades, although communities and socio-environmental organizations have denounced the negative impacts caused by energy mega-projects in our territories, these projects continue to advance rapidly.[5] In the last ten years, more than 1,200 projects have been approved in Chile's energy sector, greatly increasing non-conventional renewable energy (NCRE)[6] sources. Solar and wind went from representing less than 0.6% of the national electricity generation in 2012, to 28% in 2022.[7] Today there are more than 400 NCRE projects under construction, mainly solar and wind, without factoring in transmission system expansion projects.[8] This increase in NCRE has not translated into a lower use of hydrocarbons; in fact, the year 2021 registered the highest import of hydrocarbons in the last ten years.[9] As of today, fossil fuels continue to represent around 65% of the primary energy matrix.[10]

Although there is a decarbonization plan, which considers the closure and reconversion of thermoelectric power plants, it only focuses on coal-fired plants and does not include a clear policy to significantly reduce emissions from other hydrocarbons, such as gas or diesel. All this makes us wonder if we are going in the right direction: what is the point of having a large number of NCRE projects if this does not coincide with the reduction of the use of hydrocarbons, especially

5 We use the term "megaproject" to refer to large-scale projects with a huge investment and, therefore, considerable socio-environmental impacts.

6 "Non-conventional renewable energy" is a term used to refer to renewable energy in general, encompassing sources such as wind, solar, geothermal, biomass, and small hydroelectric plants.

7 Of that 28%, 17% corresponds to photovoltaic solar energy and 11% corresponds to wind energy. Source: Ministerio de Energía, Gobierno de Chile. *Anuario Estadístico de Energía 2023.* 46.

8 División de Desarrollo de Proyectos Unidad de Acompañamiento de Proyectos. "Reporte de Proyectos En Construcción e Inversión Del Sector Energía, Mes de Agosto de 2023." (Santiago, Chile: Ministerio de Energía, Gobierno de Chile, August 2023). https://energia.gob.cl/sites/default/files/documentos/reporte_de_proyectos_-_agosto_2023.pdf.

9 Comisión Nacional de Energía, and Gobierno de Chile Ministerio de Energía. "Anuario Estadístico de Energía 2021: Balance Nacional de Energía." Santiago, Chile: Comisión Nacional de Energía, Ministerio de Energía, Gobierno de Chile, 2022. https://www.cne.cl/wp-content/uploads/2022/07/AnuarioEstadisticoEnergia2021.pdf.

10 Ministerio de Energía, Gobierno de Chile. *Anuario Estadístico de Energía 2023.* 159.

Biobio Wind Power: Potential and Harnessed

considering that the energy sector is the main one responsible for greenhouse gas (GHG) emissions? Despite the large increase in renewable energy over the last ten years, emissions produced by the energy sector have remained constant, accounting for between 76–77% of Chile's total GHG emissions.[11] Ironically, the projects that should lead us toward another energy model and contribute to reduced GHG emissions are only increasing the impacts on the climate, while hydrocarbon imports continue unperturbed in the face of an energy transition that is going nowhere.

Throughout the Global South we see this imposed energy transition mortgaging the possibilities of thinking about other forms of energy generation, distribution, and use that respond to the needs of local communities, the well-being of ecosystems and the urgencies posed by the climate crisis. This hegemonic installation of the energy transition denies the possibility of thinking of alternatives and ways out of the current crisis. Although the landscape of renewable energy technologies considers a broad spectrum of elements and value chains—from mining for so-called "critical minerals" to exporting energy—[12]it ultimately reproduces the logic of extractivism, but now in a context of climate crisis for which it bears significant responsibility.

While the winds of change are blowing, each day we see that it is only an illusion full of uncertainty and frustration. We invite you to discover more about the winds "without" change that are felt strongly in the Biobío region, a territory that has experienced firsthand the impacts of extractivism from forestry, energy, and mining threats. Today, Biobío is one of the regions in Chile with the highest number of socio-environmental conflicts related to energy, increasingly driven in recent years by the rapid expansion of the wind energy market.

The Commodified Wind

Air is an indispensable element, but despite all its life-sustaining qualities, we don't often think much about it. The wind is perhaps the most evident form in which its presence is manifested. It is precisely this wind that fulfills important ecosystemic functions, such as pollination, the movement of seeds, the dispersion of fungal spores, and temperature regulation. Amongst all these qualities, it also allows electricity generation. Under the logic of the current model, the wind has lost its natural value, it has lost its infinite presence to become merely its economic value—a commodity. How is it possible that every element of nature, including our own lives, end up being privatized? Today there are hundreds of private wind energy projects scheduled to be installed throughout the country, thus continuing the story of the privatization of wind.

11 Ministerio de Energía, Gobierno de Chile. *Anuario Estadístico de Energía 2023.* 170.
12 Regarding the export of electricity, we recommend reviewing the "Antipodes" project, a promise of the government of Sebastián Piñera based on a submarine cable that would connect Chile with Asia to mobilize up to 600 GW of solar electricity, a project presented in 2021 at the Asia-Pacific Economic Cooperation Leaders' Summit. See Arroyo Olea, Javier. "'Antípodas': El Bochornoso Proyecto Energético Presentado Por Piñera Ante La APEC Que Nunca Fue Ejecutado." Latin American Observatory of Environmental Conflicts, December 14, 2023. https://olca.cl/articulo/nota.php?id=110424.

Through subsidies and credits to facilitate the installation of energy companies, laws to promote NCRE projects, or the provision of fiscal land, the State has played an important role in this process. As a result, we have seen that wind projects have increased rapidly throughout the country, bolstered by the delivery of public policies for their validation. In 2016, the Ministry of Energy and the Ministry of National Assets under the administration of Michelle Bachelet's second government announced that about 8,000 hectares of public land would be made available to entrepreneurs in the area of Taltal, a city located in northern Chile, where the desert meets the sea. The argument? The importance of wind energy, its promotion by the State, and the myth of sources of jobs and growth.[13]

A year later, in 2017, the government announced that Colbún S.A. would build the "Largest Wind Farm in Latin America:"[14] a project located in Taltal, in the northern region of Antofagasta, projecting a power of 607 MW. However, echoing the politicization of wind energy projects, it was in 2021 when the administration of then President Sebastián Piñera announced the project again, but with a name and new figures:[15] "Parque Eólico Horizonte," with a capacity of 778 MW, 140 wind turbines and an investment of U.S.$850 million. In this sense, the government took advantage of a previous management facilitated by a private company to take water for its own mill: a practice that has been reiterated by the Chilean political class in faithful alliance with national and transnational businessmen.

Thus, the commodified wind is part of what the State has called the "Chilean energy revolution,"[16] a concept that has been iterated since 2006, when an environmental assessment prepared by the Chilean government and the German Agency for Technical Cooperation (GTZ) pointed to energy security as a challenge for the country.[17] In other words, there was political projection, which has deepened over the years.

The wind—in itself—is not the only commodity that the business community has sought to pocket; wind projects are squarely embedded within the very framework of energy extractivism. Thus, we are not only talking about the exploitations that follow the construction of wind turbines imported by Chile, nor are we merely referring to the aforementioned commodified wind, but also to its relevance for the corporate energy transition that is driven by the race to produce Green

---

13 Ministerio de Energía. "Nueva Licitación de Terrenos Fiscales Para Energía Eólica." *Ministerio de Energía,* November 25, 2016. https://energia.gob.cl/noticias/antofagasta/nueva-licitacion-de-terrenos-fiscales-para-energia-eolica.

14 Ministerio de Energía. "Chile Tendrá El Parque Eólico Más Grande de Latinoamerica." *Ministerio de Energía,* September 27, 2017. https://energia.gob.cl/noticias/nacional/chile-tendra-el-parque-eolico-mas-grande-de-latinoamerica.

15 Ministerio de Energía. "Ministro Jobet Encabezó Anuncio Del Parque Eólico Más Grande de América Latina." *Ministerio de Energía*, September 21, 2021. https://energia.gob.cl/noticias/nacional/ministro-jobet-encabezo-anuncio-del-parque-eolico-mas-grande-de-america-latina.

16 Ministerio de Energía. "Los 10 Parques Eólicos Más Grandes de Chile." *Ministerio de Energía*, October 24, 2017. https://energia.gob.cl/noticias/nacional/los-10-parques-eolicos-mas-grandes-de-chile.

17 A collaboration between the Chilean government and the German Agency for Technical Cooperation. See: Comisión Nacional de Energía. "Guía Para La Evaluación Ambiental de Proyectos Eólicos En Chile." Santiago, Chile: Ministerio de Energía, October 2006. https://energia.gob.cl/documentos/proyectos-eolicos.

Hydrogen.[18] The latter point is well recognized by the population whose lives are most affected by wind projects in the Biobío region: "our authorities are proposing these machines," one resident notes, "as the solution to climate change to produce green hydrogen, but green is only the word for the affectations of this infrastructure."[19] There is a degree of deception to these projects that are sold to the world as "green" and harmless to the environment, but in reality maintain practices that harm nature and local communities. Green Hydrogen is the hand that props up commodified wind as a pillar of extractivism, becoming a key piece in an imposed energy transition, all the while dragging behind it the forced participation of communities, regional oversaturation of mega-parks, and impacts to existing industry. All this, omitted under the discourse of decarbonization from the business community and the State.

Biobío: The Region that Energy Businessmen are Trying to Pocket

Biobío is one of the sixteen regions of Chile. Located in the central south of the country, it is home to vast beaches, where the sound of birds and the sea breeze awakens our senses, with docks and coves in hidden places between the cliffs, and huge ports. But we also find hills, forests, animals hidden in the fog, and an impressive coastal mountain range, home to unique species, such as the monito del monte (*Dromiciops gliroides*), a marsupial endemic to the temperate rainforests of Chile and Argentina; the queule (*Gomortega keule*), a tree endemic to Chile, declared a Natural Monument; or the canelo (*Drimys winteri*), a tree endemic to Chile and Argentina, considered sacred to the Mapuche people. Bordering Argentina is the Andean mountain range, one of the lowest areas in the country, with an average altitude of 2,000 meters above sea level. The Biobío region has been historically impacted by different extractive activities: monoculture forestry has devastated much of the native forests, replacing it with pine and eucalyptus plantations while mining for coal has caused the dispossession of local, rural communities, and the destruction of nature. But these are not the only sectors that have impacted the region, there are also several energy projects, mainly hydroelectric, which have also caused serious transformations in ecosystems and communities.

These projects have been installed invasively, without prior consultation with communities and are causing serious impacts at the local level, which are affecting the lives of people and ecosystems in the region. All this is generating concern, especially after the Ministry of Energy itself pointed out in 2016 that the Biobío region "will become the second most important in the country in energy generation through wind farms, in investments that will exceed 4 billion dollars."[20]

---

18 To obtain hydrogen, it must be separated from a compound containing the hydrogen molecule. Depending on how it is obtained, it is assigned different colors. In the case of Green Hydrogen, it is the hydrogen generated from electrolysis of the water molecule, using electricity generated with non-conventional renewable energies as a source of energy. Therefore, the production of green hydrogen technically emits no greenhouse gases.

19 Interview with resident of El Agro, commune of Negrete, September, 2023.

20 Ministerio de Energía. "Biobío Se Convertirá En La Segunda Región Con Más Generación de Energía Eólica." *Ministerio de Energía*, February 2, 2016. https://energia.gob.cl/noticias/biobio/biobio-se-convertira-en-la-segunda-region-con-mas-generacion-de-energia-eolica.

This projection has clearly materialized: in recent years, media reporting has echoed the uptick of wind projects in the region, with one newspaper stating that "a notorious trend of wind-based electricity generation initiatives is consolidating, currently totaling 14 projects in operation, with a contribution to the National Electric System of 360 MW of installed power, and in five years an additional 1,500 MW are projected to come into operation, with an investment of 2.2 billion dollars."[21]

In 2021, under the idea of "accelerating the energy transition in the region," the University of Concepción launched the project Green Hydrogen Strategic Alliance for Biobío, financed with state money through funds from the Regional Government, aiming to prepare regional actors to "take advantage of the opportunities that will open up for the development of the green H_2 industry in the Biobío Region and the entire south-central macro-region of our country."[22] This initiative assembles a wide array of actors from public and private sectors: executed by a partnership between a private university and the Chilean Hydrogen Association, it is bolstered by state funding and brings together about 30 companies. Among them are AES Andes—a perpetrator of several socio-environmental conflicts related to wind energy projects, the National Petroleum Company (ENAP), the Arauco and CMPC forestry companies, GasSur, Abastible, Siderúrgica Huachipato, and Minera Aclara, which seeks to exploit rare earth minerals as a key element of the energy transition. All this is in addition to a strong network of think tanks and "civil society entities," mainly linked to the representation of the business community, such as the Chamber of Production and Commerce (CPC), the Fishing Industry Association, and Corporación Chilena de la Madera (CORMA), a Chilean trade association representing the wood and forestry sectors. This alliance amongst business interests reflects a political interpretation of the region and its connection to the wind energy industry and its offshoots. Treating it as a "privileged condition" due to the existence of industrial ports that mobilize forestry, industrial, and fishing cargo, in addition to the prevalence of other areas of production such as oil refining and steel, the companies gathered here highlight that "the region has a wind energy potential of 5,000 MW with an average plant factor of 33% (with areas at Arauco level that can reach 42%),"[23] positioning the Biobío region within the places with the greatest potential for wind energy production.

Thus, the Biobío H2V Alliance has stated that this will position the region "as a productive and technological pole of green H_2, generating clean fuels for transportation, low-emission solutions for industry, development of new products and services of high technological level."[24] That is to say, discursively, the

---

21 Ochoa, Javier. "Región Del Biobío Tendrá Este Año 16 Parques Eólicos En Operación." *Diario Concepción*, March 18, 2022. https://www.diarioconcepcion.cl/economia/2022/03/18/region-del-biobio-tendra-este-ano-16-parques-eolicos-en-operacion.html.
22 Alianza Hidrógeno Verde. "UdeC Lanza Proyecto Que Busca Desplegar El Potencial En H2 Verde de La Región Financiado Por El Gore Biobío." Gobierno Regional I Region del Biobio. *Alianza H2V Biobío*, May 7, 2021. https://www.ah2vbiobio.cl/evento/udec-lanza-proyecto-h2-verde/.
23 Alianza Hidrógeno Verde. "Región del Biobío." Gobierno Regional I Region del Biobio. *Alianza H2V Biobío*. Accessed June 4, 2025. https://www.ah2vbiobio.cl/region-del-biobio/.
24 Alianza Hidrógeno Verde. "Región del Biobío."

prevailing attitude is that the landing of projects of this nature in the region is inevitable, as though the die are already cast, when at the same time the region's fate is totally imposed without a transversal openness to dialogue.

Below, we will take a closer look at three cases of socio-environmental conflicts in the Biobío region through the eyes and voices of the territories themselves, who invite us to better understand what the explosively sudden installation of wind projects has meant in their lives. We will visit Negrete, a commune located next to the Biobío river, where agriculture and animal husbandry are part of life; El Ciruelo Sur, a rural community located in the commune of Los Ángeles, capital of the eponymous province, considered by many people as "the entrance to the south of Chile"; and Arauco,[25] a commune in the province of Arauco in Biobío's south that has been historically impacted by monoculture plantations of pine and eucalyptus, and today, the proliferation of wind energy megaprojects.

Knots in the Air, the Earth, the Waters, and the Life of Communities

There is a progressive increase in the criticism of wind projects in different territories, based on the experiences already lived in areas that have been deeply affected in multiple dimensions. Those who used to live in rural areas dedicated to a family economy linked to agriculture and livestock, now see their way of life diminished by the generation of energy from wind.

The wind, the most evident expression of the air that surrounds us—and one that we do not always consider or stop to think about—is becoming entangled in the business of energy companies and the conflicts they are generating. This entanglement has led to knots in the air, land, and water, knots in the throats of those who inhabit these territories and, of course, knots of conflict that have developed in the provinces of Arauco and Biobío.

Wind Farms Landing, the Ghost of "Progress" and the Lie of Labor

In its search for integration into the territory, the energy business has fallen into questionable procedures and positions in relation to local communities. In the case of the Vientos del Pacífico project, one inhabitant of the Tralca community points out that the company "came with brochures and models presenting the project to all of us in the sector, and in reality with deceit, because it turns out that the reality we have is different."[26] In this sense, the community points out that the companies are unaware of the knowledge and ways of life of the local communities, and instead move to impose an absolute knowledge based on technicalities, accompanied by the discourse of progress and investment.

However, these are myths that are progressively failing, along with superficial promises linked to the world of labor. By overtaking the reality lived by hundreds of people, the energy business has pretended to base its insertion in the

25 *Arauco* can be translated from Mapudungun as "muddy water" or "clayey water."
26 Interview with resident of Tralca, commune of Arauco, January, 2023.

territories through pillars such as investment, decarbonization, and labor supply, at the same time crushing the historical forms of work sustained by the communities themselves.

Thus, the corporate haste took over. "It was all very rushed,"[27] said one resident of Tralca, a commune in the municipality of Arauco. After going through participation processes related to wind projects, she has determined that companies "come with lies." Another resident from Arauco echoes this statement: "things are done the way they say and not the way we say here, even though you know the area where you live and they come in with other things."[28]

Progressively, more voices have begun questioning the location of these projects in the rural sectors of the Biobío region, concerned that "they are passing over us without respecting the rights of everyone who lives in the countryside,"[29] deconfiguring life models to impose new logics that respond to the energy market and that commercialize something as essential for life as air. Thus, the false idea of "progress" conflicts with the defense of the land, creating serious tensions: "they come with the promise of generating sources of employment—which is not even the case, because they bring specialists from their areas."[30]

In this sense, despite the promises of job creation, in practice the opportunities for new jobs have been limited. While employment is a key part of the investment discourse promoted by the companies, and one that resonates with a significant portion of the population, in reality, "they give work to those who 'plant the flag,' and that lasts just a couple of months. They deceive people with promises of the future, of economic support, but that support that is promised in the construction of headquarters or a project is nothing in relation to what is lost."[31]

This uneven tradeoff has not gone unnoticed. This characterization has led to the prevalence of an idea that is being increasingly spread today in the face of potential wind energy projects, as new voices are rising every day against the deception of the supposed winds of change. One of those affected has stated it well: "if they are cultivating mutual respect with the communities and neighbors, there is no real change, there will not be any benefit."[32]

a. Threats

In the context of broader socio-environmental conflicts in Chile, a response based on intimidation has been increasingly established against those who choose to defend the territory against the permanence of any project that affects nature and/or the population. Wind farms are no exception to this phenomenon, reproduced as a subsistence strategy of extractivism.

These forms of intimidation do not rely solely on physical violence, but also on actions and tactics aimed at discouraging communities from continuing their

27 Interview with resident of Tralca, commune of Arauco, January, 2023.
28 Interview with resident of the municipality of Arauco, January, 2023.
29 Interview with resident of El Agro, commune of Negrete, September, 2023.
30 Interview with resident of El Ciruelo Sur, commune of Los Angeles, January and December, 2023.
31 Interview with resident of El Ciruelo Sur, commune of Los Angeles, January and December, 2023.
32 Interview with resident of El Agro, commune of Negrete, September, 2023.

Energy Infrastructure

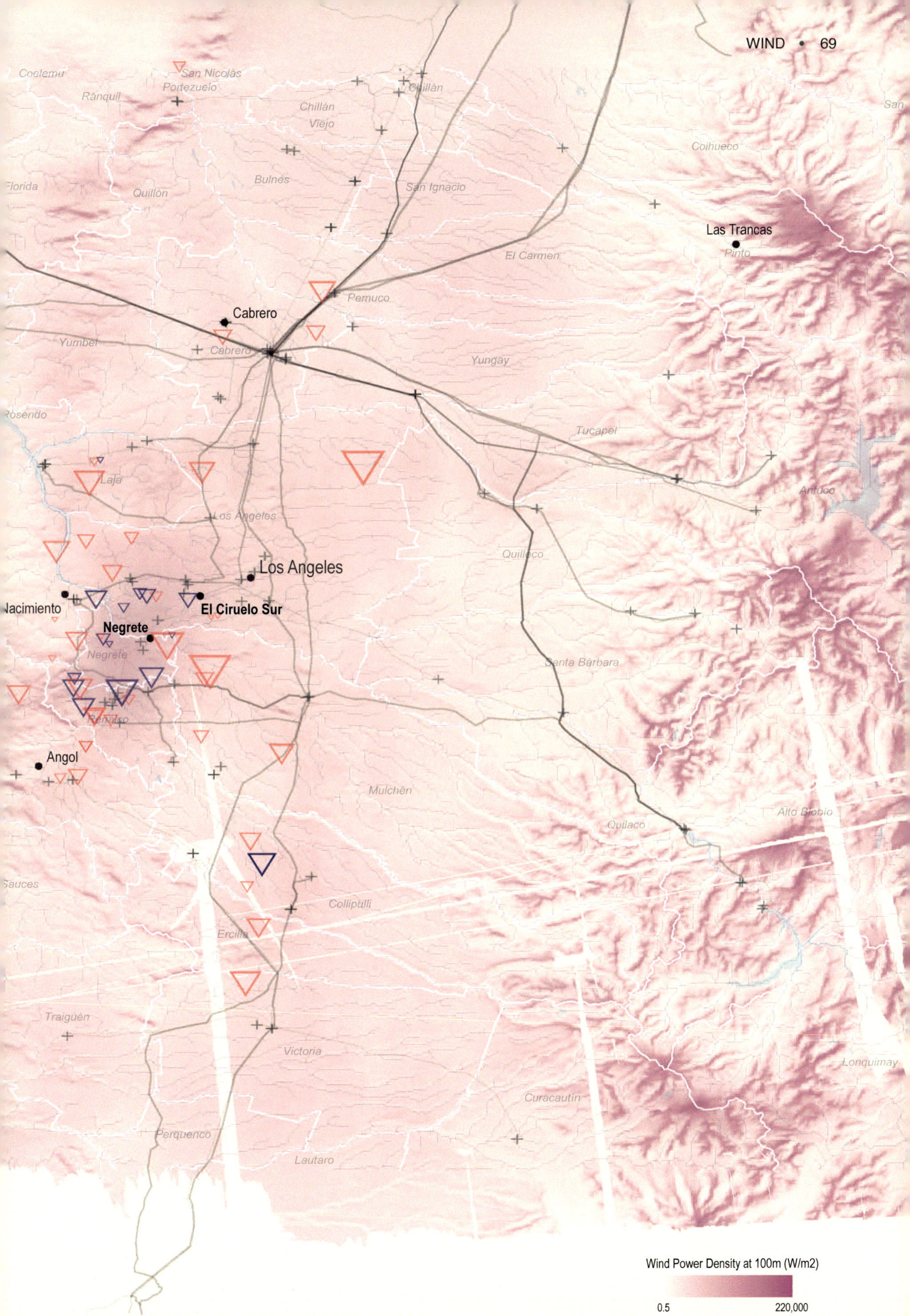
Coelemu
San Nicolás
Portezuelo
Ránquil
Chillán
Chillán
Viejo
Coihueco
Florida
Quillón
Bulnes
San Ignacio
Las Trancas
Pinto
El Carmen
Pemuco
Cabrero
Yumbel
Cabrero
Yungay
Tucapel
Antuco
Laja
Los Angeles
Los Angeles
Quilleco
El Ciruelo Sur
Negrete
Negrete
Santa Bárbara
Renaico
Angol
Mulchén
Quilaco
Alto Biobío
Collipulli
Ercilla
Traiguén
Victoria
Lonquimay
Curacautín
Perquenco
Lautaro
Wind Power Density at 100m (W/m2)
0.5
220,000

territorial defense, whether through fear, disruption, distrust, or other means of dissuasion. In this sense, processes have been developed in which even the role of Carabineros de Chile—the country's law enforcement—has swayed towards favoring wind projects, as in Campo Lindo, an initiative of AES Andes in Los Ángeles. In this case, inhabitants of the village Las Trancas accused the project of destroying an easement road protection center in the context of mobilization for damages after the company intervened in irrigation channels.[33]

Similarly, those who defend the territory, whether through public denunciation, awareness campaigns or other actions, have observed that it is not just companies generating pressure on communities: "Authorities also exert underhand pressure, because they do a dirty job of convincing the communities that it is very important and necessary for the company to be installed, and that there is no other option."[34]

In this context, the role of women in socio-environmental conflicts has come to the foreground, as they are often "the ones who defend the territory the most, who are persecuted and face restrictions in their work opportunities as a result."[35] In the case of women, the impacts occur in a very particular way, since they are the ones who often assume the care work[36] and, therefore, suffer an overload in contexts of environmental conflicts that affect the health of their families and Nature.[37] But they are also stigmatized in their role as defenders, and in some cases, "when working outside, we are branded as extremists, as if we were guilty of not allowing progress."[38] This has also led to the creation of complaints from the population for harassment and surveillance, which despite being denounced have not made much progress in terms of investigation by the Chilean justice system.[39]

These facts have led to a harsh outlook for the affected communities, who point out that "the State is not interested in the cost of living, in what is lost in the value of the landscape, in culture, in autonomy. The State is not interested in autonomy, it is interested in keeping the people submissive."[40]

33 This fact can be further explained in the following source: Arroyo Olea, J. "Acusan que en presencia de Carabineros se destruyó centro de resguardo de camino rural en Las Trancas para favorecer proyecto eólico." *Resumen*, September 7, 2022. https://resumen.cl/articulos/acusan-que-carabineros-destruyo-centro-de-resguardo-de-camino-rural-en-las-trancas-para-favorecer-proyecto-eolico.

34 Interview with resident of El Ciruelo Sur, commune of Los Angeles, January and December, 2023.

35 Interview with resident of El Ciruelo Sur, commune of Los Angeles, January and December, 2023.

36 Considering the sexual division of labor, where women have been relegated to domestic and care work.

37 Ardiles, Karen, et al. *Women and Environmental Conflicts: Differentiated Impacts*. Latin American Observatory of Environmental Conflicts (OLCA), 2023.

38 Interview with resident of El Ciruelo Sur, commune of Los Angeles, January and December, 2023.

39 A clear example of this situation has been experienced in the sectors of El Ciruelo Sur and Las Trancas, both located in the commune of Los Angeles. For more information on this, we recommend the following article: Moreno Caro, Branco. "Dirigenta Socioambiental Denuncia Hostigamientos en Medio de Conflicto entre Comunidades Rurales de Los Ángeles y AES Andes." *Resumen*. Accessed June 4, 2025. https://resumen.cl/articulos/dirigenta-socioambiental-denuncia-hostigamientos-en-medio-de-conflicto-entre-comunidades-rurales-de-los-angeles-y-aes-andes.

40 Interview with resident of El Ciruelo Sur, commune of Los Angeles, January and December, 2023.

b. Insights into Impacts on Nature and Communities

The effects of the wind power market in the Biobío region are a true reflection of what people do not want to see from energy transitions. Under the argument of decarbonization, public policy and private business have tried to caricature the demands of communities affected by new wind tower projects. However, the experience of the population as wind projects are installed in the region marks an important turning point.

The installation of wind infrastructure has profoundly disrupted the way of life in the countryside, suffocating the activity of local families. One resident from outside of Tralca speaks for her community when she explains that "most of us live off the land here, from gathering medicinal herbs. All that is going to be affected [...] we are the ones affected here, we live here, we rely on the herb collection, we live from what Nature gives us. Things have not been done well here."[41] But it seems that deaf ears prevail.

Thus, the economic activities that have lasted for years are also affected, where communities have accused companies of coming "to reap the money at our expense," because they spoil the whole environment we have here from agriculture." This negatively impacts practices related to family farming for those who "were born and raised by agriculture, and the *animalito*[42] that we also raise to sustain our lives to live. And then, the companies come to install these wind farms, to damage the land, the water."[43]

In this sense, the installation and continued operation of wind projects in the countryside of the Biobío region "has meant a total loss of quality of life." Much of this is due to where these projects are erected: in many cases, the turbines are incredibly close to the existing rural homes, nearly "on top" of them. "These people cannot sleep; the animals have seen their way of life completely altered."[44] The land has also been affected, but there has been no greater awareness into the irony of "killing the land at the hand of a discourse of renewable and clean energy that is being produced for export at the cost of our sacrifice."[45]

Water is also impacted. Though companies installing new energy projects made empty promises to communities that "the water would not be affected," the projects are in turn sited in critical hydrological areas: "where the stream that feeds us every day, where the animals of the region drink, has its source."[46] This shows us once again the deceitful course of these projects, where the communities in Biobío have stated that "with these machines, our water tables have been altered, it is no longer possible to plant in the same areas."[47] Likewise, experiences show that "the well that had been there for 10 years without any

41 Interview with resident of Tralca, commune of Arauco, January, 2023.

42 In Chile, the population is accustomed to call "animalito" (little animal) as an expression of affection to different animals, whose size may differ from one another. In rural areas, it is related to horses, cows, chickens, pigs, or others.

43 Interview with resident of the municipality of Arauco, January, 2023.

44 Interview with resident of El Ciruelo Sur, commune of Los Angeles, January and December, 2023.

45 Interview with resident of El Ciruelo Sur, commune of Los Angeles, January and December, 2023.

46 Interview with resident of Tralca, commune of Arauco, January, 2023.

47 Interview with resident of El Ciruelo Sur, commune of Los Angeles, January and December, 2023.

problem dried up, and it began to dry up a year after the wind farm began operations."[48]

In the village of Las Trancas rains in June catalyzed huge losses after the company AES Andes intervened in existing canals and irrigation channels during the construction of the Campo Lindo Wind Farm: the canal flooded catastrophically, resulting in the loss of hectares of corn plantations and rendering interior roads impassable.[49] This case shows aspects that are seldom considered in the analysis of the effects of wind projects on communities: how the wind energy business affects the water tables vital for human, animal, and irrigation consumption, in addition to the ecosystem. Likewise, it makes visible political turning points that impact the daily lives of communities near these rural wind projects: the control and facilitation of the use of easements. While legislation prioritizes extractive industries and resource extraction for export, making it difficult for citizens to stop projects, more recently, the idea that easements are mandatory for public interest has begun to be questioned. [50]

These projects not only affect nearby human communities, but also other forms of life that share the habitat. There are reports of impacts on animals living on land adjacent to the wind towers, which "alter the quality of life of the animals as well." The shadow effect, which refers to the phenomenon that occurs when the blade of a wind turbine intermittently interferes with sunlight as it rotates, causing intermittent periods of light and shadow, "stresses the animals, and the birds and bees leave."[51]

This same shadow effect has measurable impacts on mental, physical, and ecological health. In communities that are undertaking territorial defense against wind projects, residents are disrupted by the impact of the vast turbines on natural light availability, contending that "the shadow effect is something aberrant to health. Imagine being with constant power cuts,[52] it turns on and off every four seconds, that does a lot of damage."[53] Through the shadow effect, the visual impact of wind turbines, and their associated health risks, are found to alter daily life, both at the family and community level.[54] In addition, noise and vibrations are generated, which also affect people's health, especially at night, as they directly impact the quality of sleep.

---

48 Interview with resident of El Agro, commune of Negrete, September, 2023.

49 More information on this case available through the following source: Contreras Jara, Juan. "AES Chile Pretende Desalojar con Carabineros Camino Interior Resguardado por Agricultores/as Pese a Haber Inundado sus Cultivos e Intervenido Canal de Regadío." *Resumen*, July 4, 2022. https://resumen.cl/articulos/aes-chile-pretende-desalojar-con-carabineros-camino-interior-resguardado-por-agricultores-as-pese-a-haber-inundado-sus-cultivos-e-intervenido-canal-de-regadio.

50 In the area of electric energy business, easements have become a turning point, especially in rural areas. In the case of electric concessions, the easement strips correspond to the land where the company will have the right to transit and through which an electric line would pass. This element can be explored in more detail in the following paper: Observatorio Latinoamericano de Conflictos Ambientales (OLCA). "Una Linterna Para El Oscuro Mundo de Las Concesiones Eléctricas." (Santiago, Chile: OLCA, April 2023): 1–20. https://olca.cl/oca/informes/Linterna.pdf.

51 Interview with resident of El Ciruelo Sur, commune of Los Angeles, January and December, 2023.

52 The term "power outages" refers to periods without sunlight due to the alternation of light and shadow generated by the movement of the wind tower blades (shadow effect or flickering shadow).

53 Interview with resident of El Ciruelo Sur, commune of Los Angeles, January and December, 2023.

54 García-Bernarl, N. "Impacto Ambiental de Aerogeneradores: Campos Eólicos Situados En Tierra." *Asesoría Técnica Parlamentaria* (April 2021): 1–10.

A resident in Negrete who lives 300 meters from a wind turbine illustrates the impact of the noise and shadow effect on nearby families: "The towers arrived and we began to suffer the noise, the flickering shadow effect, the stress on the animals—because they interpret it as a threat—and milk production decreased, the bees began to die as soon as the wind towers began to operate [...] there are days when you cannot sleep, you cannot stand the noise. It starts to hurt your head, your ears."[55]

This threat is further deepened with a large pipeline of projects in the works: "At least ten of these projects propose the installation of more than 500 wind turbines [wind towers] in the province of Biobío. These projects arouse high rejection in the community because they strongly damage rural landscapes [...] they produce annoying noise and constant vibrations, as well as intermittent shadows on the houses that alter daily life."[56]

These impacts are a true reflection of the discourse of renewable energies and the so-called "just transition." However, as one inhabitant of El Ciruelo Sur states, "this is not fair, it cannot be normalized in rural or urban areas."[57] A resident of El Agro, a commune of Negrete, echoes this, declaring that "there is no respect for anything, our human rights are being violated by placing a tower 300 meters from my home, from my father's home, from that of other neighbors who were not even asked if they wanted it."[58]

As companies establishing wind projects are faced with the opposition coming from within communities, one resident considers it "a shame how they present the projects, no one cares to make a real study that includes the communities and their state [...], we see how our local autonomy is being undermined; the landscape value and culture are lost living an injustice because the authorities do not care that the population has a welfare."[59]

However, despite the harm caused, the resistance and defense of the territories has continued. This is a defense in which "our idea is to be able to fight against these companies so that they do not come to harm us and let us live, as we lived naturally here from the animals, the milk, the cheese, the maqui berries, the potato, that is how we live here."[60] In this way, the communities continue advocating for the defense of the land, spreading awareness in various spaces about the impacts that are often unseen and unspoken in these so-called "clean" energy projects. Awareness and advocacy campaigns continue, in these territories, in international spaces, and through social networks. Protest activities and actions also continue, as well as spaces for meetings and exchanging experiences, which nourish and provide the necessary strength to resist these strong winds that blow through Biobío.

55 Interview with with resident of El Agro, commune of Negrete, September, 2023.
56 Fundación Pongo. "Invasión de Máquinas Eólicas." In *Movimientos del FioFío: De la Costa a la Cordillera.* (Región del Biobío, Chile: Fundación Pongo, 2023): 8.
57 Interview with resident of El Ciruelo Sur, commune of Los Angeles, January and December, 2023.
58 Interview with resident of El Agro, commune of Negrete, September, 2023.
59 Interview with resident of El Ciruelo Sur, commune of Los Angeles, January and December, 2023.
60 Interview with resident of the municipality of Arauco, January, 2023.

Projections

The winds blowing in the Biobío do not bring changes, neither for the region nor for the country, but rather continue to deepen an economic model that generates serious territorial impacts, altering the way of life of the communities, destroying ecosystems and undermining local economies, all as part of a strategy that seeks to impose a hegemonic economic activity. All of this is supported by the State and by public and private institutions to the detriment of the local population.

On the other hand, the same countries that are the main causes of climate change and the ecological crisis are today at the forefront of the so-called energy transition. Today, only ten countries are responsible for almost 70% of CO_2 emissions worldwide[61] and although countries like Chile contribute very little to CO_2 emissions, it is also considered one of the most vulnerable to the impacts of the climate crisis. Thus, to achieve their own "transition," but without changing their ways of life and consumption, the countries of the global north impose mega-projects in the global south under a cloak of "green energy," ignoring local realities and deepening extractivism in our territories. It is precisely this development model, with its dependence on fossil fuels, intensive exploitation of nature, and unlimited economic growth, that has led us to the current ecological crisis. The transition proposed so far is not addressing the underlying problem. On the contrary, the historical asymmetries between the global north and south continue to be perpetuated, as well as the internal inequalities in each country, where it is precisely the impoverished sectors who are most affected by the invasion of energy projects. All this could be summarized in the following phrase: an energy transition for some at the cost of sacrificing the lives of many territories.

All this has been framed under the argument of the climate crisis and the need to decarbonize, which is an undeniable urgency, but in practice what we see is an energy expansion and not a transition. An expansion that seeks to maintain the model of growth and capitalist accumulation. In this sense, it is not enough to increasingly incorporate renewable energy into the energy matrix of our countries; it is necessary to reduce energy consumption, especially by the sectors that consume more energy, such as the industrial, mining, and transportation sectors. On the other hand, it should be the territories themselves who decide what type of transition they need, according to their local ways of life, while respecting the quality of life and health of people and Nature.

Today, more than ever, it is important to ask ourselves for whom and for what is the energy produced? Is it for the rural communities that are living day by day the impacts of wind turbines? Is it the indispensable energy for a just and dignified life? Or is it a new business of national and transnational corporations, but this time measuring their carbon footprint? Is this the transition we need? Is it possible to imagine a different form of development, a development that is in harmony with ecosystems and that responds to the urgency of the climate crisis? These questions invite us to think about the direction we want to move in as a society.

---

61 Ritchie, Hannah, and Max Roser. "CO_2 Emissions." Published online at *Our World in Data*, June 10, 2020. https://ourworldindata.org/co2-emissions.

We need to move toward a transition with justice, where the rights of local communities and workers, the wellbeing of ecosystems, people's health, and local economies are considered. A transition where communities are key actors in decision making, project planning, and implementation. A transition that is not based on the export of energy, but focused on solving local needs without compromising the health of people and Nature. For this it is necessary to break with the capitalist idea of development and with the prevailing extractivism in our countries, it is necessary to look toward new possibilities, toward post-extractivist ideas, which abandon the predatory and exporting logic of nature, where it is nothing more than a commodity. Only a new way of seeing and understanding the world that surrounds us and of which we are a part will allow us to truly advance toward a socio-ecological transition.

Currently, in different parts of the world there are experiences of energy development at the community level, which, through cooperatives,[62] community networks, neighborhood councils, schools or universities, have managed not only to generate electricity to supply themselves, but also to generate alternative proposals with other ways of generating and conceiving energy, such as, for example, that the energy comes from NCRE sources, is small-scale, collectively owned, with low environmental impact, according to the needs of the community, and contextualized to the local culture and ecosystem.

Some of these experiences are: the solar energy project in rural territories in Mali, coordinated by a women's organization in Sirakorola, where through the installation of solar panels they produce energy for homes and agriculture;[63] the solar energy project for a community nursery in Colombia, implemented by the Collective of Peasant Community Reserves of Santander (CRCCS) and the Municipal Association of Peasant Women of Lebrija (AMMUCALE), focused on enhancing the work done by women in the territory; the Solar Bakery in a rural area in northeastern Brazil, which is managed by the Associação de Mulheres do Grupo Bolo das Oliveiras;[64] and the case of the "Popular Energy" project of the Municipality of Recoleta in Chile, which consists of installing solar panels in homes in the municipality and using the surplus energy in other buildings of the Municipality.[65] Although these experiences are still few in Chile, they represent a very important step toward a transition with justice, since they are bets that put life, people's lives, their local practices, and ecosystems at the center.

The invitation is to continue reflecting on and asking ourselves all the necessary questions, always remembering that the answers lie at the heart of the communities, where nature and society intertwine to write a different story.

---

62 A cooperative is an association of people who unite under the idea of mutual support. In Chile there are worker, service, consumer, peasant, and fishing cooperatives.

63 Other initiatives present in the article by Eduardo Blanco: "Diez Experiencias Mundiales de Uso Comunitario de las Energías Renovables" *Observatorio Petrolero Sur*, June 14, 2020. https://opsur.org.ar/2020/06/14/diez-experiencias-mundiales-de-uso-comunitario-de-las-energias-renovables/.

64 Cases present in Furtado, Flávio, and Edson Soldateli. *Energy in Latin America: From Business to Common.* (Berlin: Rosa Luxemburg Foundation, 2019).

65 CNN Chile. "'Energía Popular': Recoleta Impulsa Plan con Paneles Solares para que Vecinos No Paguen Cuentas de la Luz." *CNN Chile*, December 9, 2019. https://www.cnnchile.com/pais/energia-popular-recoleta-paneles-solares-cuentas-luz_20191209/.

BIBLIOGRAPHY

Alianza Hidrógeno Verde. "Región del Biobío." Gobierno Regional I Region del Biobio. *Alianza H2V Biobío*. Accessed June 4, 2025. https://www.ah2vbiobio.cl/region-del-biobio/.

Alianza Hidrógeno Verde. "UdeC Lanza Proyecto Que Busca Desplegar El Potencial En H2 Verde de La Región Financiado Por El Gore Biobío." Gobierno Regional I Region del Biobio. *Alianza H2V Biobío*. May 7, 2021. https://www.ah2vbiobio.cl/evento/udec-lanza-proyecto-h2-verde/.

Ardiles, Karen, et al. Women and Environmental Conflicts: Differentiated Impacts. Latin American Observatory of Environmental Conflicts (OLCA). 2023.

Arroyo Olea, Javier. "Acusan que en Presencia de Carabineros se Destruyó Centro de Resguardo de Camino Rural en Las Trancas para Favorecer Proyecto Eólico." *Resumen*. September 7, 2022. https://resumen.cl/articulos/acusan-que-carabineros-destruyo-centro-de-resguardo-de-camino-rural-en-las-trancas-para-favorecer-proyecto-eolico.

Arroyo Olea, Javier. "'Antípodas': El Bochornoso Proyecto Energético Presentado Por Piñera Ante La APEC Que Nunca Fue Ejecutado." Latin American Observatory of Environmental Conflicts, December 14, 2023. https://olca.cl/articulo/nota.php?id=110424.

Blanco, Eduardo. "Diez Experiencias Mundiales de Uso Comunitario de las Energías Renovables." Observatorio Petrolero Sur. June 14, 2020. https://opsur.org.ar/2020/06/14/diez-experiencias-mundiales-de-uso-comunitario-de-las-energias-renovables/.

CNN Chile. "'Energía Popular': Recoleta Impulsa Plan con Paneles Solares para que Vecinos No Paguen Cuentas de la Luz." December 9, 2019. https://www.cnnchile.com/pais/energia-popular-recoleta-paneles-solares-cuentas-luz_20191209/.

Comisión Nacional de Energía. "Guía Para La Evaluación Ambiental de Proyectos Eólicos En Chile." Santiago, Chile: Ministerio de Energía, October, 2006. https://energia.gob.cl/documentos/proyectos-eolicos.

Comisión Nacional de Energía and Gobierno de Chile Ministerio de Energía. "Anuario Estadístico de Energía 2021: Balance Nacional de Energía." Santiago, Chile: Comisión Nacional de Energía, Ministerio de Energía, Gobierno de Chile, 2022. https://www.cne.cl/wp-content/uploads/2022/07/AnuarioEstadisticoEnergia2021.pdf.

Contreras Jara, Juan. "AES Chile Pretende Desalojar con Carabineros Camino Interior Resguardado por Agricultores/as Pese a Haber Inundado sus Cultivos e Intervenido Canal de Regadío I Resumen.cl." *Resumen*. July 4, 2022. https://resumen.cl/articulos/aes-chile-pretende-desalojar-con-carabineros-camino-interior-resguardado-por-agricultores-as-pese-a-haber-inundado-sus-cultivos-e-intervenido-canal-de-regadio.

División de Desarrollo de Proyectos Unidad de Acompañamiento de Proyectos. "Reporte de Proyectos En Construcción e Inversión Del Sector Energía, Mes de Agosto de 2023." Santiago, Chile: Ministerio de Energía, Gobierno de Chile, August 2023. https://energia.gob.cl/sites/default/files/documentos/reporte_de_proyectos_-_agosto_2023.pdf.

Energy Partnership Chile-Germany. "Facts & Figures." Accessed June 4, 2025. https://energypartnership.cl/energy-facts/?utm_source=chatgpt.com.

Fundación Pongo. "Invasión de Máquinas Eólicas." In Movimientos del FioFío: De la Costa a la Cordillera. Región del Biobío, Chile: Fundación Pongo, 2023.

Furtado, Flávio, and Edson Soldateli. *Energy in Latin America: From Business to Common*. Berlin: Rosa Luxemburg Foundation, 2019.

García-Bernarl, N. "Impacto Ambiental de Aerogeneradores: Campos Eólicos Situados En Tierra." *Asesoría Técnica Parlamentaria (April 2021): 1–10.*

International Energy Agency. "Chile – Countries & Regions." Accessed June 4, 2025. https://www.iea.org/countries/chile/efficiency-demand.

Ministerio de Energía. "Anuario Estadístico de Energía 2022." Santiago, Chile: Ministerio de Energía, Gobierno de Chile, 2022.

Ministerio de Energía. "Anuario Estadístico de Energía 2023." Santiago, Chile: Ministerio de Energía. Gobierno de Chile, 2023.

Ministerio de Energía. "Biobío Se Convertirá En La Segunda Región Con Más Generación de Energía Eólica." Ministerio de Energía. February 2, 2016. https://energia.gob.cl/noticias/biobio/biobio-se-convertira-en-la-segunda-region-con-mas-generacion-de-energia-eolica.

Ministerio de Energía. "Chile Tendrá El Parque Eólico Más Grande de Latinoamerica." Ministerio de Energía. September 27, 2017. https://energia.gob.cl/noticias/nacional/chile-tendra-el-parque-eolico-mas-grande-de-latinoamerica.

Ministerio de Energía. "Los 10 Parques Eólicos Más Grandes de Chile." Ministerio de Energía. October 24, 2017. https://energia.gob.cl/noticias/nacional/los-10-parques-eolicos-mas-grandes-de-chile.

Ministerio de Energía. "Ministro Jobet Encabezó Anuncio Del Parque Eólico Más Grande de América Latina." Ministerio de Energía. September 21, 2021. https://energia.gob.cl/noticias/nacional/ministro-jobet-encabezo-anuncio-del-parque-eolico-mas-grande-de-america-latina.

Ministerio de Energía. "Nueva Licitación de Terrenos Fiscales Para Energía Eólica." Ministerio de Energía. November 25, 2016. https://energia.gob.cl/noticias/antofagasta/nueva-licitacion-de-terrenos-fiscales-para-energia-eolica.

Moreno Caro, Branco. "Dirigenta Socioambiental Denuncia Hostigamientos en Medio de Conflicto entre Comunidades Rurales de Los Ángeles y AES Andes." *Resumen.* April 12, 2023. https://resumen.cl/articulos/dirigenta-socioambiental-denuncia-hostigamientos-en-medio-de-conflicto-entre-comunidades-rurales-de-los-angeles-y-aes-andes.

Moreno Caro, Branco. "Dirigenta Socioambiental Denuncia Hostigamientos en Medio de Conflicto entre Comunidades Rurales de Los Ángeles y AES Andes." *Resumen.* Accessed June 4, 2025. https://resumen.cl/articulos/dirigenta-socioambiental-denuncia-hostigamientos-en-medio-de-conflicto-entre-comunidades-rurales-de-los-angeles-y-aes-andes.

Observatorio Latinoamericano de Conflictos Ambientales (OLCA). "Una Linterna Para El Oscuro Mundo de Las Concesiones Eléctricas." Santiago, Chile: OLCA, April 2023. https://olca.cl/oca/informes/Linterna.pdf.

Observatorio Petrolero Sur. "Chile | OLCA Lanza Instructivo Para Orientar La Gestión Comunitaria Ante Fiebre Energética." April 25, 2023. https://opsur.org.ar/2023/04/25/chile-olca-lanza-instructivo-para-orientar-la-gestion-comunitaria-ante-fiebre-energetica/.

Ochoa, Javier. "Región Del Biobío Tendrá Este Año 16 Parques Eólicos En Operación." *Diario Concepción* March 18, 2022. https://www.diarioconcepcion.cl/economia/2022/03/18/region-del-biobio-tendra-este-ano-16-parques-eolicos-en-operacion.html.

Osorio-Aravena, Juan Carlos, Lorenzo Reyes-Chamorro, Jannik Haas, Jorge Aguilera, Rodrigo Palma-Behnke, and Christian Breyer. "Insights for Informing Energy Transition Policies – Are Decision Makers Listening to Science? The Case of Chile." *Energy Strategy Reviews* 58 (March, 2025): 101644. https://doi.org/10.1016/j.esr.2025.101644.

Ritchie, Hannah, and Max Roser. "CO_2 Emissions." *Our World in Data*, June 10, 2020. https://ourworldindata.org/co2-emissions.

Interviews

Resident of El Agro sector, commune of Negrete. In person, September 2023.

Resident of Comunidad Ciruelo Sur, commune of Los Ángeles. In person, between January and December 2023.

Resident of Community of Tralca, Municipality of Arauco. In person, January 2023.

Resident of the Arauco commune. In person, January 2023.

IMAGE CITATIONS + CREDITS

Listed by Page Number

54 Photo: Javier Olea Arroyo.
Map sources: Esri, Maxar, Earthstar Geographies, and the GIS User Community.

56 Map sources: Leah Temper, Daniela del Bene and Joan Martinez-Alier. 2015. Mapping the frontiers and front lines of global environmental justice: the EJAtlas. Journal of Political Ecology 22: 255-278. https://journals.librarypublishing.arizona.edu/jpe/article/id/1932/. Olson, D. M. and E. Dinerstein. 2002. The Global 200: Priority ecoregions for global conservation. (PDF file) Annals of the Missouri Botanical Garden 89:125-126. The Nature Conservancy, USDA Forest Service and U.S. Geological Survey, based on Bailey, Robert G. 1995. Description of the ecoregions of the United States (2nd ed.). Misc. Pub. No. 1391, Map scale 1:7,500,000. USDA Forest Service. 108pp. The Nature Conservancy (2003), based on Wiken, E.B. (compiler). 1986. Terrestrial ecozones of Canada. Ecological Land Classification Series No. 19. Environment Canada, Hull, Que. 26 pp. + map.Esri, USGS.

58 Map sources: "Add/Edit GEO Systems," accessed April 25, 2025, https://globalenergyobservatory.org/select.php?tgl=Edit; "Argentina Guarantees Natural Gas Supply to Chile Through 2024 as Vaca Muerta Shale Takes Off," September 11, 2023, https://naturalgasintel.com/news/argentina-guarantees-natural-gas-supply-to-chile-through-2024-as-vaca-muerta-shale-takes-off/; "Brazil Crude Oil: Imports, 1980–2024 | CEIC Data," accessed April 25, 2025, https://www.ceicdata.com/en/indicator/brazil/crude-oil-imports; "Chile Geoportal," accessed May 2, 2025, https://www.geoportal.cl/geoportal/catalog/35541/Industria%20Forestal%202016; "Colombia Coal Exports Keep Rebounding | Hellenic Shipping News Worldwide," accessed April 25, 2025, https://www.hellenicshippingnews.com/colombia-coal-exports-keep-rebounding/; "Colombian Coal Exports | The Coal Hub," accessed April 25, 2025, https://thecoalhub.com/colombian-coal-exports-2.html; "Colombian Rail Blockade Halts All Cerrejon Coal Shipments | Montel," Montel, n.d., https://www.montelnews.com/news/1194134/colombian-rail-blockade-halts-all-cerrejon-coal-shipments-; Editor, "Colombian Coal Exports | The Coal Hub," May 15, 2024, https://thecoalhub.com/colombian-coal-exports-2.html; "Energy Industry in Chile," accessed April 25, 2025, https://aenert.com/countries/america/energy-industry-in-chile/; Javier Treboux, "Low Level of Paraná River to Cost Near US$ 315 Million over 6-Month Period to Argentinian Agribusiness Export Complex and Farmers," Bolsa de Comercio de Rosario, June 22, 2021, http://www.bcr.com.ar/es/mercados/mercado-de-granos/news/low-level-parana-river-cost-near-us-315-million-over-6-month-period; "IDE ENERGÍA," accessed May 2, 2025, https://ide-energia.minenergia.cl/portal/apps/webappviewer/index.html?id=5c526a138b1449458e0667b2235d2b19; IEA, "Chile - Countries & Regions," accessed April 25, 2025, https://www.iea.org/countries/chile/oil; INDH, "Conflictos mediosambientales INDH," Mapa de conflictos, accessed May 2, 2025, https://mapaconflictos.indh.cl/#/; "International - U.S. Energy Information Administration (EIA)," accessed April 25, 2025, https://www.eia.gov/international/overview/country/CHL

61 Photo: Javier Olea Arroyo.
Map sources: Neil N. Davis, Jake Badger, Andrea N. Hahmann, Brian O. Hansen, Niels G. Mortensen, Mark Kelly, Xiaoli G. Larsén, Bjarke T. Olsen, Rogier Floors, Gil Lizcano, Pau Casso, Oriol Lacave, Albert Bosch, Ides Bauwens, Oliver James Knight, Albertine Potter van Loon, Rachel Fox, Tigran Parvanyan, Søren Bo Krohn Hansen, Duncan Heathfield, Marko Onninen, Ray Drummond; The Global Wind Atlas: A high-resolution dataset of climatologies and associated web-based application; Bulletin of the American Meteorological Society, Volume 104: Issue 8, Pages E1507-E1525, August 2023, DOI: https://doi.org/10.1175/BAMS-D-21-0075.1; Data obtained from the Global Wind Atlas version 3.3, a free, web-based application developed, owned and operated by the Technical University of Denmark (DTU). The Global Wind Atlas version 3.3 is released in partnership with the World Bank Group, utilizing data provided by Vortex, using funding provided by the Energy Sector Management Assistance Program (ESMAP). For additional information: https://globalwindatlas.info; Global Energy Monitor. "Global Wind Power Tracker," February 11, 2025. https://globalenergymonitor.org/projects/global-wind-power-tracker/. Esri, NASA, NGA, USGS.

68 Map sources: "Geoportal de Chile." Accessed May 6, 2025. https://www.geoportal.cl/geoportal/catalog/36419/SEIA%20Central%20E%C3%B3lica. Neil N. Davis, Jake Badger, Andrea N. Hahmann, Brian O. Hansen, Niels G. Mortensen, Mark Kelly, Xiaoli G. Larsén, Bjarke T. Olsen, Rogier Floors, Gil Lizcano, Pau Casso, Oriol Lacave, Albert Bosch, Ides Bauwens, Oliver James Knight, Albertine Potter van Loon, Rachel Fox, Tigran Parvanyan, Søren Bo Krohn Hansen, Duncan Heathfield, Marko Onninen, Ray Drummond; The Global Wind Atlas: A high-resolution dataset of climatologies and associated web-based application; Bulletin of the American Meteorological Society, Volume 104: Issue 8, Pages E1507-E1525, August 2023, DOI: https://doi.org/10.1175/BAMS-D-21-0075.1; Data obtained from the Global Wind Atlas version 3.3, a free, web-based application developed, owned and operated by the Technical University of Denmark (DTU). The Global Wind Atlas version 3.3 is released in partnership with the World Bank Group, utilizing data provided by Vortex, using funding provided by the Energy Sector Management Assistance Program (ESMAP). For additional information: https://globalwindatlas.info; Global Energy Monitor. "Global Wind Power Tracker," February 11, 2025. https://globalenergymonitor.org/projects/global-wind-power-tracker/.

76 Photo: Javier Olea Arroyo.

ALUMINUM
Smelting with Iceland's Melting Glaciers

by Theodore Teichman + Daniel Carmelo

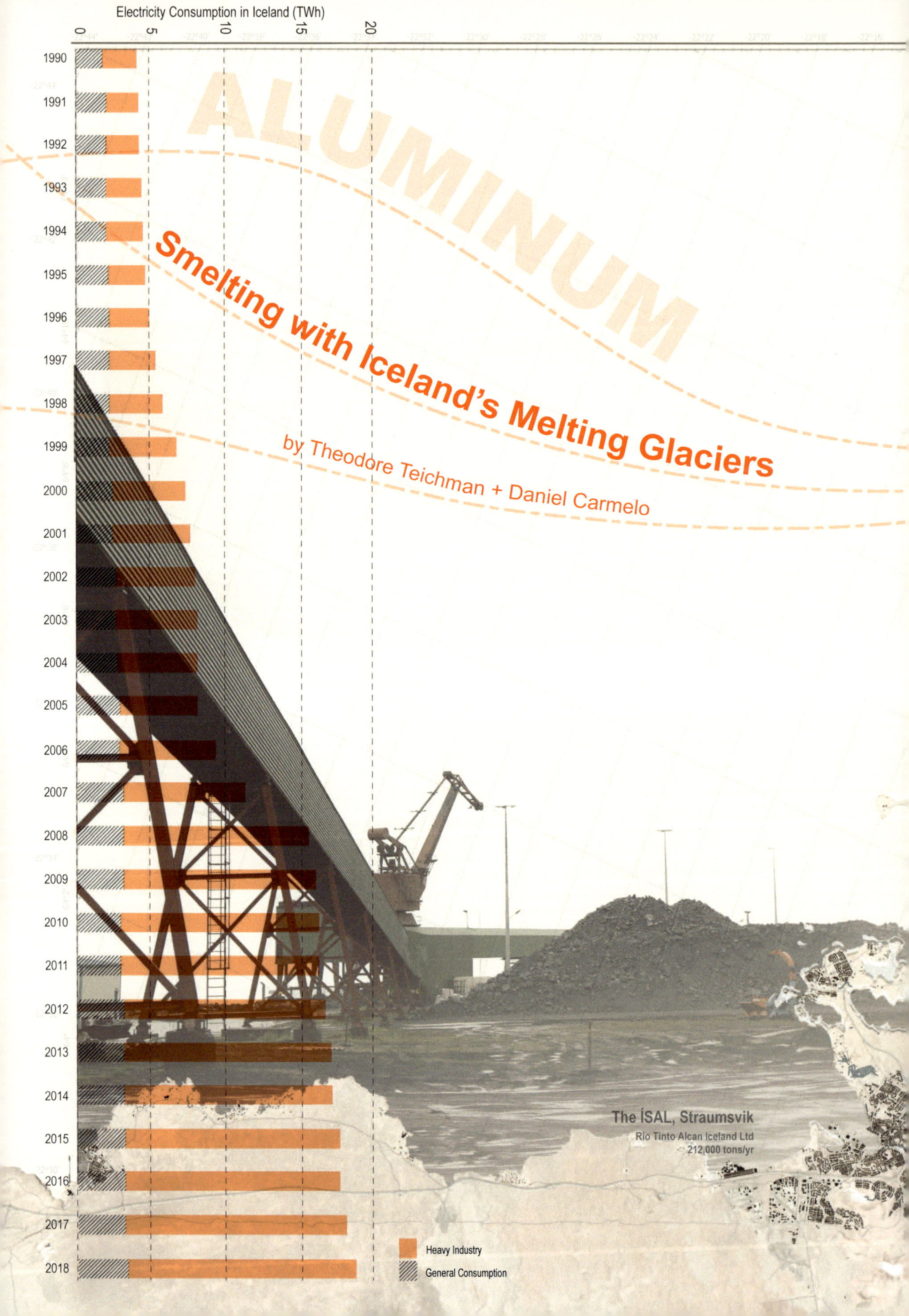

Akrane
Hvalfjörður
Norðurál, Grundartangi
Century Aluminum
317,000 tons/yr
REYKJAVIK

Green Energy Transition(s), or: Aluminum Dreams

The year is 1998, the beginning of a new epoch: "global warming" has entered the zeitgeist—buzzing with a growing swarm of green technologies, materials, and industries—and Iceland is trying to capitalize on the demand for green.[1] The national power company of Iceland, Landsvirkjun, is speculating on extensive power plant construction to fuel new aluminum smelters with green energy.[2] The place where this all began is merely 20 minutes from where I lived in 2021. Weekly, I drove past Grundartangi, the site of the smelter that established Iceland as a green energy economy.[3] Flash forward to 2011, and the aluminum bubble in Iceland has burst, dead in the northern fishing town of Húsavík where Alcoa has just canceled long-held plans for a smelter claiming the already heavily subsidized electricity costs were too high.[4] In between are ten years of conflict, ecological damage, and terraforming on a massive scale.[5] While the aluminum

1 Iceland's first smelter was the Alcan (now Rio Tinto Alcan) smelter built in Straumsvik, just south of Reykjavik. Construction began in 1966 and the smelter opened in 1969. In the 1960s, the Ministry of Industry's newly formed "Industrial Development Committee" was specifically tasked with finding ways to sell (and export) Iceland's robust low-cost electricity. This is what drew the first smelter and what drew the subsequent ones as well.

2 Pálmason, Sigurður Gísli. *Dreamland* (*Draumalandið*). Directed by: Þorfinnur Guðnason, Andri Snær Magnason. Based on the book: *Dreamland: A Self-Help Manual for a Frightened Nation* by Andri Snær Magnason. Ground Control Productions, 2009. https://dreamland.is; Guðmundsdóttir, Hrönn, Wim Carton, Henner Busch, and Vasna Ramasar. "Modernist Dreams and Green Sagas: The Neoliberal Politics of Iceland's Renewable Energy Economy." *Environment and Planning E: Nature and Space* 1, no. 4 (2018): 590.

3 Pálmason, Sigurður Gísli. *Dreamland* (Draumalandið). I (Theodore) had my first opportunity to live in Iceland for a year between 2021 and 2022 to conduct a year of fieldwork concerning the cultural landscapes of ecological restoration, sponsored by a Fellowship from the Leifur Eiriksson Foundation and in association with the Agricultural University of Iceland. Daniel came first to Iceland in the summer of 2023 as part of the Wright-Ingram Institute's Field Stations program, learning about ecological and cultural change and resilience through Iceland's landscapes. Iceland has a close hold on us both as a place and as a network of friends and colleagues that keep us coming back.

4 While this year marked a symbolic end to that smelter and dam-building boom, it is also the moment when industrial consumption would rapidly expand (and continue to stay high, even without new facilities being constructed); Alcoa's proposed plant at Bakki was presented as innovative and renewable based on the proposed plan to run the plant off of a newly constructed geothermal power plant. However, it appeared that based on the proposed energy needs of the smelter, new hydroelectric plants would be constructed, even if not directly adjacent to the smelter, then as part of the larger grid to handle increased demand; "Alcoa Drops Iceland Smelter Plan I Reuters." Reuters, October 17, 2011. https://www.reuters.com/article/alcoa-iceland-smelter-idUSL5E7LH3MR20111017/; Sheller, Mimi. *Aluminum Dreams: The Making of Light Modernity.* MIT press, 2014: 209 ; Guðmundsdóttir, Hrönn, Wim Carton, Henner Busch, and Vasna Ramasar. "Modernist Dreams and Green Sagas; Skúlason, Jón Björn, and Roger Hayter. "Industrial Location as a Bargain: Iceland and the Aluminium Multinationals 1962–1994." Geografiska Annaler: Series B, *Human Geography* 80, no. 1 (1998): 29–48 ; Breitschopf, Barbara, and Lin Zheng. Rep. "Electricity Costs of Energy Intensive Industries in Iceland: A Comparison with Energy Intensive Industries in Selected Countries". Department of Energy, Industry and Business affairs, Ministry of Industries and Innovation, Iceland. November 2020. 6–7; Krater, Jaap. "Bakki Impact Assessment Should Include Dams." *Saving Iceland,* August 22, 2008. https://www.savingiceland.org/2008/08/bakki-impact-asessment-should-include-dams/.

5 The Kárahnjúkar hydroelectric project is an infamous example and a site of massive public controversy in Iceland. In order to generate the promised electricity output, this hydropower plant required the creation of three immense reservoirs through the construction of five massive dams on the glacial rivers, Jökulsá á Dal and Jökulsá í Fljótsdal. These reservoirs not only flood valleys at higher elevations, but the process of damming changes water and sediment flow systems affecting wetland health downstream. By altering wetland systems both up and downstream of the dam, this act of terraformation has fragmented habitats; disrupted migration

fervor of the pre-2008 era has dissipated, aluminum production is likely to continue to surge. Smelters are the largest consumer of electricity in Iceland, and the small island nation is still the tenth largest exporter of aluminum in the world.[6] According to the Aluminum Association, the demand for sustainable transportation in the form of electric cars over the next decade will increase demand for aluminum: infinitely recyclable, lightweight, strong and cheap.[7]

---

routes and nesting grounds; and reduced the ecological health of the wetland soil system, releasing large quantities of greenhouse gasses from the earth. The environmental impacts of smelting in Iceland extend beyond the consequences of earthworks for dam-building. The smelters themselves, while operating on green energy, are still sources of chemical pollution, such as the gaseous releases of excess fluoride that affected hay and sheep. These releases provoked outcry from local farmers outraged about their land and sheep being exposed to toxic chemicals and worried about the risks of living in the windshed of an industrial facility. Beyond the ecological consequences, this project was seen as destructive of the landscape because it altered and destroyed unique geological features and a landscape significant to the environmental imaginary, attracting attention and support locally and internationally. Even the Icelandic Planning Agency ruled (at the time) that the project should not be conducted because Landsvirkjun's Environmental Impact Assessment had too little environmental data and the project would have limited economic benefits especially compared to the irreversible negative impact on the natural environment. Likewise, environmental nongovernmental organisations (NGOs) compiled their own set of environmental data about the negative impacts of the project on the environment in a legally protected conservation area. In a controversial decision, Iceland's Minister of the Environment, Siv Friðleifsdóttir removed protections for the area to be flooded and overturned the Icelandic Planning Agency's decision to halt the project. Not only was this project controversial within the halls of government, but it became a major cultural force within Iceland that garnered international attention. Massive protests (between 8,000–13,000 people) were gathered in Reykjavik and a group of activists camped out on barricades at the construction site. Major Icelandic artists and musicians openly voiced their opposition and outrage at the project and released songs in cultural protest, such as Björk's "Náttúra", Valgeir Sigurðsson's "Grýlukvæði", and Sigur Rós's "Vaka". The band Sigur Rós even staged a performance at the protest camp on the site of the dam's construction to demonstrate cultural resistance to the hydropower project. The conflict surrounding the construction of the Kárahnjúkar hydroelectric project remains prominent in the cultural consciousness, epitomizing the conflict between the socio-technical imaginary of green energy and the environmental imaginary of a thriving natural landscape. While lagging somewhat behind the pace of industrial investment in the 1990s and 2000s, Iceland has been developing a unique master planning process to balance energy infrastructure development and protection of natural and cultural sites ("The Master Plan for Nature Protection and Energy Utilization." *Rammaáætlun*. Accessed December 19, 2024. https://www.ramma.is/english/. The plan uniquely recognizes that energy infrastructures inherently are questions of landscape form and function. This plan and its process help us to envision what green energy landscapes of the future might look like and the institutional frameworks that might facilitate that transition and novel land management regime. However, the plan neglects to address the non-technical, political dimensions of land use change that are contingent on shared or divergent imaginaries of socio-technical and environmental form and function that any construction project will enlist in collaboration or conflict ; Rekow, Lea. "Hydropower in Iceland: A Review of the Kárahnjúkar Project. Wright Ingraham Institute. March 18, 2023. https://wright-ingraham.org/hydropower-in-iceland-a-review-of-the-karahnjukar-project/; Fontaine, Andie Sophia. "From Iceland – Landowners Demand End to Smelter Pollution." *The Reykjavik Grapevine*, September 23, 2014. https://grapevine.is/news/2014/09/23/landowners-demand-end-to-smelter-pollution/; Benediktsson, Karl. "Conflicting Imaginaries in the Energy Transition? Nature and Renewable Energy in Iceland." *Moravian Geographical Reports* 29, no. 2 (2021): 93; "The Master Plan for Nature Protection and Energy Utilization." Rammaáætlun. Accessed December 19, 2024. https://www.ramma.is/english/; Benediktsson, Karl, and Edda Ruth Hlín Waage. "Power Without Politics? Nature, Landscape and Renewable Energy in Iceland." *Annales de geographie*, no.6 (2020): 107–130. Cairn/Publilog, 2020.

6 And it is the largest export commodity of Iceland; "Raw Aluminium in Iceland." *The Observatory of Economic Complexity*. Accessed December 7, 2024. https://oec.world/en/profile/bilateral-product/raw-aluminium/reporter/isl; Breitschopf, Barbara, and Lin Zheng. Rep. "Electricity Costs of Energy Intensive Industries in Iceland." 36. https://www.stjornarradid.is/library/01--Frettatengt---myndir-og-skrar/ANR/ThKRG/Report%20Iceland-FINAL.pdf; "Raw Aluminium in Iceland." *The Observatory of Economic Complexity*. Accessed July 10, 2024. https://oec.world/en/profile/bilateral-product/raw-aluminium/reporter/isl.

7 In a life cycle analysis, Sujit reports that building lighter weight vehicles with aluminum, reducing vehicle mass by 25%, also reduced life cycle energy consumption and CO_2 emissions by 25%. However, 90% of this reduction is from the use phase—and thus unaffected by the lower CO_2 emissions of hydropower-fed aluminum smelting operations. Through historical analysis, Kawajiri and colleagues argue that merely substituting lightweight materials like aluminum for steel does not necessarily result in lighter vehicles. Over the past 30 years, consumer demand has driven an increase in vehicle weight. The authors argue that the reduction in greenhouse gasses from vehicle efficiency does not create large enough shifts in total greenhouse gas emissions, and we should prioritize overall reduction of greenhouse gas-intensive materials; Das, Sujit. "Life Cycle Energy and Environmental Assessment of Aluminum-Intensive Vehicle Design." *SAE International Journal of Materials and*

Global Bauxite Mines + Aluminum Smelting

Aluminum remains a key actor in the global green energy economy and the imaginary of a green future—etched into the landscapes of Iceland where these smelters forge together distant pasts and all too near futures.[8]

Smelting Landscapes, or: There Are No Mines

Iceland is a mineral economy, but there are no mines.[9] Instead, this economy lives in places like Grundartangi, set in a pristine-looking fjord.[10] I pass the road

---

Manufacturing 7, no. 3 (2014): 588-595; Kawajiri, Kotaro, Michio Kobayashi, and Kaito Sakamoto. "Lightweight Materials Equal Lightweight Greenhouse Gas Emissions?: A Historical Analysis of Greenhouse Gases of Vehicle Material Substitution." *Journal of Cleaner Production* 253 (2020): 119805; "Auto Aluminum Growth in an Era of Electrification." *The Aluminum Association*. Accessed December 19, 2024. https://www.aluminum.org/auto-aluminum-growth-era-electrification.

8 Benediktsson summarizes that an imaginary is "a common mindset that has been gradually articulated by a large collectivity." He quotes Jasanoff and Kim to define specifically socio-technical imaginaries as a "collectively imagined forms of social life and social order reflected in the design and fulfilment of nation-specific scientific and/or technological projects [that] at once describe attainable futures and prescribe futures that states believe ought to be attained". He further expands, quoting Peet and Watts that the environmental imaginary is "a way of imagining nature, including visions of those forms of social and individual practice which are ethically proper and morally right with regard to nature." Benediktsson argues that the landscapes of energy production in Iceland are indelibly tied. These landscapes are created by, and create, multiple, competing imaginaries that hybridize the socio-technical and environmental by their very function as infrastructural landscapes; Benediktsson, Karl. "Conflicting Imaginaries in the Energy Transition? Nature and Renewable Energy in Iceland." *Moravian Geographical Reports* 29, no. 2 (2021): 89–90; JASANOFF, S., KIM, S-H. (2009): "Containing the Atom: Sociotechnical Imaginaries and Nuclear Power in the United States and South Korea." Minerva 47, no. 2 (2009): 119–146; Peet, R., and M. Watts. "Liberation Ecology: Development, Sustainability, and Environment in an Age of Market Triumphalism." In Peet, R., and M. Watts (eds.), *Liberation Ecologies: Environment, Development, Social movements* (pp. 1–45). London, Routledge.

9 Iceland does not have many mineral deposits of note. This is due to the young geology of the island which has not had the time required to form minerals of commercial/industrial interest. There are some mineral resources, however, these are not being mined commercially. The exception is the pumice, which is mined directly from the volcanic mountain, Hekla; Pokornı, Richard, and Helgi Torfason. *Mineral Resources in Iceland Volume II: Ores*. Cambridge Scholars Publishing, 2024. https://pubs.usgs.gov/myb/vol3/2019/myb3-2019-iceland.pdf

10 Specifically it is called the "Grundartangi Eco-Industrial District." Industrial activity began at Grundartangi at the Icelandic Iron and Steel Association's smelter in 1979, which today produces silicon metal (which, according to the Grundartangi Development Association's website, is used in high efficiency home appliances, wind turbines, and electric cars). However, activity at Grundartangi entered the current "eco-industrial" epoch in the 1990s in step with the emergence of the "green economy." While early uses of hydropower for aluminum were driven by demand for the material, the 1990s marked a change in this paradigm. After 1995, this changed. The hydropower itself began to be seen as the resource to be cultivated and sold. Supply has begun manufacturing demand. Grundartangi is where, in 1998, this bubble began. This smelter—built by the American company, Century Aluminum—would kick off this era of aluminum in Iceland. While the aluminum smelting in Hvalfjordur only began in 1997, energy extraction in this site spans centuries. While Hvalfjordur's name literally translates to "whale fjord," it is not actually likely there were ever whales in the fjord itself. In fact, it was more likely named for the whaling station built at the far shore of the fjord. The fjord does not actually have whales, merely facilities the processing of them. Just the same, Hvalfjordur does not have aluminum itself, but just the facilities to process it. Today, the whaling station is still active and uses the whale oil as fuel for the whale ships themselves despite its economic unviability. Whaling was a circuitous—rather than circular—resource economy: whales are hunted so that whaling ships can have fuel to hunt whales. In this glacial valley, this theme is being repeated with but now with aluminum and hydropower. The whaling station, the only one in Iceland at the time, was built in 1948 on the remains of the U.S. military base built during the period of U.S. occupation of Iceland (with the British), from July 1941 to June 17, 1944 (when Iceland declared itself a republic, which was intended to forestall Iceland from being invaded by Germany). The former base was opportune for the whaling station as the American military had already constructed a bridge, barracks, and other buildings that could be adapted and reused for this new purpose. The whaling station was established at the site of the WWII era U.S. naval base in Hvalfjordur (using the same structures). The base was built in 1940 after first the British then the Americans seized Iceland as a protectorate to prevent the German military from claiming Iceland after annexing Denmark (which owned Iceland), due to Iceland's strategic position in the north atlantic. Following this, Iceland gained independence but was still held closely in the U.S. sphere of influence with a large military presence as a key NATO base buffering the U.S. from the USSR. This period however was marked by massive protests, first for Independence from Denmark and then to resist joining NATO and the construction of the U.S. military base. Iceland's aluminum industry and the US military industrial complex are intricately connected; for more information see the 2009 doc-

sign that marks the turn for the smelter Norðurál. The simplicity of the sign belies the extent of the processing complex at the end of this road. Hidden behind this entire landscape sprawls the vast constellation of global supply chains, terraforming infrastructures, and geologies of places far away in space and time. The smelter reveals itself in glimpses along my drive, unsettling my illusions of untouched nature. In the dark days of Icelandic winter, traces of the smelter shine in a halo of lights in the sky and reflect onto the water. I realize that the lights never turn off and begin to understand that aluminum smelters do not either. The pylons surging across the mountain line make me curious where all this energy comes from.[11] Smelting 4 kg of bauxite into merely 1 kg of aluminum requires 13—15 kWh, the same energy consumed in an hour by 12 average U.S. homes.[12] In Iceland, this immense energy demand is not a problem; it makes the aluminum industry desirable. Iceland has a surplus of energy, not embodied in mineral resources but in *potential energy* held in reserve as glacial ice waiting to thaw and flow downhill. Aluminum smelting translates this potential energy into a unit of exchange, liberating it from the confines of power lines and geography.[13] As aluminum, energy can be extracted from Iceland's glacial landscape and shipped around the world. While Iceland may not be mining the earth, it is mining its energy and the natural systems that produce it. In this paradigm, aluminum is merely the proxy.

Aluminum Histories, or: Economies of Infinite Growth

The first hydropower plant in Iceland was constructed in 1937 in Grímsnes municipality.[14] By 1960, Iceland had built five more plants, and its industrial authorities realized that since most domestic energy consumption was now satisfied by hydropower, they could expand into other markets.[15] In 1961, the Icelandic Ministry of Industry formed the Industrial Development Committee to find and bring large energy consumers to Iceland.[16] Their initial success was

umentary, *Draumalandið*; "The Beautiful Hvalfjörður." Your Friend in Reykjavik. April 7, 2023. https://yourfriend-inreykjavik.com/the-beautiful-hvalfjordur/; "History of the U.S. and Iceland." U.S. Embassy in Iceland. Accessed July 10, 2024. https://is.usembassy.gov/our-relationship/policy-history/io/; "Whaling in Iceland." Whale & Dolphin Conservation USA. Accessed May 29, 2024. https://us.whales.org/our-goals/stop-whaling/whaling-in-iceland/; Ćirić, Jelena. "Whaling Has Little Economic Impact on Iceland." *Iceland Review.* August 22, 2023. https://www.icelandreview.com/nature-travel/whaling-has-little-economic-impact-on-iceland/; Pálmason, Sigurður Gísli. *Dreamland (Draumalandið).*

11 The pylons themselves have aluminum conductors and cables that allow high voltages to travel far distances: in this case from hydropower plants up in the rocky highlands down to the smelter nestled between wetlands and the shore. Aluminum smelting is highly energy-intensive due to the stronger chemical bond between aluminum and oxygen compared to iron and oxygen. Consequently, a much greater amount of energy is necessary to break the bond and form the metal.

12 The Hall-Héroult process; "13 KW Solar Battery". SunWatts. Accessed July 10, 2024. https://sunwatts.com/13-kwh-solar-battery/; Kvande, Halvor, and Per Arne Drabløs. "The Aluminum Smelting Process and Innovative Alternative Technologies." *Journal of Occupational and Environmental Medicine* 56 (2014): S23–S32.

13 Sheller, Mimi. "Aluminum Dreams."

14 There was actually an earlier one constructed in the municipality of Hafnarfjörður in 1904. However, we are not counting that as the origin point as it was much smaller than the kind we see today, powering 15 houses and 4 street lamps.

15 "Landsvirkjun Power Stations." Landsvirkjun. Accessed December 19, 2024. https://www.landsvirkjun.com/powerstations.

16 Skulason, Jon Bjorn. "Foreign Investment and the Bargaining Process: Case Study of the Aluminum Industry in Iceland." University of Iceland, 1992: 6

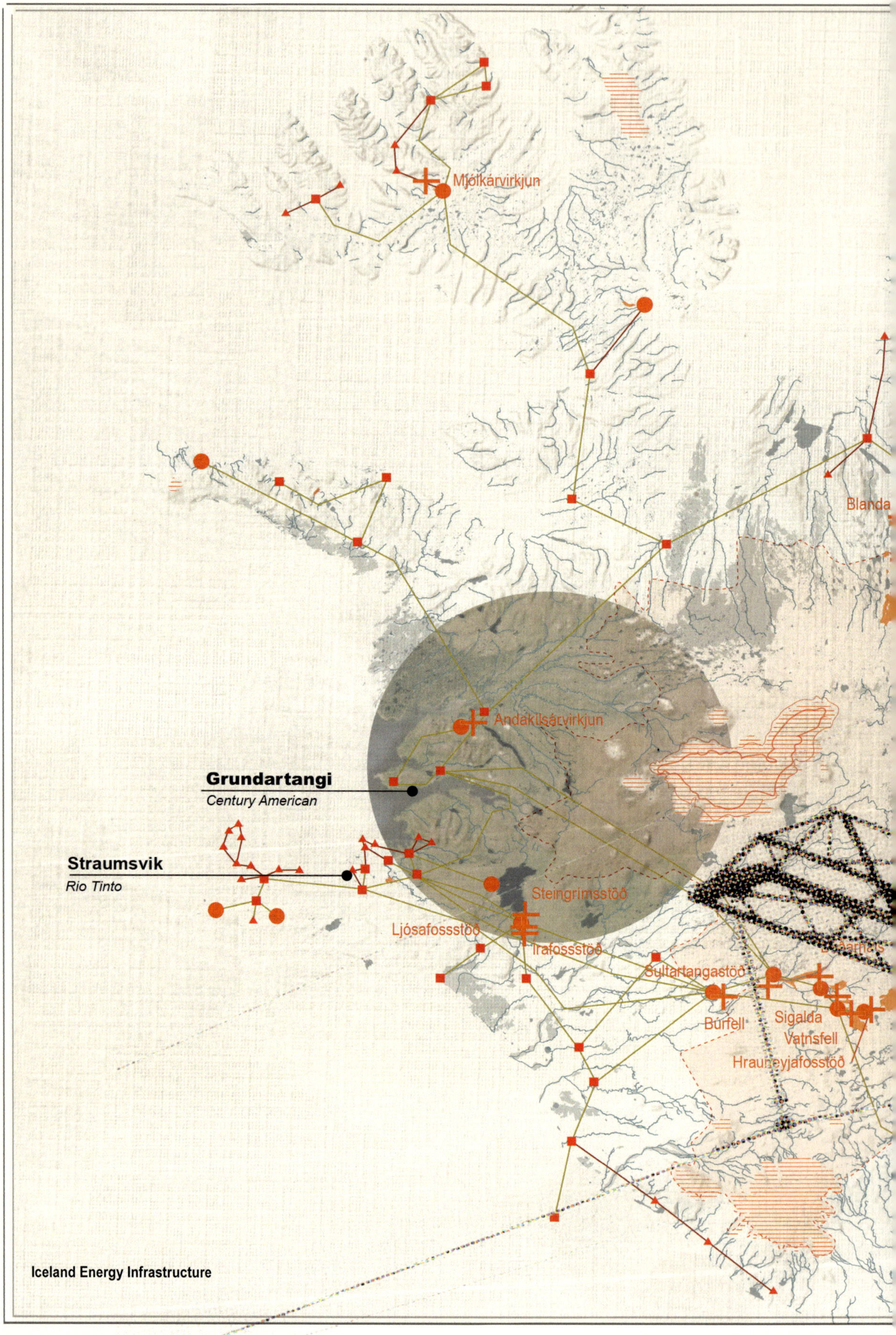
Mjólkárvirkjun
Blanda
Andakílsárvirkjun
Grundartangi
Century American
Straumsvik
Rio Tinto
Steingrímsstöð
Ljósafossstöð
Írafossstöð
Sultartangastöð
Búrfell
Sigalda
Vatnsfell
Hrauneyjafosstöð
Iceland Energy Infrastructure

Power Station
Main Substation
District Substation
Transmission Line
Distribution Line
Central Highlands
Reservoir
Glacier
2018 Snowline
1973 Snowline
Waterway
Wetland
Water
Aluminum Smelter
Hydroelectric Power Plant
Laxárstöðvar
Lagarfoss
Kárahnjúkar
Fjarðaál - Alcoa

Iceland's first smelter, the Alcan smelter in Straumsvik, which began operation in 1969 powered by a new hydropower plant at Búrfell.[17] Iceland's aluminum industry had taken hold. However, nearly thirty years passed between the construction of this smelter and two more in 1998 and 2007, the Century Aluminum smelter in Grundartangi and the Alcoa smelter in Reydarfjordur, respectively. This return of interest in aluminum processing was historically contingent. In a neoliberal experiment, Iceland sought to reduce high unemployment rates in the late 1980s and early 1990s through investments from multinational corporations.[18] In the 1990s, manufacturers began to replace steel car bodies with aluminum to reduce weight and improve efficiency, which drove demand for primary aluminum production.[19] Aluminum expansion projects became attractive and more profitable in the decade afterward because from 1990 to 2003 Icelandic legislation reduced the corporate tax from 50 percent to 18 percent in an effort to attract large industries to reduce the rise in unemployment.[20] Iceland became an aluminum economy because of cheap electricity in an expanding market and the government's ideological shift to neoliberalism.[21] While smelters in Iceland market themselves as carbon-friendly *replacements* to coal-powered aluminum production elsewhere in the world, global aluminum companies increased overall production instead.[22] As the consumer vehicle market grew, most fossil fuel-powered aluminum plants continued to operate and expand outside of Iceland. Between 2000 and 2019, the average carbon footprint of aluminum production actually increased by 38%, despite ostentatious investments in "low carbon" aluminum smelters in places like Iceland. In Iceland, the aluminum landscapes appear calm, pristine, and green. However, globally, from bauxite mine to aluminum ingot, 12–17 metric tons of CO_2-equivalents are released for each ton of aluminum.[23]

A Landscape of Pass-Through Places, or: The Hero's Journey

Stacks of shipping containers and industrial cranes line the Grundartangi deepwater port that allows heavy alumina-laden cargo ships access to the Norðurál smelter.[24] The aluminum processing complex is defined not by what

17 The smelter is called ISAL. The company is now Rio Tinto Alcan; "Isal." Rio Tinto. Accessed December 19, 2024. https://www.riotinto.com/en/operations/iceland/isal.

18 Skulason, Jon Bjorn. "Foreign Investment and the Bargaining Process: Case Study of the Aluminum Industry in Iceland." University of Iceland, 1992: 7.

19 Fridlyander, I. N., V. G. Sister, O. E. Grushko, V. V. Berstenev, L. M. Sheveleva, and L. A. Ivanova. "Aluminum Alloys: Promising Materials in the Automotive Industry." Metal Science and Heat Treatment 44, no. 9 (2002): 365–370.

20 Ólafsson, Stefán. "Welfare Trends of the 1990s in Iceland." *Scandinavian Journal of Public Health* 31, no. 6 (2003): 401–404.

21 Benediktsson, Karl. "Conflicting Imaginaries in the Energy Transition? Nature and Renewable Energy in Iceland." *Moravian Geographical Reports* 29, no. 2 (2021): 93.

22 Saevarsdottir, Gudrun, Thordur Magnusson, and Halvor Kvande. "Reducing the Carbon Footprint: Primary Production of Aluminum and Silicon with Changing Energy Systems." *Journal of Sustainable Metallurgy* 7, no. 3 (2021): 848–857.

23 Saevarsdottir et al. "Reducing the Carbon Footprint." 848-857

24 The port at Grundartangi was constructed in 1977 and is now one of the largest ports in the country. The largest ships that come to this port are those that unload alumina to the Norðurál smelter.

Jamalco Bauxite Mine, Jamaica
JAMALCO
Port Esquivel
Port Kaiser
Grundartangi
First shipment of bauxite 1963
First shipment of alumina 1972
1973
482,000 tonnes 1980
471,000 tonnes 1981
237,000 tonnes 1982
420,000 tonnes 1983
1984
1986
1989
1991
1998
1999
Capacity increased from 500,000 tonnes to 1,250,000 tonnes per year 2003
Capacity increased to 1,400,000 tonnes per year 2007
2009
2019
2023
17°54'30"N
17°54'N
17°53'30"N
17°53'N
17°52'30"N
17°52'N
17°51'30"N
17°51'N
77°16'30"W
77°16'W
77°15'30"W
77°15'W
77°14'30"W
77°14'W
77°13'30"W
Global Density of Commercial Shipping Vessels

exists here but by where its materials come from and where they are going. This landscape has the strange temporality of pass-through places: nothing stays here—or it isn't supposed to.[25] In both myths and material economies, islands are not the destination, they are landscapes of transformation where the hero—or in this case, material—becomes themselves before re-entering the world. In "The Vital Materiality of Aluminum: Light Modernity and the Global Atlantic," Mimi Sheller writes:

> *Following other materials to get at the material basis of their geo-energetic sources usually leads outside any one state's borders, not only because production chains, labor forces and companies are transnational, but also because the earth's geology distributes things like oil, coal, rivers, and metals in far-flung places, along with the "wastes" left over during their production and consumption.*[26]

As I write this, back in Virginia, I have just returned home from a work trip in a company van, the first electric hybrid vehicle I have driven. The body for this aluminum van started as red soil underneath a seasonal tropical forest or savannah woodland in a place like Jamaica's Cockpit Country where the winter is dry and the summer and fall are hot and wet.[27] After trees were cut and the topsoil removed, the bauxite was extracted, cleaned, loaded onto ships, and

---

25 The Grundartangi Eco-Industrial Park project is an attempt to change this. The Grundartangi Development Association emphasizes the focus of the project (in addition to renewable energy utilization) is "circularity." For example, the Alur Aluminum Processing produces commercial aluminum from the aluminum slag produced by Norðurál.

26 Sheller, Mimi. "The Vital Materiality of Aluminum: Light Modernity and the Global Atlantic." *Atlantic Studies* 11, no. 1 (2014): 67–81. doi:10.1080/14788810.2014.872331.

27 Bauxite is not an ore but an ancient type of soil—a regolith. Bauxite forms through soil formation and leaching. It can only form over rocks that are high in aluminum and poor in silica, such as clay-rich limestone. Bauxite requires strong weathering over a long period of time, which leaches anything soluble out of the soil and leaves behind only insoluble aluminum hydroxides. Importantly, this weathering process depends on location-specific climate factors and cycles: tectonics, geomorphology, vegetation cover, microbial activity, and atmospheric chemistry. One of the largest factors is a mean annual temperature greater than 22 degrees Celsius and mean annual precipitation of greater than 1,200 mm. Areas with wet and dry season cycles enable weathering and leaching processes through the drop in the water table during dry seasons. Vegetation is an important factor in weathering and bauxite formation. Most bauxite formations mined today are found in areas with seasonal tropical forests and savanna-type woodlands, which experience dry seasons more favorable to bauxite formation than tropical rainforests. However, we find bauxite even in areas not considered to have tropical climates due to tectonic shifting during the Cretaceous periods; tectonic plates moved through tropical climates and stayed there long enough for bauxite formation to occur. Jamaica is the main country of origin for the alumina that comes to the Century Aluminum plant at Grundartangi. The second largest supplier is Guinea, which also has its share of corruption, controversies, unequal distribution of benefits, and even riots in the bauxite industry. Cockpit Country is the center of Jamaica's biodiversity, source of freshwater, and home to the indigenous Maroon people. Although a 1738 treaty with the British granted the Maroons the right to settle and manage the region, in 2021, Jamaica's Minister of Transport and Mining revised the laws governing the land use in this area to open the lands up to bauxite mining. Bauxite mining in this important ecosystem destabilizes its resilience by increasing pollution, downstream flooding, and habitat loss for sensitive endemic species. Cordua, Bill. "Bauxite and Greenhouse Earth." Minds@UW (blog). University of Wisconsin, 2007. https://minds.wisconsin.edu/bitstream/handle/1793/11537/Bauxite.pdf?sequence=1&isAllowed=y; Yu, Wenchao, et al. "Climatic and Hydrologic Controls on Upper Paleozoic Bauxite Deposits in South China." *Earth-Science Reviews* 189 (2019): 159–176; Korneliusdóttir, Anna Margrét. "A Cradle-to-Gate Life Cycle Assessment of Primary Aluminium Production at Norðurál." PhD diss. University of Iceland, 2014; "'What Do We Get out of It?'" Human Rights Watch. March 28, 2023. https://www.hrw.org/report/2018/10/04/what-do-we-get-out-it/human-rights-impact-bauxite-mining-guinea; Taylor, Gladstone. "Jamaica's 'Cockpit Country' Faces Growing Threats from Mining Interests." Pulitzer Center, November 21, 2022. https://pulitzercenter.org/stories/jamaicas-cockpit-country-faces-growing-threats-mining-interests; Widengård, Marie. "The Mining-Conservation Nexus in Jamaica's Cockpit Country." *Edge Effects*. April 2, 2024. https://edgeeffects.net/cockpit-country-mining-conservation/; "Climate." *Cockpit Country*. Accessed July 10, 2024. https://cockpitcountry.com/climate.html.

brought to the Sherwin Alumina Company, LLC in Gregory, Texas.[28] There, bauxite was processed into powdery alumina and hoisted aboard a cargo ship bound for an Icelandic fjord, where the aluminum smelters stretch along the shore like the glacial mountains of the fjord itself.[29] One of these smelters transformed the white powder into shining ingots, the building blocks of the aluminum economy. This raw aluminum traveled onward to Europe for its next transformation into consumer products like my company van, which was shipped to the U.S. to tempt climate-conscious buyers like my employers.[30] The aluminum economy that brought me this car creates relationships between people and places, but the distance between mines and Icelandic smelters thins the sense of connection between these materials and each extractive step in their processing.[31] Companies like Alcoa, Rio Tinto Alcan, and Century Aluminum—

28 Unlike the base metal ores, most of the bauxite mined does not require complex processing to improve its grade before use. Its quality can be improved by relatively simple and inexpensive processes for removing clay, known as "beneficiation," which include washing, wet screening, and mechanical or manual sorting. Once the bauxite is loosened into manageable pieces, it is loaded into trucks, railroad cars or conveyors and transported to crushing and washing plants or to stockpiles before being shipped to alumina refineries. These refineries are usually located close to bauxite mines due to the inconvenience of shipping raw bauxite. However, while bauxite processing is inexpensive, it carries massive risk from a residue called "red mud," which is toxic to humans and ecosystems. This mud is stored indefinitely in industrial lakes and reservoirs. As aluminum production continues, the "red mud" continues to pile up, which is not just a waste of resources, but as demonstrated by the industrial accident in Ajka, Hungary in 2010 (cited below), poses significant risks to communities and ecosystems downstream of these industrial sites. "TENORM: Bauxite and Alumina Production Wastes." EPA. Accessed August 11, 2024. https://www.epa.gov/radiation/tenorm-bauxite-and-alumina-production-wastes; Rosenthal, Elisabeth. "Hungary's Red Sludge Spill: The Media and the Eco-Disaster." *Yale E360.* October 21, 2010. https://e360.yale.edu/features/hungarys_red_sludge_spill_the_media_and_the_eco-disaster; Donoghue, A. Michael, Neale Frisch, and David Olney. "Bauxite Mining and Alumina Refining: Process Description and Occupational Health-Risks." Journal of Occupational and Environmental Medicine 56 (2014): S12-S17; Korneliusdóttir, Anna Margrét. "A cradle-to-gate life cycle assessment of primary aluminium production at Norðurál."

29 Iceland's geomorphology is characterized by a coastline of fjords, deep channels with steep mountains, carved out by glaciers over thousands of years. The fjords are not only visually striking, but also opportune for shipping infrastructure as they can be used for sea-faring vessels yet are also sheltered from the open waters of the sea. Beyond aluminum industries taking advantage of these landscapes, the German company Bremen Ports has announced plans to build a deep sea port in Finnafjord in the North of Iceland, which they intend to serve as an intermediary port built in anticipation of the economies of a changing arctic. As the ice sheets covering Greenland and the Arctic Sea melt, Bremen Ports intends the Finnafjord port to provide a hub for the shipping of mineral resources from Greenland to China's manufacturers (especially for electronics) along the Northern route, which is anticipated to be key to trade between China and Europe in the coming century; Lo, Chris. "Finnafjord: A New Port for the Melting Arctic." *Ship Technology.* July 4, 2019. https://www.ship-technology.com/features/arctic-shipping-finnafjord-port-project/.

30 "Raw Aluminium in Iceland."

31 Aluminum economies reinforce the exploitative relationship between the Global North and the Global South. Iceland's shiny aluminum economy hides an underbelly of exploitation, ecological destruction, corporate malfeasance, political destabilization, environmental risk, and community displacement. These effects fall largely on Indigenous communities in the Global South. The way bauxite is extracted at an industrial scale is an intensive and destructive process. Bauxite is usually extracted by open cast mining since it is almost always found near the surface. This means that before mining can commence, the land needs to be cleared of timber and vegetation. Even if ecological rehabilitation occurs, the mining process can still result in habitat loss, groundwater disturbance, flooding, soil erosion, and soil degradation. This is what happened in the Jarrah Forest region of western Australia. In addition to the effects open cast mining has on non-humans, it also harms human communities by displacing inhabitants from the land and from their long-standing forms of livelihood and ecological stewardship. This is happening to the Maroons in Jamaica's Cockpit Country. Extraction processes also have broader effects on local governance and society. For example, in April and September 2017, uprisings in Boke, Guinea were led by local communities motivated by a lack of water and local services as mining companies continue to expand. Locals in Boke are experiencing displacement from their farmlands and the loss of a means to support themselves because of an absence of clear legal protections for customary land rights. Mining also degrades water systems significantly, and water pollution and scarcity impact the health and survival of local inhabitants. Because the mining companies have an outsized influence on local politics, locals find their concerns go unheard; they are not fairly compensated for their losses, and they are not provided alternative sources of livelihood or clean water. "A Thousand Cuts." WAFA, Wilderness Society and Conservation Council of Western

respectively headquartered in Pittsburgh, Montreal, and Chicago—unintentionally link the people and places of Iceland to Jamaica and to me, the consumer. The invisibility of these relationships allows corporations and governments to commodify landscape systems, and this commodification of land disrupts the ability of inhabitants to receive and respond to feedback from these systems. Extraction can hide in beauty. In Grundartangi, water laps gently onto the dark-pebbled shore, sheep graze peacefully in the hills, and the electric glow of the smelter glinting off the clouds points to the global web of industry. Following the powerlines away from the smelter leads us to the looming glaciers, impressive in size even at this distance.

Smelting the Unit of Exchange, or: Selling Electricity as an Island Nation

The aluminum fervor of the 1990s has cooled and given way to hot pursuit of new energy-intensive ventures like data centers, silicon production, and underwater transmission lines.[32] Each of these bubbles has grown and burst. Aluminum is only one of many proxies for the real commodity: hydropower. [33]Almost all the largest hydroelectric plants in the world exist for aluminum production, which accounts for 3% of global electricity use. As electricity costs rose in Europe and the U.S. during the late 20th century, the aluminum industry shifted its new

---

Australia. April 2022. https://www.wilderness.org.au/images/resources/A-Thousand-Cuts-NJF-Report.pdf; Taylor, Gladstone. "Jamaica's 'Cockpit Country' Faces Growing Threats from Mining Interests."; "'What Do We Get out of It?'" Human Rights Watch.

32 After aluminum industry expansion fell through, data centers came to take their place. In 2008 Rio Tinto Alcan cancelled the expansion of the ISAL plant in Straumsvik, even after the construction of the new Búdarháls by Landsvirkjun meant to power it. Shortly after, Verne Holdings announced plans to construct a data center using the electricity generated that was originally going to go to aluminum. Likewise, when aluminum industries began to falter and withdraw from Icelandic projects, Landsvirkjun turned its attention to Silicon smelters to use hydropower electricity to produce the raw components for the solar panels that are to be a key component of much of the world's green energy transition. After aluminum investments pulled out, Iceland saw a wave of investments from silicon companies: United Silicon plant in Reykjanes (starting operations in 2016), the PCC BakkiSilicon smelter in Húsavík, and the long planned Silicor smelter at Grundartangi. While the premise of using clean hydropower to make these solar panels is a compelling utilization of Iceland's abundant resources to power a larger scale transition, these plants have been plagued by financial and environmental issues, raising concerns about the neoliberal approach to the green energy transition. Adalbjornsson, Tryggvi. "Iceland's Data Centers Are Booming – Here's Why That's a Problem." *MIT Technology Review*. August 22, 2024. https://www.technologyreview.com/2019/06/18/134902/icelands-data-centers-are-booming-heres-why-thats-a-problem/; "Rio Tinto Alcan Cancels Smelter Project in Iceland." Iceland Review. December 15, 2008. https://www.icelandreview.com/news/rio-tinto-alcan-cancels-smelter-project-iceland/; "Verne Global – Iceland Data Center in Keflavik." Data Center Map. Accessed December 19, 2024. https://www.datacentermap.com/iceland/keflavik/verne-global/; Dalrymple, Colin Arnold. "From Iceland – Silicon Folly: The Failure of the Latest Industrial Trend." *The Reykjavik Grapevine*. September 10, 2018. https://grapevine.is/news/2018/09/10/silicon-folly-the-failure-of-the-latest-industrial-trend/; "Silicor Materials Iceland Ehf." SisAl Pilot. September 3, 2020. https://www.sisal-pilot.eu/partners/silicor-materials-iceland-ehf/; Kelly, Jemima, and Nerijus Adomaitis. "Giant Iceland-UK Power Cable Plan Seen Facing Brexit Delay." *Reuters*. October 21, 2016. https://www.reuters.com/article/business/giant-iceland-uk-power-cable-plan-seen-facing-brexit-delay-idUSKCN12L1O4/.

33 There is current research that makes it fully explicit that smelting is not even about the production of goods but about transferring hydropower electricity into a form that can be easily transferred and moved out of the local system for consumption elsewhere. Batteries cannot be used at large scale because they have low energy density and discharge prematurely, meaning they are inefficient for long term energy storage. Aluminum has the characteristics that would make it a viable energy carrier in lieu of batteries because it is relatively abundant, its material is recyclable, and it has high-energy density and high cycle efficiency (40%). When used for energy storage, aluminum will not self-discharge. When energy is needed, it can be released by oxidizing the aluminum with water to produce hot-hydrogen steam. Not only does this mean the aluminum can be used for energy transfer, but this function can be incorporated into larger aluminum material cycles, moving between structural use and energy use as the material is, in theory, infinitely recyclable; Trowell, K. A., S. Goroshin, D. L. Frost, and J. M. Bergthorson. "Aluminum and Its Role as a Recyclable, Sustainable Carrier of Renewable Energy." *Applied Energy* 275 (2020): 115112.

smelter production to sites in the Caribbean, Africa, and Australia already mining bauxite and refining alumina. In smaller countries, smelters consume a third or more of the national power supply.[34]

Hydropower is not a new and "green" technology for powering smelting; it has always been part of aluminum's story. Not only has aluminum always needed hydropower, it has driven the development of some of the largest hydropower projects we might not even recognize as being part of the aluminum industrial web. While Iceland lacks mineral wealth, it abounds with electricity.[35] Like the smelter at Grundartangi, the ravenous smelters across Iceland feast on cheap hydropower. Iceland's National Energy Company, Landsvirkjun, proudly proclaims that electricity generation in Iceland is 100% renewable.[36] Although Iceland is one of the few countries with robust geothermal energy, 73% of its electricity comes from hydropower, and the majority of these plants are dependent on glacial rivers.[37] Iceland's rivers run through steep, easily dammable valleys perfectly suited for hydropower infrastructure. However, the Langjökull, Hofsjökull, and Vatnajökull glaciers are the island's most unique hydropower resource. Under stable conditions, they function as rechargeable batteries in the water cycle by holding the winter rain as ice and releasing thawed meltwater in the spring. Beyond making electricity, hydropower plants turn the ephemeral process of the glacial water cycle into a commodity. Beginning in 2007, a green energy credit trading system was implemented for Icelandic power producers to sell Guarantees of Origin to power suppliers in the European Union / European Economic Area.[38] This allowed European suppliers to call their energy "renewable" because their use of fossil fuels is offset by Icelandic energy. Yet the Icelandic power grid is not physically connected to that of continental Europe, so this energy is not actually being exported, transferred, and utilized in Europe. Instead, buyers of this certificate purchase an electronic document that only represents green energy–not the electricity itself. These certificates illustrate poignantly what all high energy industries in Iceland are attempting: to commodify hydropower into a unit of exchange and, thus, a valuable export.[39]

34 Sheller, Mimi. "The Vital Materiality of Aluminum."

35 Landsvirkjun as an institution is deeply imbricated with power-intensive industries. According to Grande, "Landsvirkjun, the national power company of Iceland, which was initially founded in order to provide electricity to foreign-owned power intensive industries"; Grande, Daniel. "Rethinking Aluminum in Iceland's Economy: Balancing Risk and Reward." *Perspectives on Business and Economics* 29 (2011). https://doi.org/10.18275/pbe-v029-002.

36 Sveinsson, Óli Grétar Blöndal, Úlfar Linnet, and Halldór Björnsson. "Hydropower Expansion and Improved Management in Response to Increased Glacier Melt in Iceland." *Climate Adapt*. February 6, 2019.

37 And geothermal is part of Iceland's charismatic image of what a green energy landscape looks like, which is severely contrasted to hydropower, which has a worse reputation for environmental costs. However, geothermal power plants still have massive infrastructural and environmental costs and considerations that lie hidden in the fantastical sociotechnical imaginary of "limitless" clean energy and bypasses the real, wicked challenges of co-constructing the complex infrastructural-natural landscapes of the future; Benediktsson, Karl. "Conflicting Imaginaries in the Energy Transition?"; Nature and renewable energy in Iceland." *Moravian Geographical Reports* 29, no. 2 (2021): 94; Sveinsson, Óli Grétar Blöndal, Úlfar Linnet, and Halldór Björnsson. "Hydropower Expansion and Improved Management in Response to Increased Glacier Melt in Iceland."

38 "General Information." Umhverfisstofnun. Accessed December 4, 2024. https://ust.is/english/air-climate/eu-ets/aviation/general-information/; Adalbjornsson, Tryggvi. "Iceland's Data Centers Are Booming-Here's Why That's a Problem."

39 Energy is produced and used in Iceland, and carbon-intensive energy is still produced and consumed at the plants where demand is being met. In theory, this system does mean a trade wherein Icelandic consumers in-

Climate Mining, or: Prospecting on Glacial Decay

Glaciers are a water battery, holding seasonal rains in their icy grasp and releasing them with the spring thaw. In the North Atlantic, the Gulf Stream brings the precipitation that recharges these batteries and keeps retreat and expansion in balance. But this pattern of recharge and release is not as constant as it once was. These icy behemoths are currently in a phase of retreat—the third retreat in the last 100 years.[40] Iceland's glaciers have retreated consistently since 1995.[41] One glacier, Okjökul, has already disappeared, and its location is now marked with only a plaque and traces left on maps made before 2014.[42] A changing climate means changes to hydropower infrastructure in Iceland and, thus, its aluminum economy. As temperatures rise, glaciers melt. When the glaciers die, these energy resources will also vanish. But before then, a brief window during increasing meltwater creates a potential glut of energy. Hydropower production is not merely incidentally affected by climate change–in many ways now, it depends on it. Landsvirkjun is looking to climate models to understand how climate change will alter meltwater flows (and potential hydropower resources) in the coming decades.[43] Their 2017 report describes how they used this data to justify the expansion of the Burfell Hydropower Plant, increasing capacity by 30 MW.[44] [45] Similarly, the capacity of the new Búðarháls Hydropower Plant was "increased

---

directly purchase and consume the burden of fossil fuel energy being produced and used elsewhere in Europe. This system is inherently confusing because of the dissociation of the physical and accounting dimensions of electricity production and consumption. The aluminum companies claim to be the greenest smelters in the world relying on Iceland's image of green energy and the reality that there are (physically) no fossil fuel energy producers in Iceland. However, in reality, because of the European electricity market, these smelters are not buying renewable electricity, but instead, in majority, buying fossil fuel energy. Yet electricity in Iceland is the least expensive and with the lowest taxes for industrial buyers compared to peer countries, effectively subsidizing the utilization of energy that is not actually green. Adalbjornsson, Tryggvi. "Iceland's Data Centers Are Booming-Here's Why That's a Problem."; Breitschopf, Barbara, and Lin Zheng. Rep. Electricity Costs of Energy Intensive Industries in Iceland." A Comparison with Energy Intensive Industries in Selected Countries. Department of Energy, Industry and Business affairs, Ministry of Industries and Innovation, Iceland, November 2020, 36.

40 Although current climatic and oceanic warming is unequivocally human-driven, the complexity of the whole system means the climate and glaciers may react unpredictably. The glaciers are projected to increase melting as a result of rising temperatures, but climate changes may cause local cooling or increases in precipitation that may actually expand the glaciers. Climate is a dynamic system and one ill-suited for resource speculation; Chandler, Benjamin M.P., David J.A. Evans, and David H. Roberts. "Recent Retreat at a Temperate Icelandic Glacier in the Context of the Last~ 80 Years of Climate Change in the North Atlantic Region." *arktos* 2 (2016): 1–13.

41 Aðalgeirsdóttir, Guðfinna, Eyjólfur Magnússon, Finnur Pálsson, Thorsteinn Thorsteinsson, Joaquín M.C. Belart, Tómas Jóhannesson, Hrafnhildur Hannesdóttir, et al. "Glacier Changes in Iceland from ~1890 to 2019." *Frontiers in Earth Science* 8 (2020): 523646.

42 Technically it hasn't disappeared completely. Ice still remains on the mountain, but the mountain was downgraded and no longer meets the criteria for a glacier; Kenny, Stuart. "The Story of Okjökull: The Glacier That Disappeared." *Much Better Adventures Magazine*. April 25, 2023. https://www.muchbetteradventures.com/magazine/okjokull-glacier-history-iceland-disappeared/.

43 They base their study on simulations by the ECHAM4 and HadCM2-coupled OAGCM models, which are founded on the IPCC SRES B2 emission scenario. For more information on the utilization of IPCC scenarios for climate prediction Wigley, T. M. L., and S. C. B. Raper. "Implications for Climate and Sea Level of Revised IPCC Emissions Scenarios." *Nature* 357, no. 6376 (1992): 293-300; Sveinsson, Óli Grétar Blöndal, Úlfar Linnet, and Halldór Björnsson. "Hydropower Expansion and Improved Management in Response to Increased Glacier Melt in Iceland."

44 By our calculations, 30MW is sufficient to smelt 2,000-2,300 kg of aluminum or to drive the average electric car for 157,894 kilometers/98,110.78 miles; "EV Database." 2019. https://ev-database.org/cheatsheet/energy-consumption-electric-car.

45 Sveinsson, Óli Grétar Blöndal, Úlfar Linnet, and Halldór Björnsson. "Hydropower Expansion and Improved Management in Response to Increased Glacier Melt in Iceland."

from the originally planned 80 MW to 95 MW in response to climate change."[46] The designed capacity of the proposed Hvammur Hydropower Project was also increased by 13 MW. By using climate models as prospecting tools for energy resources, Iceland is prototyping a new form of mining: climate mining.[47] In this case study, Landsvirkjun reduces climate change to quantifiable temperature and precipitation changes materialized as glacial melt rates.[48] This prospecting maneuver equates climate change to projected megawatts of electricity. While the "green" aluminum smelters of Iceland do not directly use coal or petroleum, they are indirectly and ironically dependent on continued use of glacier-melting fossil fuels to meet their speculated potential. Someday, however, the landscapes created by the smelters and dams will all be post-glacial.

Encounters with the Glacier, or: Cronch-Cronch-Cronch

Within the next couple decades, the lower elevations of the three largest glaciers in Iceland will melt away, and their last vestiges will vanish completely in the next 100–200 years.[49] In the lurching drive across the plains approaching Vatnajökull, the largest glacier in Iceland, my sense of scale and time shakes loose. It never feels any closer even as we progress through the landscape of glacial till and moraines, witnessing the glacier's previous expansions etched into the topography. As the glacier retreats, vegetative cover increases—but slowly. These plains are mostly sparse, and even areas that have been post-glacial for 100 years are still highly dynamic ecosystems.[50] The glacier is just an abstract presence until I begin my crampon-clad climb across the wet cones of dust and pitted ice interspersed with glinting streams.[51] Moving across this surface requires more stagger than walk, and my balance reorients to adapt. The constant trickling reshapes the ice, revealing and carrying away histories of ash. Fluctuations are quick enough that no landmarks last, and when I try to retrace my steps, the speckled expanse traps me for an hour in a directionless wander. A new cave opened up this week, and I creep through a narrow dripping crevasse into glowing blue light. Meltwater runs in sheets down all sides and joins the trickle of the streams above and the barely perceptible groaning beneath. The walls are sharper than they look and sting my hands as I slip along the edge, and icy water runs into my boots to numb my feet as I exit. On a ledge where the

46 "The runoff change may be important for the design and operation of hydroelectric power plants and other utilisation of water"; Thorsteinsson, Thorsteinn, and Halldór Björnsson, eds. Climate Change and Energy Systems.. Nordic Council of Ministers, 2012: 57. This implies retrofitting dams and reservoirs to handle larger capacity and expanding power plant generator facilities to make use of greater flow. This is framed from an hydro-engineering perspective but neglects the whole system. It raises the question, are these also designed to change when the runoff decreases? The answer is most likely no as it is not mentioned in the document and that seems to be indicative of epistemic bias.

47 Black, Megan. "Prospecting the World: Landsat and the Search for Minerals in Space Age Globalization." *Journal of American History* 106, no. 1 (2019): 97–120.

48 Thorsteinsson, Thorsteinn, and Halldór Björnsson, eds. *Climate Change and Energy Systems.*

49 Jóhannesson, Tómas, G. Adalgeirsdóttir, Helgi Björnsson, Finnur Pálsson, and O. Sigurdsson. "Response of Glaciers and Glacier Runoff in Iceland to Climate Change." Nordic Hydrological Conference 2004 (NHC-2004), 2004: 651–660.

50 Glausen, Taylor G., and Lawrence H. Tanner. "Successional Trends and Processes on a Glacial Foreland in Southern Iceland Studied by Repeated Species Counts." *Ecological Processes* 8 (2019): 1–11.

51 *CRONCH-CRONCH-CRONCH.*

expanse is white and dustless, I lie flat. Bumps in the pocked ice press into my spine and send faint vibrations through my chest, transmitting from the shifting depths and gushing hollows below. The meltwater from disappearing glaciers forms ephemeral rivers across the treeless expanse of the dark sand floodplain, creating unique microbial communities and nutrient systems where it meets groundwater.[52] At the base of the glacier, my boots stick and slip as I trudge muddily away from the evidence of inexorable change in a living giant. Iceland is at the edge of habitability, and this boundary between white and green has ebbed and flowed many times. The whole landscape echoes with the ghosts of glaciers past and present. The glacier lives on as a remnant through this ecosystem by creating the conditions for lichens, bacteria, and mosses to slowly build new soil and new worlds.

Time-Machines, or: What Happens when the Water Dries Up

The present must tend to the conditions for the future. However, energy that leaves a system and does not return to it cannot build its future. The value must be reinvested. When the glacial melt that feeds them diminishes and the glut of hydropower runs dry, what will remain beyond a landscape littered with defunct dams, useless hydropower plants, and empty reservoirs? If hydropower is a resource that is running out, then the present resource needs to be directed to future needs: needs to be given back to the land, designed for decomposition, succession, and transition.[53] Our recycling challenge extends beyond the material scale of aluminum to the entire landscape—and the energy needs of the people in it. The constellation of the global aluminum industry alienates the material from the labor of the earth systems—and peoples—that make it possible. The aluminum smelter is a time machine: in an instant, it reverses millions of years of the weathering and geological processes that create bauxite. Time can be undone—all it takes is an immense amount of energy. The value of aluminum is often conceptualized through its material properties instead of the geologic and climatic processes that create that value. Aluminum is made valuable through the energy poured into its smelting. This energy does not come from nowhere; it comes from the specific properties and histories of these glacial landscapes. In Iceland, dams like Kárahnjúkar Hydropower Plant are constructed on the highland commons land, but the wealth generated from this energy enriches private corporations and their shareholders.[54] In 2012, the intergovernmental group Norden wrote in their report "Climate Change and Energy Systems,"

> *While the companies obviously aim to minimise negative impacts, it must be remembered that it is also possible that some features of renewable*

52 Purkamo, Lotta, Brighid Ó Dochartaigh, Alan MacDonald, and Claire Cousins. "Following the Flow—Microbial Ecology in Surface and Groundwaters in the Glacial Forefield of a Rapidly Retreating Glacier in Iceland." *Environmental Microbiology* 24, no. 12 (2022): 5840–5858.

53 This is not a material investment in Iceland *or* Jamaica but both simultaneously. Right now, each stage of material processing is extractive, and each phase should be redesigned to invest collaboratively in the health of the places affected by it.

54 Ögmundardóttir, Helga. *The Shepherds of Þjórsárver: Traditional Use and Hydropower Development in the Commons of the Icelandic Highland.* Uppsala: Department of Cultural Anthropology and Ethnology, 2011.

> *resources may create new opportunities for the power plants in the future. ... Since the life-time of power plant investments is usually less than 40 years, there is seldom a need for a longer planning period in an economic study. Private investors also tend to focus more on the near future because of the interest rate and because of the larger uncertainty surrounding the distant future.*[55]

This report was written for investors and shareholders, and it hints that medium-term climate change speculation may provide economic value for savvy industrialists. However, the earth is a complex system of which an economic understanding alone can barely grasp a fraction. Surviving the uncertain future requires imagination beyond economic priorities, and it needs more than what we are currently giving it. Although economic arguments for future adaptation planning and circular economies have a place, I will not cede the discursive territory to economic prioritization. Framing the challenges of climate change through only an economic and energy lens gives these factors undue emphasis. The land is so much more, and a future with it requires connections across scales and times.

Adopted Ecologies, or: Making Kin in a Changing World

In the 2018 Icelandic film *Woman at War (Kona fer í stríð),* climate change disrupts the traditional family and requires new bonds of kinship across bloodlines, cultures, and ecologies—and explicitly across scales of environmental destruction.[56] The film tells the story of Halla, a matronly choir teacher whose desire to adopt an orphan from Ukraine and become a mother is imperiled by her simultaneous decision to become an eco-terrorist. She decides to fight in Iceland not to preserve a pristine landscape while the rest of the world is beset by climate change, but because she believes the smelters connect to a larger system of environmental destruction and exploitation.

> *Vandalism against…nature which causes global warming is a crime against humanity and life on earth. The system of these multinationals is an infernal power that democracy is obviously unable to stop. There are higher laws than those of men, ancestral laws which will not be obsolete tomorrow. It's our duty to protect life and the health of future generations. We are the generation that is the most powerful that has ever existed and also the last one to be able to stop the war against Mother Earth. Our children and grandchildren cannot change anything. For them it will be too late. Today is the moment, it's [up to] us to act.*[57]

---

55 Keränen, Jaana, Riitta Molarius, Jari Schabel, Jenny Gode, Edward James-Smith, Noora Veijalainen and Kirsti Jylhä. "The Effects of Climate Change on Power & Heat Plants—Assessing the Risks and Opportunities" in *Climate Change and Energy Systems: Impacts, Risks and Adaptation in the Nordic and Baltic Countries*, Thorsteinn Thorsteinsson and Halldór Björnsson, eds. Nordic Council of Ministers, Copenhagen. 2012: 195.

56 Erlingsson, Benedikt. *Woman at War* (*Kona fer í stríð*). Directed by: Benedikt Erlingsson. Written by: Benedikt Erlingsson & Ólafur Egill Egilsson. Starring: Halldóra Geirharðsdóttir as Halla / Ása, 2018. https://www.imdb.com/title/tt7279188/?ref_=ttfc_fc_tt.

57 Erlingsson. *Woman at War.*

Glacial Retreat and Lichenous Growth

Although Halla's activism threatens her dreams of family, her concern for the environment is an extension of the same desire for parenthood. Halla is a mother in a much broader sense—a post-human sense. A strength of the film is that these acts of care are not selfless, but they entwine her deeper into relationships. In the first scene, she disrupts the operation of an aluminum smelter by bringing down its power line with only a bow and arrow. While fleeing from the authorities across the mossy terrain of the highland commons, she is discovered by a sheep farmer. He decides to protect her based on the justification that her family also comes from this part of the country. Specifically, he claims that since his grandfather was promiscuous they might actually be cousins, reifying this bond of place into an unusual form of family. This interaction and the subsequent escalation of his willingness to help her demonstrate that shared care for a place creates kinship.

The film further argues this kinship extends beyond human ties to bonds with the land itself. Halla is pursued by authorities after attempting to leave the country to adopt her daughter, and she makes a final run from helicopters and police across the mossy expanse. When she realizes she has nowhere left to run, Halla makes a symbolic burial of her dream of family by entrusting her photo of her daughter into the care of the moss. Her dream of connection with her daughter dies, but while she did not get to embrace her daughter, she is embraced by the land instead—as a mother would a child. As the film resolves, even Halla's relationship with her sister—her only biological relationship depicted in the film—develops an expanded definition of family. Halla's dream of motherhood rekindles when her identical twin sister swaps places with her in prison and Halla unites with her daughter in Ukraine at last.

All of Halla's kinships are irregular; her daughter is adopted, a stranger risks his freedom to help a "cousin," and her sister takes on her identity in prison so Halla can become a mother. In the final scene, the bus carrying Halla and her daughter is caught in extreme flooding, and the passengers must disembark and wade to dry ground. Halla and her daughter become family against a backdrop of climate change as she carries the child into a flooded and uncertain future. This kinship does not look like the bonds of blood but who we choose and are chosen by. However, while many of the bonds in the film are between people, these are secondary to kinships between people and land.

Nutrient Cycling, or: Lessons from the Already Post-Glacial Landscapes

While the film describes this relationship of care through a heroic resistance to "vandalism against…nature which causes global warming", caring for a future is also built on the embodied, embedded, ongoing interaction with the materials and systems of a dynamic landscape.[58] Northwest of expansive highlands near the fertile Lake Myvatn lies Svartárkot, a sheep farm that has been continuously inhabited for most of Iceland's history. Iceland's story is the story of glaciers, and this landscape is already post-glacial. After a long drive across barren lava fields,

58 Erlingsson. *Women at War.*

I duck inside a tiny smoke hut where farmers Guðrún and Sigurlína Tryggvadóttir still cure meat and fish using sheep dung, and nothing is wasted. The dung is a plentiful source of not only fuel but also worldbuilding nutrients. At the edge of the farm where grass cannot grow, the sisters continue a tradition of terraforming volcanic rock with yearly applications of manure, inviting the slow succession of lichens and moss. Across the river in Holasandur, farming communities work with the Soil Conservation Service of Iceland, spreading sheep dung and blackwater on the gravely barrens to feed future grasslands and forests.[59] This comes from a sense of responsibility to place and connection to dynamic systems beyond the individual scale. Hulda Brynjólfsdóttir describes that, as a sheep farmer, she feels responsibility to the nutrient cycle that extends beyond her own lifespan:

> *The sheep have fed me all these years, it is my duty to feed them when I die by feeding the soil with my body. We have been ruining the soil by using fertilizers. We have broken this circle by being buried in the ground. We ingest nutrients and then let it out, not back to the land. If we have healthy organic food coming into us, our waste can and should be used to fertilize. We should spread the nutrients we get from the food we eat back into nature.*[60]

Full new ecosystems take generations to emerge, but this nutrient cycle is both longitudinal and balanced: some fuel for food preservation in the present, some food for microbial co-conspirators building the future. Extractivism removes material and energy from a system, leading to its demise. In contrast, this model of nutrient cycling guides a framework for energy consumption that extends beyond farm life. The lifecycle of infrastructure can also continue not only after, but ideally through, its death. When the glacial melt that feeds them diminishes and the glut of hydropower runs dry, what will remain beyond a landscape littered with defunct dams, useless hydropower plants, and empty reservoirs? If hydropower is a resource that is running out, then the present resource needs to be directed to future needs: needs to be given back to the land, designed for decomposition, succession, and transition. Although climate change is an ongoing catastrophe, the energy potential from the next fifty years of glacier melt may be used to build habitability for new futures, not just speed present collapse.

Outwitting Trolls, or: Dynamic Phenomena Require Dynamic Solutions

However, the next 50 years are unpredictable. Not only is the climate changing, the way climate change is happening is changing also. Climate systems dependent on a cooler planet are becoming unstable under globally warmer conditions. Despite predictions by Landsvirkjun that glacier meltwater rates would increase, the winter of 2023–2024 was cold and dry, and the rivers had lower than anticipated flows.[61] At some stations, the flows were the lowest on record.

---

59 Teichman, Theodore S. Interview with Magnus Johansson. Personal, January 29, 2022.

60 Teichman, Theodore S. Interview with Hulda Brynjólfsdóttir. Personal, January 30, 2022.

61 "Climate Change Affecting Electricity Production in Iceland!" arcticportal.org. April 15, 2024. https://arcticportal.org/.

The authors of the 2012 Norden report on climate claim that the uncertainty in their climate predictions is due to coarse spatial resolution and differences between general circulation and regional climate models.[62] This explanation may be accurate, but it also neglects the possibility that some aspects of climate change and natural systems may be inherently unpredictable and outside the bounds of what is modelable. Researcher Patricia Carralero warns that modelers' neglect of this uncertainty can skew interpretations of model outputs, which has effects on sustainability decisions by policy-makers.[63] "Higher quality information involves understanding that knowledge is not simply a body of objective facts, but that it includes questions of framing, interpretation, narration, and testimony, and thus forefronts modalities of experience in relation to knowledge," she writes.[64] Standalone facts can also be insufficiently contextualized within models. Tolerance for uncertainty can also be an asset when relating to a nonlinear system like climate. High-quality information about a system describes the kind of relationships we have with the subjects of dynamic models. This kind of data is not the kind that can be line-itemed in spreadsheets; it is embedded in people, embedded in place—and sometimes, in the stories they tell about it.

According to a local story, the valleys near Myvatn are home to the troll woman Kraka.[65] To punish the locals for evading her amorous advances, she animated the river to be difficult to control. Traditionally the farmers were dependent on temporary wood, stone, and turf river dams to introduce nutrients to the fields and prevent erosion on the riverbanks, and Kraka expected that the locals would eventually succumb to the river's unpredictability. This story is a unique case study because Icelandic farmers are intimately dependent on climate to manage agriculture, and the potential for catastrophe is high because the growing season is short. The Kraka myth reveals local anxieties around survival and resource management. However, its most important function is as a technology that encodes a material and infrastructural framework for responding to dynamic environmental systems. Not only must the community manage the river and its sediment but they must also steward the turf, stone, earth, and trees that they depend on for dam construction. If the farmers overbuild, these resources will quickly run out, and Kraka wins. The myth provides a framework for living in relationship to the river and also gives the community a common purpose: do not let the troll win. The whims of Kraka—the whims of the river—represent changes of climate, which locals manage as both risk and asset with which to co-create.

Iceland has always been terrain with slim survival margins, and its culture emerged from practices of dwelling amid environmental instability. In an increasingly unstable global climate system, extreme weather events are

62 Thorsteinsson, Thorsteinn, and Halldór Björnsson. *Climate Change and Energy Systems: Impacts, Risks and Adaptation in the Nordic and Baltic Countries*. 2012.

63 Carralero, Pamela Alicia. "Uncertainty Discourse: Climate Models, Gender, and Environmental Literature in the Anthropocene." Ph.D. diss. Purdue University, 2019: 23.

64 Carralero, Pamela Alicia. "Uncertainty Discourse." 25.

65 Sigurðardóttir, Ragnhildur, Anthony J. Newton, Megan T. Hicks, Andrew J. Dugmore, Viðar Hreinsson, Astrid E.J. Ogilvie, Árni Daníel Júlíusson et al. "Trolls, Water, Time, and Community: Resource Management in the Mývatn District of Northeast Iceland." *Global Perspectives on Long Term Community Resource Management* (2019): 77–101.

narrowing survival margins in the rest of the world too. A community can both create and emerge around a shared story of human-ecological relationships, and a unified purpose—like outsmarting the troll—can provide structure for solving local challenges in a way that meets both social and technical needs. Dynamic systems, like temporary dams moved across seasons, are suited for dynamic phenomena like the river. Partially through lack of technical specificity, shared myths can describe relationships between communities and dynamic climate systems in a way that is flexible enough to adapt to both cultural and ecological change. However, not all stories about dynamic relationships are in the mythological past; they extend into the future also.

Future Stories, or: Life-Spanning at the End of the World

My relationship with the glacier is superficial; we have met only once. My relationship to the glacier is profound; its life affects mine through a global network of climate impact cause-and-effect. All the miles I have already driven are etched into the retreating crevices and crags of the glacier.[66] This relationship extends across time as well as place. I stand at the edges of the meltwater lagoon and look behind me to the moraines, the piles of glacial till marking this furthest edge. These moraines are the ghosts of the glacier's past expanse. It strikes me that—while past moraines might expect to be moved again by future glacial expanse—these are the last. From this point on, the glacier will move further away from its till piles until it disappears fully. In the end, what is left of the glacier is not the hydropower but the moraines. What lies between the world of the glaciers and their last moraines is millions of tons of meltwater.

The soggy ground beneath my feet is rich in volcanic andosols, one of the richest soils for aluminum compounds.[67] Iceland does not have mineral resources—yet. But a hundred thousand to a few million warm humid years will weather these andosols down into bauxite, the progenitor of aluminum. This tropical climate is far from today's Iceland of glaciers and hydropower, but this future is growing likely as we blow past increasingly urgent carbon emission reduction

66 This is not about guilt, it is about love. I save aluminum foil not out of guilt but because I respect its lifespan and want it to have many futures beyond me. It, like the lichens, deserves better. Likewise, even if intangible, it is better to love even the greenhouse gasses and strive to treat them better than we have.

67 Gow, Neil N., and Gian Paolo Lozej. "Bauxite." *Geoscience Canada* 20, no. 1 (1993): 12.

commitments. The journey of Jamaican bauxite to Iceland for aluminum smelting helps create the warm post-glacial future in which Iceland itself will hold the conditions necessary for bauxite formation. The Intergovernmental Panel on Climate Change (IPCC) emissions and temperature projections estimate that by 2050, the time when glacier meltwater will reach its peak, Iceland's temperatures may reach 0.86–4.64 degrees Celsius warmer under current climate agreements—and the rate of increase is also increasing.[68] Before the climate can work soil into bauxite under a humid forest canopy, stone and glacial till must be transformed into soil. But that is tens of thousands of years in the future. The glacial till around me is near-barren gravel now, but the pioneering lichens dotting this rocky rubble are beginning the soil formation work that will feed future ecosystems. The estimated lifespan of some arctic lichens stretches 5,000 years, making these colorful symbiotes charismatic mediators between the far past and future.[69] On the perception of time and glacial landscapes, Icelandic author Andri Snær Magnason describes overlapping human lifespans to his daughter,

> *Imagine that. Two hundred and sixty-two years. That's the length of time you connect across. You'll know the people who span this time. Your time is the time of the people you know and love, the time that molds you. And your time is also the time of the people you will know and love. The time that you will shape. You can touch two hundred and sixty-two years with your bare hands. Your grandma taught you, you will teach your great-granddaughter. You can have a direct impact on the future, right up to the year 2186.*[70]

Magnason is talking about human family relationships: time as *who* you know, not as a unit. These trans-temporal intimacies extend to more than human relationships. In 1947, the power plant Andakílsárvirkjun was built to dam the Andakíl river that feeds the wide grassy wetlands of the estuary outside my

68 "Trends & Variability – Projection." World Bank Climate Change Knowledge Portal. Accessed December 18, 2024. https://climateknowledgeportal.worldbank.org/country/iceland/trends-variability-projections.

69 Armstrong, Richard A. "Adaptation of Lichens to Extreme Conditions." *Plant Adaptation Strategies in Changing Environment* (2017): 1–27.

70 Magnason, Andri Snær. *On Time and Water.* Translated by Lytton Smith. Rochester, NY: Open Letter, 2021: 12–13.

Norðurál Smelter

window 70 years later. On the slopes of this reservoir, I gathered angelica seeds in the spring and planted birch seedlings in the fall. Up in the highlands two years later, I admired rainbows in the mists below the thundering Karahnjukar dam, built in 2009 to last over 100 years—but not long enough to witness the end of the Vatnajökull glacier that will take twice as long. The oldest lichens spreading across those moraines are from the early 19th century, and the timespan I have touched stretches over 200 years into the past. If these ancestral lichens survive the warming climate, I may have touched thousands of years into the future also.[71] These overlapping, interwoven timescales are different from time encoded in a climate projection because they are defined intimately, not abstractly. Time happens relationally in this landscape: through the lichen, through the soil, and through aluminum—not only in smelters and cars but bound in the compounds of fresh glacial till.

Although they are our connection to this future landscape, paradoxically, these lichens are unlikely to exist in it. Lichens would need one million years to adapt to a temperature increase of just 1 degree Celsius, which is the *minimum* anticipated warming by 2100.[72] However, nothing about the coming years is as certain as a projection might anticipate. Even if they survive only a hundred years more, the lichens I love here will at least survive me. In two hundred years, they will witness worlds I will never see: the power plants shut down, the dams broken, and the last glacier's rocky remains.

Our future, too, emerges in what we leave behind.

---

71 Decaulne, Armelle. "Lichenometry in Iceland, Results and Application." *Géomorphologie: Relief, Processus, Environnement* 22, no. 1 (2016): 77–91.

72 Wetzel, Corryn. "Lichens Are in an Evolutionary Race against Climate Change." Smithsonian.com. February 17, 2022. https://www.smithsonianmag.com/smart-news/lichens-are-in-an-evolutionary-race-against-climate-change-180979587/#:~:text=Their%20work%20revealed%20that%20lichens,increase%20of%20one%20degree%20Celsius.

Acknowledgements

To Viðar Hreinsson, for his introduction to Icelandic climate activism and contagious enthusiasm for scholarship, stories, and sheep.

To Þorvarður Árnason, for his introduction to the glacial landscape.

To Guðrún and Sigurlína Tryggvadóttir, for their hospitality at Svartarkot and many stories.

To the Wright-Ingraham Institute Field Stations program and facilitators.

To Magnus Johansson at the Soil Conservation Service of Iceland for his generosity with his expertise.

To Hulda Brynjólfsdóttir at Uppspunni for her time, hospitality, and poignant insights.

To Matthew Seibert for being a patient and supportive editor in this research journey.

To Samaneh Nickayin for her advice and support.

To the Agricultural University of Iceland for its support and foundation.

To my friends Eysteinn Eyjólfsson, Davíð Bjarki Jónsson, and Beatrice Mazarin for coming with me on reconnaissance expeditions to the dams and smelters.

BIBLIOGRAPHY

Aðalgeirsdóttir, Guðfinna, Eyjólfur Magnússon, Finnur Pálsson, Thorsteinn Thorsteinsson, Joaquín MC Belart, Tómas Jóhannesson, Hrafnhildur Hannesdóttir, et al. "Glacier Changes in Iceland from ~1890 to 2019." *Frontiers in Earth Science* 8 (2020): 523646.

Adalbjornsson, Tryggvi. "Iceland's Data Centers Are Booming – Here's Why That's a Problem." MIT Technology Review, August 22, 2024. https://www.technologyreview.com/2019/06/18/134902/icelands-data-centers-are-booming-heres-why-thats-a-problem/.

The Aluminum Association. "Auto Aluminum Growth in an Era of Electrification." Accessed December 19, 2024. https://www.aluminum.org/auto-aluminum-growth-era-electrification.

Andri Snær Magnason. On Time and Water. Translated by Lytton Smith. Rochester, NY: Open Letter, 2021: 12–13.

arcticportal.org. "Climate Change Affecting Electricity Production in Iceland!" April 15, 2024. https://arcticportal.org/.

Armstrong, Richard A. "Adaptation of Lichens to Extreme Conditions." *Plant Adaptation Strategies in Changing Environment* (2017): 1–27.

Benediktsson, Karl. "Conflicting Imaginaries in the EnergyTransition? Nature and Renewable Energy in Iceland." *Moravian Geographical Reports* 29, no. 2 (2021): 89–90, 93–94.

Benediktsson, Karl, and Edda Ruth Hlín Waage. "Power Without Politics? Nature, Landscape and Renewable Energy in Iceland." *Annales de Geographie*, no.6 (2020): 107–130.

Black, Megan. "Prospecting the World: Landsat and the Search for Minerals in Space Age Globalization." *Journal of American History* 106, no. 1 (2019): 97–120.

Breitschopf, Barbara, and Lin Zheng. Rep. "Electricity Costs of Energy Intensive Industries in Iceland: A Comparison with Energy Intensive Industries in Selected Countries." Department of Energy, Industry and Business affairs, Ministry of Industries and Innovation, Iceland, November 2020: 6-7, 36. https://www.stjornarradid.is/library/01--Frettatengt---myndir-og-skrar/ANR/ThKRG/Report%20Iceland-FINAL.pdf.

Carralero, Pamela Alicia. 2019. "Uncertainty Discourse: Climate Models, Gender, and Environmental Literature in the Anthropocene." Ph.D. diss., Purdue University, 23.

Chandler, Benjamin MP, David JA Evans, and David H. Roberts. "Recent Retreat at a Temperate Icelandic Glacier in the Context of the Last ~80 Years of Climate Change in the North Atlantic Region." *arktos* 2 (2016): 1–13.

Ćirić, Jelena. "Whaling Has Little Economic Impact on Iceland." *Iceland Review*. August 22, 2023. https://www.icelandreview.com/nature-travel/whaling-has-little-economic-impact-on-iceland/.

Cockpit Country. "Climate." Accessed July 10, 2024. https://cockpitcountry.com/climate.html.

Cordua, Bill. "Bauxite and Greenhouse Earth." Minds@UW (blog). University of Wisconsin, 2007. https://minds.wisconsin.edu/bitstream/handle/1793/11537/Bauxite.pdf?sequence=1&isAllowed=y.

Data Center Map. "Verne Global – Iceland Data Center in Keflavik." Accessed December 19, 2024. https://www.datacentermap.com/iceland/keflavik/verne-global/.

Dalrymple, Colin Arnold. "From Iceland – Silicon Folly: The Failure of the Latest Industrial Trend." *The Reykjavik Grapevine*. September 10, 2018. https://grapevine.is/news/2018/09/10/silicon-folly-the-failure-of-the-latest-industrial-trend/.

Das, Sujit. "Life Cycle Energy and Environmental Assessment of Aluminum-Intensive Vehicle Design." *SAE International Journal of Materials and Manufacturing* 7, no. 3 (2014): 588–595.

Decaulne, Armelle. "Lichenometry in Iceland, Results and Application." *Géomorphologie: Relief, Processus, Environnement* 22, no. 1 (2016): 77–91.

Donoghue, A. Michael, Neale Frisch, and David Olney. "Bauxite Mining and Alumina Refining: Process Description and Occupational Health Risks." *Journal of Occupational and Environmental Medicine* 56 (2014): S12–S17.

EPA. "TENORM: Bauxite and Alumina Production Wastes." Accessed August 11, 2024. https://www.epa.gov/radiation/tenorm-bauxite-and-alumina-production-wastes.

Erlingsson, Benedikt. *Woman at War* (*Kona fer í stríð*). Directed by: Benedikt Erlingsson. Written by: Benedikt Erlingsson & Ólafur Egill Egilsson; Starring: Halldóra Geirharðsdóttir as Halla / Ása, 2018. https://www.imdb.com/title/tt7279188/?ref_=ttfc_fc_tt.

"EV Database." 2019. https://ev-database.org/cheatsheet/energy-consumption-electric-car.

Fontaine, Andie Sophia. "From Iceland – Landowners Demand End to Smelter Pollution." *The Reykjavik Grapevine*. September 23, 2014. https://grapevine.is/news/2014/09/23/landowners-demand-end-to-smelter-pollution/.

Fridlyander, I. N., V. G. Sister, O. E. Grushko, V. V. Berstenev, L. M. Sheveleva, and L. A. Ivanova. "Aluminum Alloys: Promising Materials in the Automotive Industry." *Metal Science and Heat Treatment* 44, no. 9 (2002): 365–370: 366.

Glausen, Taylor G., and Lawrence H. Tanner. "Successional Trends and Processes on a Glacial Foreland in Southern Iceland Studied by Repeated Species Counts." *Ecological Processes* 8 (2019): 1–11.

Gow, Neil N., and Gian Paolo Lozej. "Bauxite." *Geoscience Canada* 20, no. 1 (1993): 12.

Grande, Daniel. "Rethinking Aluminum in Iceland's Economy: Balancing Risk and Reward." *Perspectives on Business and Economics* 29 (2011). https://doi.org/10.18275/pbe-v029-002.

Guðmundsdóttir, Hrönn, Wim Carton, Henner Busch, and Vasna Ramasar. "Modernist Dreams and Green Sagas: The Neoliberal Politics of Iceland's Renewable Energy Economy." *Environment and Planning E: Nature and Space* 1, no. 4 (2018): 590. 592.

Helga Dís Björgúlfsdóttir, "The Beautiful Hvalfjörður." Your Friend in Reykjavik. April 7, 2023. https://yourfriendinreykjavik.com/the-beautiful-hvalfjordur/.

Human Rights Watch. "'What Do We Get out of It?'" March 28, 2023. https://www.hrw.org/report/2018/10/04/what-do-we-get-out-it/human-rights-impact-bauxite-mining-guinea.

Iceland Review. "Rio Tinto Alcan Cancels Smelter Project in Iceland." December 15, 2008. https://www.icelandreview.com/news/rio-tinto-alcan-cancels-smelter-project-iceland/.

Jasanoff, S., S. H. Kim. (2009): Containing the Atom: Sociotechnical Imaginaries and Nuclear Power in the United States and South Korea. Minerva 47, no. 2 (2009): 119–146.

Jóhannesson, Tómas, G. Adalgeirsdóttir, Helgi Björnsson, Finnur Pálsson, and O. Sigurdsson. "Response of Glaciers and Glacier Runoff in Iceland to Climate Change." *Nordic Hydrological Conference* 2004 (NHC-2004), pp. 651–660. 2004.

Kawajiri, Kotaro, Michio Kobayashi, and Kaito Sakamoto. "Lightweight Materials Equal Lightweight Greenhouse Gas Emissions?: A Historical Analysis of Greenhouse Gases of Vehicle Material Substitution." *Journal of Cleaner Production* 253 (2020): 119805.

Kelly, Jemima, and Nerijus Adomaitis. "Giant Iceland-UK Power Cable Plan Seen Facing Brexit Delay." *Reuters*. October 21, 2016. https://www.reuters.com/article/business/giant-iceland-uk-power-cable-plan-seen-facing-brexit-delay-idUSKCN12L1O4/.

Kenny, Stuart. "The Story of Okjökull: The Glacier That Disappeared." *Much Better Adventures Magazine*. April 25, 2023. https://www.muchbetteradventures.com/magazine/okjokull-glacier-history-iceland-disappeared/.

Korneliusdóttir, Anna Margrét. "A Cradle-to-Gate Life Cycle Assessment of Primary Aluminium Production at Norðurál." PhD diss. University of Iceland, 2014.

Krater, Jaap. "Bakki Impact Assessment Should Include Dams." *Saving Iceland*. August 22, 2008. https://www.savingiceland.org/2008/08/bakki-impact-asessment-should-include-dams/.

Kvande, Halvor, and Per Arne Drabløs. "The Aluminum Smelting Process and Innovative Alternative Technologies." *Journal of Occupational and Environmental Medicine* 56 (2014): S23–S32.

Landsvirkjun. "Landsvirkjun Power Stations." Accessed December 19, 2024. https://www.landsvirkjun.com/powerstations.

Lo, Chris. "Finnafjord: A New Port for the Melting Arctic." *Ship Technology*. July 4, 2019. https://www.ship-technology.com/features/arctic-shipping-finnafjord-port-project/.

Ögmundardóttir, Helga (2011). *The Shepherds of Þjórsárver: Traditional Use and Hydropower Development in the Commons of the Icelandic Highland*. Uppsala: Department of Cultural Anthropology and Ethnology.

Ólafsson, Stefán. "Welfare Trends of the 1990s in Iceland." *Scandinavian Journal of Public Health* 31, no. 6 (2003): 401–404.

The Observatory of Economic Complexity. "Raw Aluminium in Iceland." Accessed July 10, 2024. https://oec.world/en/profile/bilateral-product/raw-aluminium/reporter/isl.

Pálmason, Sigurður Gísli. *Dreamland* (*Draumalandið*). Directed by: Þorfinnur Guðnason, Andri Snær Magnason. Based on the book: *Dreamland: A Self-Help Manual for a Frightened Nation* by Andri Snær Magnason. Ground Control productions, 2009. https://dreamland.is.

Peet, R., Watts, M. (1996): Liberation Ecology: Development, Sustainability, and Environment in an Age of Market Triumphalism. In Peet, R., and M. Watts (eds.), *Liberation Ecologies: Environment, Development, Social Movements*. London: Routledge, 1–45.

Pokornı, Richard, and Helgi Torfason. *Mineral Resources in Iceland Volume II: Ores. Cambridge Scholars Publishing, 2024*. https://pubs.usgs.gov/myb/vol3/2019/myb3-2019-iceland.pdf.

Purkamo, Lotta, Brighid Ó Dochartaigh, Alan MacDonald, and Claire Cousins. "Following the Flow—Microbial Ecology in Surface and Groundwaters in the Glacial Forefield of a Rapidly Retreating Glacier in Iceland." *Environmental Microbiology* 24, no. 12 (2022): 5840–5858.

Rammaáætlun. "The Master Plan for Nature Protection and Energy Utilization." Accessed December 19, 2024. https://www.ramma.is/english/.

Rekow, Lea. "Hydropower in Iceland: A Review of the Kárahnjúkar Project." Wright Ingraham Institute. March 18, 2023. https://wright-ingraham.org/hydropower-in-iceland-a-review-of-the-karahnjukar-project/.

Reuters. "Alcoa Drops Iceland Smelter Plan." October 17, 2011. https://www.reuters.com/article/alcoa-iceland-smelter-idUSL5E7LH3MR20111017/.

Rio Tinto. "Isal." Accessed December 19, 2024. https://www.riotinto.com/en/operations/iceland/isal.

Rosenthal, Elisabeth. "Hungary's Red Sludge Spill: The Media and the Eco-Disaster." Yale E360, October 21, 2010. https://e360.yale.edu/features/hungarys_red_sludge_spill_the_media_and_the_eco-disaster.

Saevarsdottir, Gudrun, Thordur Magnusson, and Halvor Kvande. "Reducing the Carbon Footprint: Primary Production of Aluminum and Silicon with Changing Energy Systems." *Journal of Sustainable Metallurgy* 7, no. 3 (2021): 848–857.

Sheller, Mimi. Aluminum Dreams: the Making of Light Modernity. MIT press, 2014: 209.

Sheller, Mimi. "The Vital Materiality of Aluminum: Light Modernity and the Global Atlantic." *Atlantic Studies* 11, no. 1 (2014):67–81. doi:10.1080/14788810.2014.872331.

Sigurðardóttir, Ragnhildur, Anthony J. Newton, Megan T. Hicks, Andrew J. Dugmore, Viðar Hreinsson, Astrid EJ Ogilvie, Árni Daníel Júlíusson et al. "Trolls, Water, Time, and Community: Resource Management in the Mývatn District of Northeast Iceland." *Global Perspectives on Long Term Community Resource Management* (2019): 77–101.

SisAl Pilot. "Silicor Materials Iceland Ehf." September 3, 2020. https://www.sisal-pilot.eu/partners/silicor-materials-iceland-ehf/.

Skúlason, Jón Björn, and Roger Hayter. "Industrial Location as a Bargain: Iceland and the Aluminium Multinationals 1962–1994." Geografiska Annaler: Series B, *Human Geography* 80, no. 1 (1998): 29–48.

Skúlason, Jón Björn. "Foreign Investment and the Bargaining Process: Case Study of the Aluminum Industry in Iceland." University of Iceland, 1992: 6–7.

Sveinsson, Óli Grétar Blöndal, Úlfar Linnet, and Halldór Björnsson. "Hydropower Expansion and Improved Management in Response to Increased Glacier Melt in Iceland." Climate Adapt, February 6, 2019. https://climate-adapt.eea.europa.eu/en/metadata/case-studies/hydropower-expansion-and-improved-management-in-response-to-increased-glacier-melt-in-iceland.

SunWatts. "13 KW Solar Battery." Accessed July 10, 2024. https://sunwatts.com/13-kwh-solar-battery/

Taylor, Gladstone. "Jamaica's 'Cockpit Country' Faces Growing Threats from Mining Interests." *Pulitzer Center.* November 21, 2022. https://pulitzercenter.org/stories/jamaicas-cockpit-country-faces-growing-threats-mining-interests.

Teichman, Theodore S. Interview with Hulda Brynjólfsdóttir. Personal, January 30, 2022.

Teichman, Theodore S. Interview with Magnus Johansson. Personal, January 29, 2022.

Thorsteinsson, Thorsteinn, and Halldór Björnsson, eds. *Climate Change and Energy Systems: Impacts, Risks and Adaptation in the Nordic and Baltic Countries.* 2012.

Trowell, K. A., S. Goroshin, D. L. Frost, and J. M. Bergthorson. "Aluminum and its Role as a Recyclable, Sustainable Carrier of Renewable Energy." *Applied Energy* 275 (2020): 112–115.

U.S. Embassy in Iceland. "History of the U.S. and Iceland." Accessed July 10, 2024. https://is.usembassy.gov/our-relationship/policy-history/io/.

Umhverfisstofnun. "General Information." Accessed December 4, 2024. https://ust.is/english/air-climate/eu-ets/aviation/general-information/.

WAFA, Wilderness Society and Conservation Council of Western Australia. "A Thousand Cuts." April 2022. https://www.wilderness.org.au/images/resources/A-Thousand-Cuts-NJF-Report.pdf.

Whale & Dolphin Conservation USA. "Whaling in Iceland." Accessed May 29, 2024. https://us.whales.org/our-goals/stop-whaling/whaling-in-iceland/

World Bank Climate Change Knowledge Portal. "Trends & Variability – Projection." Accessed December 18, 2024. https://climateknowledgeportal.worldbank.org/country/iceland/trends-variability-projections.

Wetzel, Corryn. "Lichens Are in an Evolutionary Race against Climate Change." Smithsonian.com. February 17, 2022. https://www.smithsonianmag.com/smart-news/lichens-are-in-an-evolutionary-race-against-climate-change-180979587/#:~:text=Their%20work%20revealed%20that%20lichens,increase%20of%20one%20degree%20Celsius.

Widengård, Marie. "The Mining-Conservation Nexus in Jamaica's Cockpit Country." *Edge Effects.* April 2, 2024. https://edgeeffects.net/cockpit-country-mining-conservation/.

Wigley, T. M. L., and S. C. B. Raper. "Implications for Climate and Sea Level of Revised IPCC Emissions Scenarios." *Nature* 357, no. 6376 (1992): 293–300.

Yu, Wenchao et al. "Climatic and Hydrologic Controls on Upper Paleozoic Bauxite Deposits in South China." *Earth-Science Reviews* 189 (2019): 159–176.

IMAGE CITATIONS + CREDITS

Listed by Page Number

78 Photo: Theodore Teichman.

Map sources: "Barbara Breitschopf and Lin Zheng. Rep. "Electricity Costs of Energy Intensive Industries in Iceland: A Comparison with Energy Intensive Industries in Selected Countries." Department of Energy, Industry and Business affairs, Ministry of Industries and Innovation, Iceland, November 2020, 36. "Kortasjá." Accessed May 16, 2025. https://kort.gis.is/mapview/?app=kort&lang=is."

80 Photo: Theodore Teichman.

83 Photo: Theodore Teichman.

Map sources: "List of Aluminium Smelters." Wikipedia, last modified June 4, 2025. https://en.wikipedia.org/wiki/List_of_aluminium_smelters; U.S. Geological Survey. "Major Mineral Deposits of the World" *U.S. Geological Survey Open-File Report 2005-1294*. Reston, VA: U.S. Geological Survey. Accessed June 8, 2025. https://mrdata.usgs.gov/major-deposits/; World Bank Group and International Monetary Fund. Global Ship Density – Commercial Vessels (AIS data January 2015–February 2021). World Bank Group, May 3, 2021. Accessed July 1, 2024. https://datacatalog.worldbank.org/search/dataset/0037580/Global-Shipping-Traffic-Density. Licensed under Creative Commons Attribution 4.0 International (CC BY 4.0).

86 Photo: Theodore Teichman.

Map sources: "Hydropower in Europe." Accessed May 19, 2025. https://www.hydropower.org/region-profiles/europe; "Hydro Power Plants." Accessed May 19, 2025. https://www.government.is/topics/business-and-industry/energy/hydro-power-plants/; "Kortasjá." Accessed May 16, 2025. https://kort.gis.is/mapview/?app=kort&lang=is; "Kort af Íslandi | Map of Iceland." Accessed May 19, 2025. http://www.map.is/.

89 Map sources: Prepp. "Indian Administrative Service – IAS Exam." Accessed May 19, 2025. https://prepp.in/news/e-492-bauxite-resources-in-india-geography-notes; ESRI; CGIAR; USGS; Maxar. World Bank Group and International Monetary Fund. Global Ship Density – Commercial Vessels (AIS data January 2015–February 2021). World Bank Group, May 3, 2021. Accessed July 1, 2024. https://datacatalog.worldbank.org/search/dataset/0037580/Global-Shipping-Traffic-Density. Licensed under Creative Commons Attribution 4.0 International (CC BY 4.0).

98 Map sources: Iceland Monitor. "Icelandic Glaciers Rapidly Retreating." Accessed May 16, 2025. https://icelandmonitor.mbl.is/news/nature_and_travel/2020/05/09/icelandic_glaciers_rapidly_retreating/

102 Photo: Theodore Teichman.

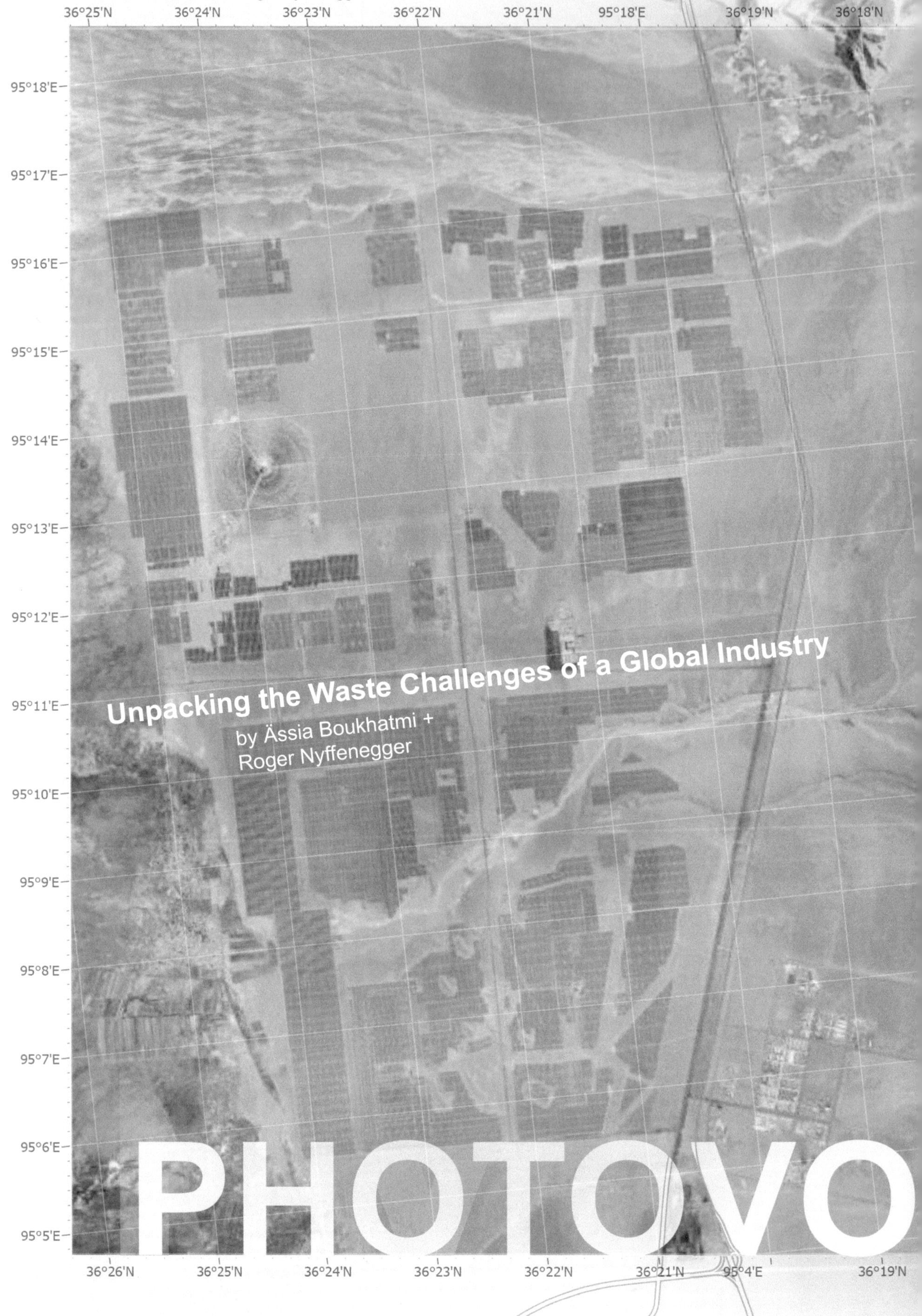

PHOTOVO

Unpacking the Waste Challenges of a Global Industry

by Ässia Boukhatmi +
Roger Nyffenegger

36°16'N
36°15'N
36°14'N
36°13'N
95°17'E
36°12'N
36°11'N
36°10'N
95°16'E
95°15'E
95°14'E
95°13'E
95°12'E
95°11'E
95°10'E
95°9'E
95°8'E
36°10'N
95°7'E
95°6'E
95°5'E
95°4'E
95°3'E
LTAICS
36°17'N
36°16'N
36°15'N
36°14'N
36°11'N

Introduction

From the dawn of civilization, the sun has been a symbol of life, power, and influence. Across cultures and epochs, humanity has revered the ethereal body of the sun, not just as a source of light and energy, but also as a sacred force that governs the rhythms of existence. The ancient Egyptians worshipped Ra, the sun god, believing he sailed across the sky in his solar barque each day, defeating the chaos of night to bring light to the world. In Hindu tradition, Surya, the sun deity, is depicted riding a chariot pulled by seven horses, symbolizing the seven colors of visible light. The ancient Incas venerated Inti, the sun god who they believed gifted life upon the earth, with the rays of his brilliance nurturing their crops and guiding their empire. These myths reveal a deep, almost universal reverence for the sun as the ultimate source of energy and life. This fascination has stood the test of time, evolving from sacred stories to scientific exploration and technological innovation.

Today, the global photovoltaic (PV) industry is a modern embodiment of this ancient worship, converting sunlight into electricity that is the fuel for our modern world. But in the wake of this change, new myths are being woven that recognize solar energy as not just a solution, but the savior of our energy- and resource-intensive lifestyles.

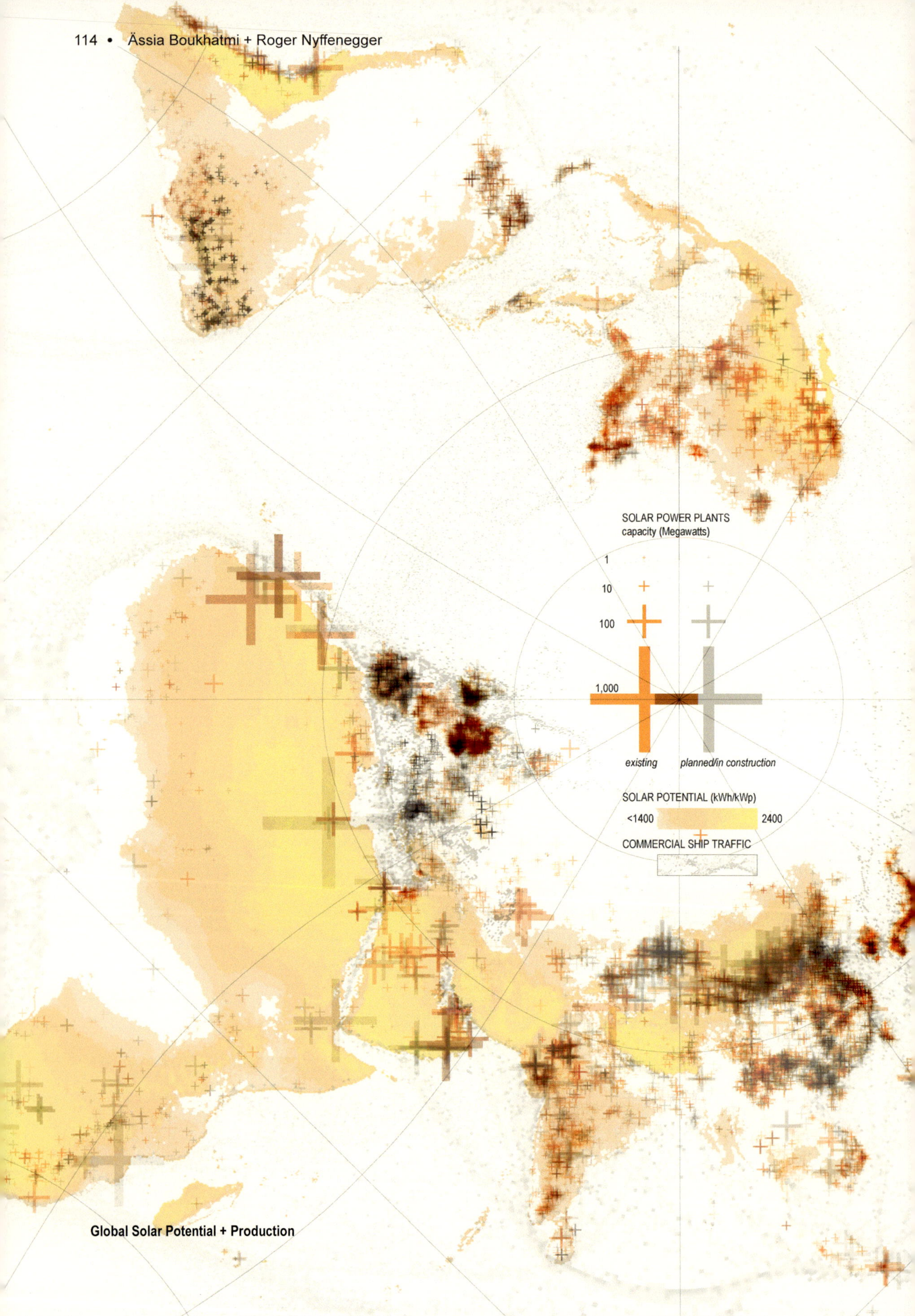

Global Solar Potential + Production

While solar energy is undoubtedly a greener alternative to fossil fuels and an important component of the climate transition, the idea that it is a flawless solution to our environmental challenges can be misleading. In our eagerness to embrace solar technology as a clean, renewable resource, we often overlook the complexities and contradictions that lie beneath the surface of our capitalist system in which resources are unconscionably extracted, utilized, and ultimately discarded. Starting at the beginning of a PV module's life cycle, production involves energy-intensive processes, scarce resources, and often harsh working conditions that can have a significant environmental and social impact. At the end of the life cycle, in turn, the disposal of PV modules poses a further challenge, contributing to the growing volumes of electronic waste—a problem that will place an increasing burden on waste management systems worldwide in the near future. Technically, a significant portion of solar modules could be reused, but the insatiable hunger of our prevailing economic model does not allow PVs to be given a second life. Instead, many of these modules, rather than being recycled responsibly in the countries where they are deployed, are often shipped to the Global South, untested and with no accountability for their end-of-life (EoL) disposal.

Just as ancient myths once attributed qualities such as omnipotence and purity to the sun, today we mythologize solar energy as a panacea for unsustainable practices. But much like these ancient tales, our modern myths can obscure the reality. To unpack these myths, we first examine the current linear system, shaped by global geopolitical price competition and growing waste streams. We then critically turn to the concept of the Circular Economy, presenting it as a transformative alternative that offers diverse strategies to combat resource depletion, overproduction, and escalating end-of-life volumes.

Global Surge in Solar Energy

In 2024, the Global Solar Council celebrated the historic achievement of reaching 2 TW of installed solar capacity worldwide.[1] For comparison, a solar capacity of 2 TW is equivalent to the annual electricity consumption of a country such as India and could potentially supply around 400 million households with electricity, which is twice the total number of households in the entire European Union.[2] This remarkable growth is largely attributable to substantial investments and technological advancements, which have made solar energy more accessible and cost-effective in recent years. Countries around the world are increasingly recognizing the potential of solar power to meet their energy needs while reducing carbon emissions to reach net zero by 2050.[3] While the global surge in solar capacity is significant, the pace and scale of photovoltaic expansion vary significantly across different regions.

1 "Global Installed PV Capacity Tops 2 TW," pv magazine International, November 13, 2024, https://www.pv-magazine.com/2024/11/13/global-installed-pv-capacity-tops-2-tw/.

2 IRENA, "IRENA – International Renewable Energy Agency," 2024, https://www.irena.org/; Pei-Hua Yu, "What Does 1 GW of Installed Solar Energy Mean? How Many Houses Can It Power?," *Medium* (blog), 2021, https://peihuayu.medium.com/what-does-1-gw-of-installed-solar-energy-mean-how-many-houses-can-it-power-bb32e-a6cc19.

3 United Nations, "The Paris Agreement I UNFCCC," 2015, https://unfccc.int/process-and-meetings/the-paris-agreement.

Africa, home to nearly one-fifth of the world's population, accounts for less than 3% of global energy-related CO_2 emissions. Yet, the continent is disproportionately affected by the negative impacts of climate change, including water scarcity, reduced food production, more frequent extreme weather events, and slower economic growth. Despite these challenges, solar installations across the continent are gaining momentum. Notably, the Benban Solar Park in Egypt, with a capacity of 1.8 GW—enough to power around 360,000 homes—stands as the largest solar farm in Africa and the sixth largest in the world.[4] By 2030, solar PV is expected to provide around 15% of Africa's power generation, which will require an increase in investment: currently, most solar projects are still primarily funded by external donors, which often lack financial sustainability and limit involvement of national governments. Thus, policy interventions by both African governments and international actors are essential to accelerate solar energy adoption across the continent.[5]

2023 marked another record year for solar PV in the European Union, with almost 60 GW installed across 27 member states, showing a doubling of market growth in only two years. Especially in the midst of the invasion of Ukraine, solar energy represents an environmentally friendly, cost-effective, and incomparably fast solution to reduce the dependence on Russian fossil fuels. One reason for this tremendous expansion of solar power is the EU-wide political measures that demand an increase in new installations. In addition, the cost of solar power decreased by 82% between 2010 and 2020, making it the most competitive electricity source in many parts of the EU[6], which is reflected in the fact that almost 17 million more European homes became powered by solar in 2023.[7] However, the EU also recognizes the potential of energy communities—collective, decentralized initiatives in which citizens, small businesses, and local entities jointly produce, manage, and consume renewable energy. Recognizing this potential, the REPowerEU Plan sets a shared goal of establishing at least one energy community in every EU municipality with a population over 10,000 by 2025.[8]

In 2023, Asia had over 840 GW of solar energy capacity, which is three times more than Europe. In the same year, China installed more solar panels than any other nation, and was responsible for 60% of the added solar capacity, versus Europe's 14%.[9] Of course, this figure is linked to the fact that China has the second-largest population in the world and tops the list of the largest solar manufacturing countries worldwide. The roots of China's renewable energy dominance can be traced back to investment decisions made in the mid-2000s, as the country's decade-long period of rapid GDP growth was drawing to a close. During the early

4 Tom Gill, "The 15 Largest Solar Farms in the World 2022," *The Eco Experts* (blog), 2022, https://www.theecoexperts.co.uk/solar-panels/biggest-solar-farms.

5 Ademola A. Adenle, "Assessment of Solar Energy Technologies in Africa-Opportunities and Challenges in Meeting the 2030 Agenda and Sustainable Development Goals," *Energy Policy* 137 (February 1, 2020): 111180, https://doi.org/10.1016/j.enpol.2019.111180.

6 European Commission, "Solar Energy," 2022, https://energy.ec.europa.eu/topics/renewable-energy/solar-energy_en.

7 Solar Power Europe, "EU Market Outlook for Solar Power 2023–2027," 2024, https://www.solarpowereurope.org/insights/outlooks/eu-market-outlook-for-solar-power-2023-2027#download.

8 European Commission, "Solar Energy."

9 Ember, "Solar Power," Ember, 2022, https://ember-climate.org/topics/solar/.

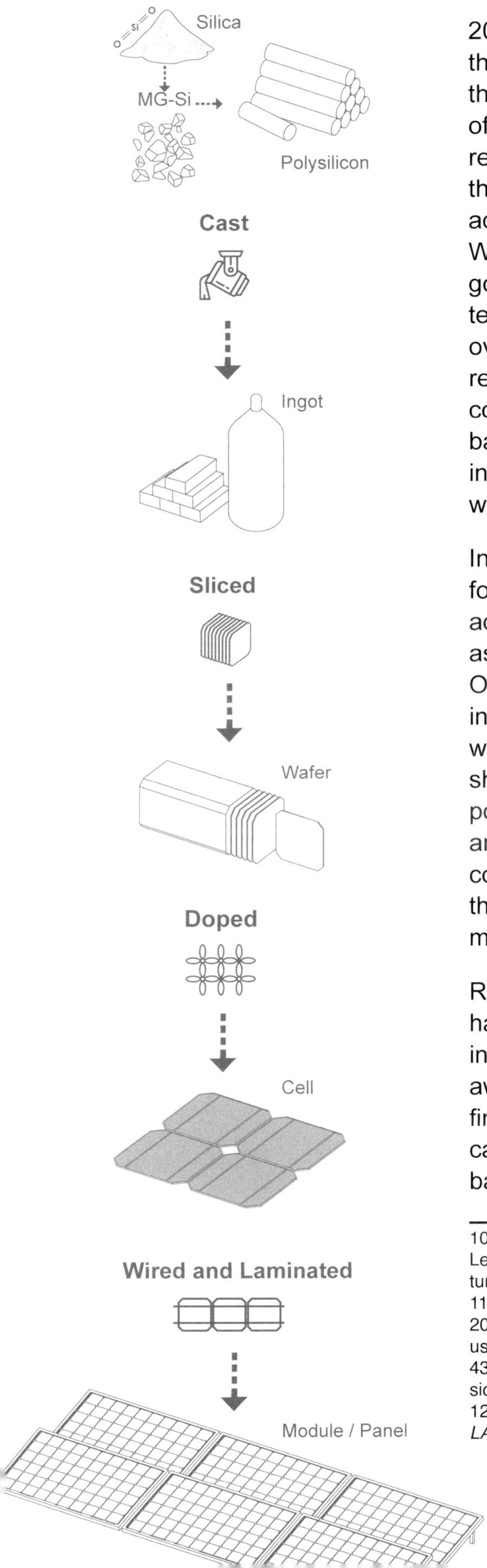

2000s, China's greenhouse gas emissions more than doubled, and by 2006, it had surpassed the U.S. to become the world's largest emitter of greenhouse gases by volume. Faced with this reality, Chinese policymakers sought investments that would pave the way for a more technologically advanced future while also curbing emissions. Within a decade, China not only achieved its goal of dominating the global production of solar technologies but also established a near-monopoly over their supply chains. The vast scale of China's renewable energy production has driven down costs worldwide, significantly lowering the financial barriers for renewable energy adoption, especially in lower-income countries.[10] Later in this chapter, we will shed light on the other side of this coin.

In the first quarter of 2024, solar energy accounted for 75% of all new electricity-generating capacity added to the U.S. grid, reinforcing its position as the leading source of new power capacity. Over the next five years, the U.S. is projected to install around 40 GW of solar capacity annually, with growth slowing due to constraints like labor shortages, high equipment costs, and trade policy uncertainties. Potential new policies and antidumping duties on Southeast Asian imports could further impact the supply chain, but at the same time provide potential to protect U.S. manufacturers.[11]

Rich in solar sources, Latin and South America have the potential to become worldwide leaders in renewable energy. The global pressure to move away from fossil fuels accompanied by vanishing financing and the effectiveness of opposition campaigns serve to enhance the attractive baseline conditions for solar development.[12]

10 Yale School of Environment, "How China Became the World's Leader on Renewable Energy," 2024, https://e360.yale.edu/features/china-renewable-energy.
11 Solar Energy Industries Association, "US Solar Market Insight," 2024, https://www.woodmac.com/industry/power-and-renewables/us-solar-market-insight/thank-you/?__FormGuid=8e2fb1e9-c50c-43ef-9601-e9e96361421a&__FormLanguage=en&__FormSubmissionId=68e0222d-13e6-4be3-b19d-10f8e7dc1c27.
12 Sophia Bauer et al., "A Race to the Top: Latin America 2023," *LATIN AMERICA*, 2023.

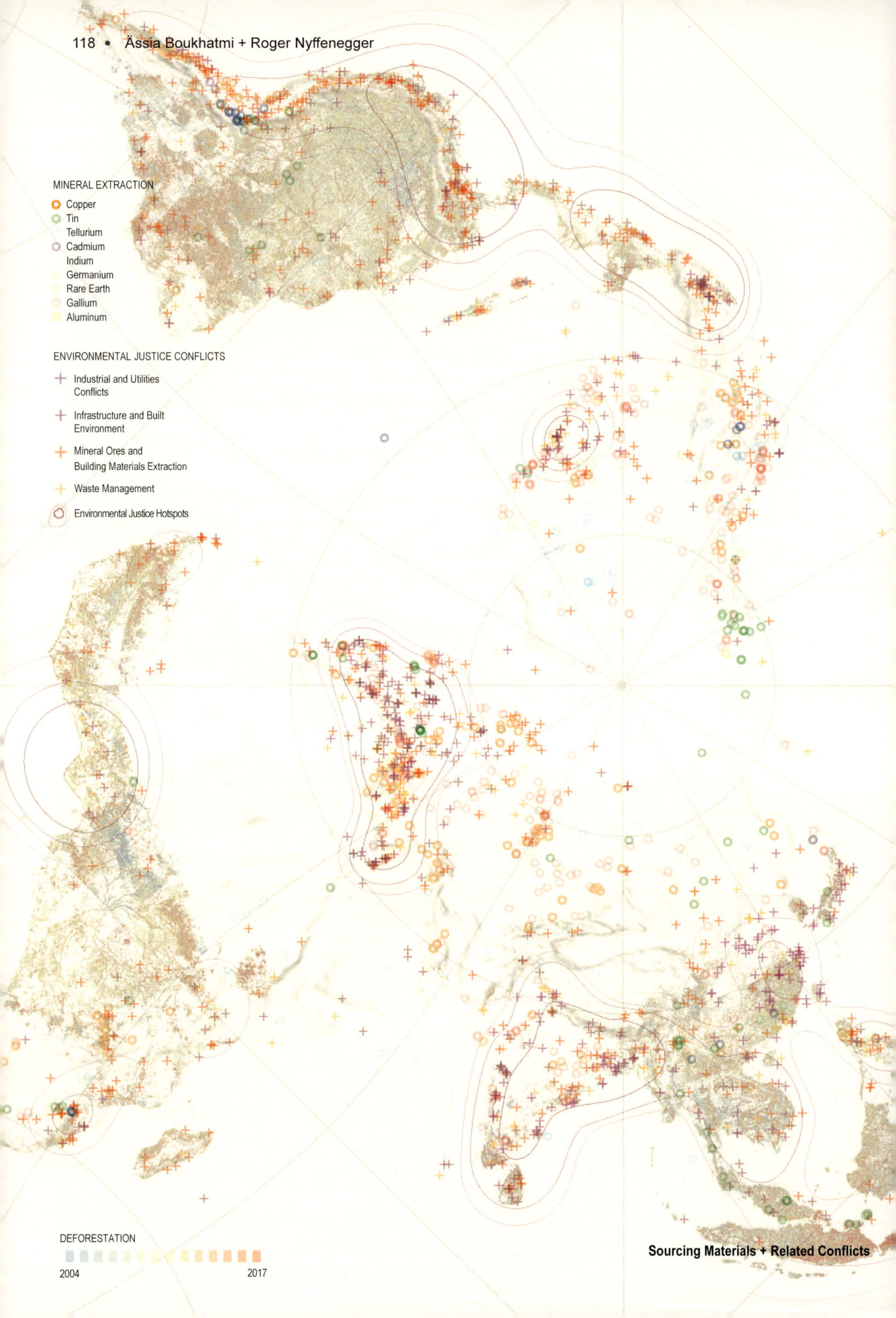

Sourcing Materials + Related Conflicts

Today, Brazil, Chile, and Colombia are at the vanguard for operating utility-scale solar, and owe their continued expansion to well-established energy auctions, openness to private investments, the decreasing costs of solar installations, and policy responses to climate change. In contrast, Mexico led the development of large-scale solar energy from 2013 to 2021, but changes to national energy policy in May 2021 have dampened growth prospects. As the region still tries to advance into the renewable energy future, these contrasting trends underscore the vital role of national policy frameworks.[13]

The rapid growth in global solar PV capacity marks a significant shift toward renewable energy, driven by diverse regions in unique ways. This transformation is shaped by an interplay of regional dynamics, policy frameworks, economic conditions, and investment strategies. From Africa's emerging solar markets and Europe's accelerated policy-driven expansion to Asia's dominance in production and America's large-scale utility deployments, solar power is now central to the global pursuit of a sustainable future. But with this unprecedented scale comes a pivotal question:

Can this transformation itself be truly sustainable?

The Myths of a Sustainable Solar Industry

The global shift toward renewable energy demands immense resources to produce new solar modules and system components. Although solar PV modules are often regarded as inherently sustainable, they actually adhere to the same linear production model as other products. To address this misconception, let us debunk this *first myth*. While it is true that the solar PV sector's linear structure has facilitated substantial growth in solar energy worldwide, as discussed above, this model brings significant challenges to true sustainability. The PV industry, like many others, not only contributes to pollution but also to the degradation of natural ecosystems, exacerbating global challenges such as climate change and biodiversity loss.[14] Materials are extracted, manufactured into PV modules using significant amounts of energy, utilized over a certain period, and ultimately discarded at the end of their life cycle.

PV modules and subsequent materials such as inverters, cables, and mounting structures are rarely used to their full potential since a linear system moves in a single direction—from raw material to waste.[15] To illustrate this, consider the amount of silver required to produce PV modules with a capacity of 1 TW, which is roughly the annual global production capacity of 2024—and approximately the

13 Bauer et al., "A Race to the Top: Latin America 2023."
14 Muhammad Tawalbeh et al., "Environmental Impacts of Solar Photovoltaic Systems: A Critical Review of Recent Progress and Future Outlook," *Science of The Total Environment* 759 (March 2021): 143528, https://doi.org/10.1016/j.scitotenv.2020.143528.
15 Roger Nyffenegger et al., "How Circular Is the European Photovoltaic Industry? Practical Insights on Current Circular Economy Barriers, Enablers, and Goals," *Journal of Cleaner Production* 448 (April 5, 2024): 141376, https://doi.org/10.1016/j.jclepro.2024.141376.

amount required to meet India's total electricity needs.[16] Maintaining business as usual could require over 20% of the current annual silver supply by 2027 and a cumulative 450–520 kt of silver until 2050, approximately 85–98% of the current global silver reserves.[17] Therefore, the myth of a sustainable solar industry can be debunked by looking at the extreme sums of material needed to support a linear model.

However, besides those negative impacts, linearity also has positive outcomes. The linear value chain not only created the conditions for making the production of PV modules more efficient, but also for scaling them up considerably in the last years. As a result, module prices fell drastically and made solar energy competitive with other energy sources, with hugely positive effects on efforts to achieve the energy transition and curb CO_2 emissions. The linear nature of the solar PV module value chain has therefore also facilitated the expansion of solar energy. This raises the critical question: *How can the increasing demand for the expansion of PV systems be met while minimizing environmental and societal impacts*?The *second myth* we will address relates to the solar PV industry's promise to free energy production from dependence on fossil-fuel-producing countries, many of which face serious human rights issues. While it is true that solar PV reduces reliance on fossil fuel and generally on import dependency—once installed, it can generate electricity for up to 30 years—there is often an overlooked aspect: China's central role in the global PV market.[18] With approximately 80% of the world's solar module production capacity, China's dominance stems from substantial government subsidies and strategic investments in solar technology since the late 2000s.[19] Under these favorable market conditions, Chinese manufacturers have achieved economies of scale that allow them to produce solar modules at exceptionally low costs. This has led to a substantial expansion of production capacity in China, but with the recent decline in China's domestic demand, the global market now faces an oversupply of PV modules. This oversupply has put even more of a downward pressure on prices, forcing non-Chinese manufacturers, particularly in Europe and the U.S., out of the market.

At the same time, amidst China's rapid expansion in the PV industry, there have been growing concerns about forced labor practices, particularly in the Xinjiang region. Reports indicate that a substantial portion of the global solar supply chain is linked to forced labor, raising ethical and human rights issues.[20] In response, the U.S. have implemented measures such as the Uyghur Forced Labor Prevention Act, which have not had a significant impact on the issue.[21] Taking the question

---

16 International Energy Agency, "Electricity 2025 – Analysis and Forecast to 2027," February 2025.

17 Brett Hallam et al., "The Silver Learning Curve for Photovoltaics and Projected Silver Demand for Net-Zero Emissions by 2050," *Progress in Photovoltaics: Research and Applications* 31, no. 6 (2023): 598–606, https://doi.org/10.1002/pip.3661.

18 Arcipowska Aleksandra et al., "Role of Solar PV in Net-zero Growth: An Analysis of International Manufacturers and Policies," *Progress in Photovoltaics: Research and Applications* 32, no. 9 (September 2024): 607–622, https://doi.org/10.1002/pip.3797.

19 International Energy Agency, *Special Report on Solar PV Global Supply Chains*, 2022, https://doi.org/10.1787/9e8b0121-en.

20 Alan Crawford and Laura T Murphy, "Uyghur Region Exposure Assessment for Solar Industry Sourcing," *Sheffield Hallam University Helena Kennedy Centre for International Justice*, November 2023.

21 "Uyghur Forced Labor Prevention Act I U.S. Customs and Border Protection," Accessed February 19, 2025, https://www.cbp.gov/trade/forced-labor/UFLPA.

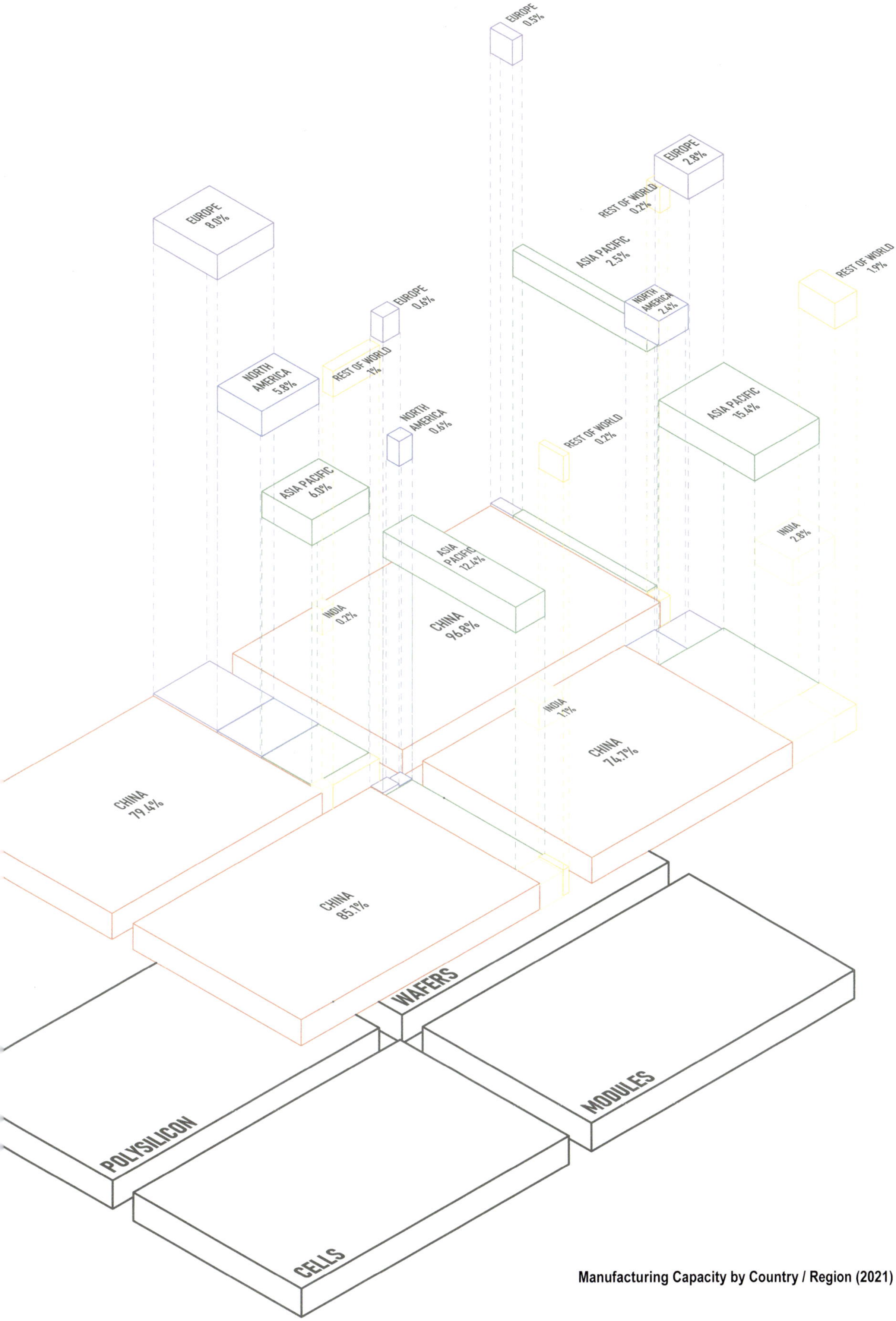

Manufacturing Capacity by Country / Region (2021)

raised above, we can now add: *How can the increasing demand for the expansion of PV systems be met while minimizing the environmental and societal impacts, particularly in a global market dominated by China?* As all good things come in threes, there is a *third myth* to consider, which is closely interlinked with the linear organization of the PV industry. There is a widespread underestimation of the huge numbers of end-of-life PV systems coming up in the next years. Initial projections estimate that global cumulative PV panel waste could reach 78 million tons by 2050,[22] with Europe alone potentially contributing up to 33 million tons, the USA 10 million tons, and China 20 million tons.[23]

22 IRENA, "End-of-Life Management Solar Photovoltaic Panels," 2016, https://www.irena.org/publications/2016/Jun/End-of-life-management-Solar-Photovoltaic-Panels.

23 Amjad Ali et al., "Policies and Regulations for Solar Photovoltaic End-of-Life Waste Management: Insights

PV Waste Comparison

To put this into perspective, managing 78 million tons of waste would require approximately 3.9 million truckloads, with each truck carrying 20 tons. Lined up end-to-end, this would form a line of trucks stretching about 78,000 kilometers—nearly twice the circumference of the Earth. Or, alternatively, a volume that dwarfs the Burj Khalifa, the tallest building in the world. Imagine the vast amount of space, equipment, energy, and financial investment needed to collect, transport, and recycle such a volume by 2050. Even in regions like Europe, where regulations such as the Waste Electrical and Electronic Equipment (WEEE) Directive[24] exist, managing end-of-life PV systems presents substantial challenges. This is primarily because such regulations only apply once a PV system is officially classified as waste.[25] Ultimately, how the PV industry addresses these waste management issues will play a crucial role in determining its future reputation and the degree to which it is seen as genuinely "green." Taking up the above raised question, we can expand upon it by asking: *How can the increasing demand for the expansion of PV systems be met while minimizing environmental and societal impacts, particularly in a global market dominated by China, and simultaneously addressing the end-of-life challenges?*

from China and the USA," *Chemosphere* 340 (November 1, 2023): 139840, https://doi.org/10.1016/j.chemosphere.2023.139840; Czajkowski et al., "Energies I Free Full-Text I Prediction of the Market of End-of-Life Photovoltaic Panels in the Context of Common EU Management System," 2022, https://www.mdpi.com/1996-1073/16/1/284.resulting in a substantial increase in the amount of end-of-life.

24 European Parliament & Council, "Directive 2012/19/EU of the European Parliament and of the Council of 4 July 2012 on Waste Electrical and Electronic Equipment (WEEE)" (Official Journal of the European Union, July 24, 2012).

25 Simone Cimadomo, "Advancing a Circular Economy for Solar Photovoltaics Exported for Reuse – Analysing the Institutional Feasibility of International Extended Producer Responsibility for EU-West Africa Transboundary Movements" Lund: Master thesis – IIIEE. Lund University, 2024.

In light of these myths, the industry's future path is increasingly locked in its current trajectory. Present investments are mainly directed toward optimizing the existing linear model, focusing on efficiency, material use, and expanding sales channels. China is anticipated to maintain its dominant position within the solar value chain, outpacing firms in other regions. Furthermore, the current linear structure of the PV industry offers limited incentives to address the end-of-life challenges associated with growing waste streams. Thus, while the industry supports the global energy transition by providing affordable solar products, it passes the burden of waste management onto society, allowing dominant players to continue reaping substantial profits.

Having set the scene, let us unpack some other essential aspects of the solar PV industry. Generally, actors within the solar ecosystem lack integration and cooperation across value chains, partly due to their geographic dispersion and limited transparency. While production-stage actors may collaborate with their immediate upstream suppliers and customers, they, for example, rarely engage with end-of-life stakeholders to facilitate access to recycled materials. Additionally, markets for secondary materials in the upstream value chain are underdeveloped, primarily due to a lack of recovery infrastructure, reverse logistics, and quality testing systems, particularly for materials like glass and silicon. Meanwhile, the availability of inexpensive virgin materials further hampers the development of secondary markets. By-products generated during the production process such as the sidewall slabs, tails, and tops from ingot squaring or kerf from wafer slicing are challenging to access and subject to supply fluctuations, exacerbating the disconnect between value chain actors who lack a mutual understanding of each other's challenges. Another significant barrier lies in the limited exchange of information across the value chain, leading to insufficient traceability of PV modules. For instance, PV module manufacturers often lack knowledge about the locations where their products are deployed. As a result, access to critical data on performance, early failures, reparability, and recyclability of PV modules is restricted, hindering efforts to improve the solar industry[26].

A New Circular Paradigm in Global PV Production?

While the rapid expansion of the global solar industry is a crucial driver of the climate transition, it also presents a range of challenges that affect every stage of the PV value chain. Starting at the upstream value chain, vast quantities of resources need to be exploited in order to meet the volumes required for the annually increasing targets for new installations. If we continue to relentlessly extract virgin materials to supply the solar industry, certain elements, like silver, could be depleted from the earth in the foreseeable future. In the midstream value chain, current legislation, such as in Europe,[27] incentivize a premature replacement of solar panels for repowering purposes, which leads to dismantling of modules

26 Nyffenegger et al., "How Circular Is the European Photovoltaic Industry?"
27 See European Union. "Directive 98/70/EC as Regards the Promotion of Energy from Renewable Sources," 2023. https://eur-lex.europa.eu/legal-content/EN/TXT/HTML/?uri=OJ%3AL_202302413.

that would still be technically functionable. At the end of the value chain, in turn, such events will increasingly reinforce the rising growth of end-of-life volumes in the future. As the number of produced and installed modules grows each year, we will inevitably face a corresponding increase in modules entering the waste stream.

So, what could be a viable solution for addressing the increasing demand for PV systems, particularly in a global market dominated by China, while simultaneously addressing end-of-life challenges? A sustainable expansion of the global solar industry would require a fundamental shift from our current linear model to a fully circular one. In an ideal circular economy, resource and energy flows are maintained in a closed loop to minimize waste and losses. Production facilities would rely less on virgin materials and repurpose production waste internally. The lifespan of solar modules would be maximized through extended minimum lifetimes, as modules would mostly be made available through product-service systems rather than pure sales. Reuse and repair activities would ensure that products only enter the end-of-life stream when they are no longer technically functional. Recycling processes would be so efficient that materials recovered from old solar panels could be seamlessly reintegrated into the production of new ones. Generally, there are nine key R-principles[28] that can be applied across different stages of industrial value chains.[29] Three of these principles are the most common and widely practiced ones: reduce, reuse, and recycle. We will look at these three principles in more detail to understand the extent to which the global solar industry has already made efforts to become more circular.

Reduce

The principle of reduce focuses on minimizing waste from the outset by making production processes more sustainable.[30] The strategies we discussed earlier—such as the re-introduction of internal and external production waste into manufacturing, the use of recycled materials, modular design with fewer material input, as well as a lower energy consumption and adherence to high environmental and ethical standards—align with this principle. The re-introduction of production waste is already common in industrial processes for economic reasons, although it is not always labelled as a circular economy practice. But how easy is it to re-introduce production waste in the PV panel production? A typical crystalline photovoltaic module has a layered sandwich structure and weighs between 18 to 25 kg (approximately 40 to 55 lb). It mainly consists of approximately 67% glass, 15% aluminium, 5% silicon, and 10% different plastics (EVA, PET, PVF), plus valuable metals such as silver and copper (<1% each).[31]

The flat glass used in PV modules is typically made from silica sand, which is abundant and can be sourced from various regions worldwide, such as China,

28 From high to low circularity: Refuse, Rethink, Reduce, Reuse, Repair, Refurbish, Remanufacture, Repurpose, Recycle, Recover.
29 José Potting et al., *Circular Economy: Measuring Innovation in the Product Chain*, 2017.
30 Potting et al.
31 Rong Deng, Yuting Zhuo, and Yansong Shen, "Recent Progress in Silicon Photovoltaic Module Recycling Processes," *Resources, Conservation and Recycling* 187 (December 1, 2022): 106612, https://doi.org/10.1016/j.resconrec.2022.106612.

North America, and Europe.[32] The glass in particular is subject to high quality requirements (purity, optical clarity, and durability) in order to protect the individual components and withstand the many years of weathering to which the module is exposed. In addition, module types are constantly changing so that no two production batches are the same. "The smallest changes in the composition of the glass would require the process parameters in the melt furnace to be adjusted, which is why the use of glass waste from internal or external sources is not possible at the moment," as one of the largest European flat glass manufacturers confirmed in an interview. The situation is different with silicon, which, despite its high material quality requirements, can be utilized from production waste in other industry sectors thanks to advanced purification processes that bring it to the level of purity required for new solar cells. Accordingly, solar module manufacturers—for example, in Europe—are already utilizing silicon kerf scraps from other industrial sectors. Recyclates are also amongst the popular external by-products, but their use is often limited due to the low material purity achieved with current recycling technologies.[33]

The use of recycled materials is not the only way to reduce resource consumption at the beginning of the value chain. Technological advancements can significantly enhance resource efficiency in solar module production. For instance, perovskite-on-silicon tandem solar cells improve upon traditional silicon panels by stacking a thin layer of perovskite, a crystalline mineral, on top of the silicon layer. This configuration not only allows for higher energy conversion efficiencies—exceeding 30% compared to the typical 20–22% of conventional silicon cells—but also results in lighter modules that require less material overall. By generating more electricity from a reduced mass, these tandem cells help decrease the demand for raw materials and support the development of a more circular solar industry.[34] Such advancements highlight the potential of technology to improve practices and promote sustainability across the solar value chain. While the solar industry has taken steps to implement "Reduce" practices, such as improving material efficiency and advancing PV technology, these efforts are often more focused on cost reduction than true sustainability. Nevertheless, PV manufacturers in Europe recognize a growing demand for ethically and environmentally friendly produced modules, which could provide a promising differentiation strategy in the future global market. As one European manufacturer reflected on in an interview, consumers have increasingly started asking questions that imply a growing consciousness about ethical practices: "Where do you get your materials from? Where are your certificates to prove they aren't from forced labor?" Although many buyers do "focus on efficiency and capacity, the criteria of being manufactured and designed in Europe is increasingly important to them."[35] But ultimately and in the near future, the reliance on virgin materials, energy-intensive manufacturing processes, and limited consideration for the full lifecycle of PV modules reveal significant gaps in adopting a genuinely circular approach. To progress further, the industry must

32 Keith Burrows and Vasilis Fthenakis, "Glass Needs for a Growing Photovoltaics Industry," *Solar Energy Materials and Solar Cells* 132 (January 1, 2015): 455–459, https://doi.org/10.1016/j.solmat.2014.09.028.
33 Nyffenegger et al., "How Circular Is the European Photovoltaic Industry?"
34 Oxford PV, 2024, https://www.oxfordpv.com/.
35 Nyffenegger et al., "How Circular Is the European Photovoltaic Industry?"

move beyond reducing material use to embrace broader circular principles after the first effective lifecycle of a PV module.

Reuse

Having shed light on reducing materials in the production process, we now turn our attention to the principle of reuse. By extending the operational life of PV modules, reuse minimizes the demand for new raw materials and reduces the energy consumption and carbon emissions associated with producing new modules. Additionally, reuse offers cost advantages by making PV modules more affordable, particularly in regions where the high upfront costs of new installations pose a significant barrier. Moreover, reuse prevents functional modules from being prematurely recycled, thereby allowing recycling technologies more time to advance toward effectively closing the loop, as module volumes enter the recycling stream later due to their extended lifetimes. This approach is especially relevant for PV modules decommissioned before their official end-of-life, often replaced by newer, more efficient models due to technological advancements and current low prices. Such repowering activities are becoming increasingly common in developed economies with large-scale PV installations, such as those in the EU, and are likely to result in future PV waste volumes that exceed current projections. Implementing effective reuse strategies will be crucial in providing suitable second-life solutions for these modules.

However, reuse currently remains niche and often informal due to a range of barriers that hinder the development of a robust reuse market for PV modules.[36] One major obstacle is the lack of a standardized protocol for preparing PV modules for reuse. Such a protocol should provide cost-effective, clear, and straightforward guidelines for testing the performance, longevity, and safety of decommissioned PV modules.[37] Additional barriers to market development include insufficient traceability, poor logistics handling, unpredictable volumes, low awareness and acceptance of secondhand PV modules, an unfavorable regulatory environment, and a lack of guaranteed performance, safety, and quality.[38] Furthermore, the current low prices of new PVs hinder the market for reused modules, as the decreasing profit margins make them less attractive. Despite these challenges, reuse remains the preferred option over recycling from a circularity perspective, provided that the PV modules still possess a remaining technical lifetime and are reused locally. According to calculations by the International Energy Agency (IEA), extending the lifetime of PV modules through reuse, rather than opting for immediate recycling, can significantly reduce environmental impact and enhance circularity, especially when the second-life installation is geographically close to the decommission site.[39]

36 Bertrand Lempkowicz and Jade Liu, "RE-USE of PV Modules, Challenges and Opportunities of the Circular Economy – Study of Re-Used PV Modules," Press release – PV Cycle & imec, August 3, 2021, https://pvcycle.be/nl/wp-content/uploads/Press-Release-Reuse-08032021.pdf.
37 Arvid van der Heide et al., "Re-Use of PV Modules: Progress in Standardisation and Learnings from a Real Case Study," 2023, https://doi.org/10.4229/EUPVSEC2023/5DO.15.6.
38 Roger Nyffenegger et al., "How Circular Is the European Photovoltaic Industry?"; G Oviedo Hernandez et al., "Trends and Innovations in Photovoltaic Operations and Maintenance," *Progress in Energy* 4, no. 4 (October 1, 2022): 042002, https://doi.org/10.1088/2516-1083/ac7c4f.
39 International Energy Agency, "Advances in Photovoltaic Module Recycling – Literature Review and Update to Empirical Life Cycle Inventory Data and Patent Review," IEA PVPS Task 12 Sustainability, 2024.

In addition to local reuse, second-life activities for PV modules in non-European countries have become increasingly common. Research indicates that countries, particularly developing nations in the West Asia North Africa region, represent a growing market for the international reuse of solar PV modules.[40] These countries have a demand for affordable off-grid energy to support local livelihoods and economic opportunities, even if the panels have lower performance. Applications include home energy, battery charging, and agricultural needs like irrigation and refrigeration. While the global market for reused solar PV is currently limited compared to new installations, it is expected to grow as the volume of decommissioned panels in the EU increases. However, this trade presents several ambiguities and risks, particularly in relation to extended producer responsibility (EPR). Unlike electrical and electronic waste, used electrical and electronic products fall outside the scope of current EPR schemes. As a result, exported volumes of used products are excluded from the estimates of waste generated in the EU, allowing producers to evade EPR fees if these are not collected upfront. Consequently, EPR schemes in the EU can be criticized for inadvertently incentivizing exports for reuse. Furthermore, exporters, through direct contracts with PV system owners, can bypass national EPR collection schemes, diverting PV module volumes and exporting used products, including non-functional equipment falsely declared as used products. This issue arises due to the lack of clear standards for assessing the functionality and reliability of used products, blurring the distinction between effective waste and used products.[41] Additionally, the Harmonized Commodity Description and Coding System (HS codes) used by the World Trade Organization does not provide a clear differentiation between waste and used products. This ambiguity complicates the identification of legal exports of used products, which are permissible under the EU Waste Shipment Regulation and the Basel Convention,[42] versus illegal exports of waste.[43] Although the current WEEE Directive specifies responsibilities for inspections and sets requirements for distinguishing between used products and waste, enforcement at exporting ports remains largely inadequate, and fraudulent activities leading to illegal shipments are prevalent.

Overall, export activities have negative implications for both Europe and non-European countries. For Europe, the export of PV modules diverts significant quantities away from local reuse and recycling initiatives. In terms of reuse, this reduces the opportunity to extend the operational life of PV modules, thereby decreasing reliance on new modules and reducing the energy consumption and

40 Cimadomo, "Advancing a Circular Economy for Solar Photovoltaics Exported for Reuse.

41 Kaustubh Thapa et al., "Transboundary Movement of Waste Review: From Binary towards a Contextual Framing," *Waste Management & Research* 41, no. 1 (January 1, 2023): 52–67, https://doi.org/10.1177/0734242X221105424.

42 The Basel Convention on the Control of Transboundary Movements of Hazardous Wastes and their Disposal was adopted on 22 March 1989 by the Conference of Plenipotentiaries in Basel, Switzerland, in response to a public outcry following the discovery, in the 1980s, in Africa and other parts of the developing world of deposits of toxic wastes imported from abroad (Accessed February, 18, 2025: www.basel.int/Home)

43 Apoorva Arya and Arpit Bhutani, "Study on Items Shipped for Reuse and Extended Producer Responsibility Fees – A Case for Extending EU EPR Fees to Cover End-of-Life Activities of Products Shipped Outside the EU," Open Society Foundation and the LIFE Programme of the European Union, May 17, 2023, https://eeb.org/library/items-shipped-for-reuse-and-extended-producer-responsibility-fees-a-case-for-extending-eu-epr-fees-to-cover-end-of-life-activities-of-products-shipped-outside-the-eu/.

carbon emissions associated with their production. Regarding recycling, exports hinder the growth of recycling facilities, as this industry relies on a consistent volume of inflow to remain economically viable. As a result, fewer secondary materials are recovered, despite the increasing demand for these materials in the production of low-carbon products. For non-European countries importing electrical and electronic equipment, the burden of end-of-life management falls on them without financial support from the original producers. Additionally, since functionality testing is typically not conducted before export, and the collection and transportation are often handled improperly, a substantial number of exported PV modules arrive damaged or broken. Consequently, many of these modules are frequently sent to hazardous e-waste processing sites and finally disposed of in landfills due to a lack of appropriate recycling facilities, with severe environmental and social consequences.

As important receiving countries, a specific focus is on Nigeria and Ghana. Recent studies have revealed that up to 30% of used electric and electronic product imports into Ghana are non-functional and, therefore, classified as waste.[44] In Nigeria, this figure is estimated to be as high as 70%. Across the African continent, estimates suggest that 150,000 to 250,000 tons of waste are shipped annually, with Nigeria and Ghana identified as primary recipients. Notably, Europe is the source of approximately 75% of these shipments. [45] In Nigeria, despite efforts to regulate e-waste imports and the establishment of an EPR scheme, enforcement remains weak. As a result, electric and electronic waste continues to enter Nigeria, often ending up in informal and hazardous e-waste processing sites like Lagos's "computer village."[46] Similarly, Ghana has implemented its Hazardous and Electronic Waste Control and Management Act and developed guidelines for e-waste management; however, the absence of an EPR scheme and the lack of ratification of the Bamako Convention[47] have resulted in ongoing challenges.[48] In both countries, the improper handling of electric and electronic waste poses significant environmental and health risks, with the World Health Organization (WHO) estimating that nearly 30% of Nigeria's health burden in 2021 was directly linked to e-waste exposure.

From a circular economy perspective, the export of PV modules is problematic. These modules, containing critical materials that were originally imported from Asia, primarily China, are often exported after use, leading to environmental and social harm. As these critical materials become increasingly scarce, it is more sustainable to retain and recycle them through proper processes. By preserving the PV module base through reuse and extending their lifespan, European countries can better supply their industries with recycled materials, thereby reducing energy

44 Arya and Bhutani. "Study on Items Shipped for Reuse and Extended Producer Responsibility Fees."
45 Arya and Bhutani. "Study on Items Shipped for Reuse and Extended Producer Responsibility Fees."
46 Cimadomo, "Advancing a Circular Economy for Solar Photovoltaics Exported for Reuse."
47 Adopted in 1991, comprising 30 articles and three annexes, The Bamako Convention, aims to regulate and minimize the transboundary movement and disposal of hazardous wastes within Africa. It reflects a collective commitment amongst member states to safeguard the continent's environmental integrity and public health (Accessed February 18, 2025: www.bamakoconvention.org).
48 Cimadomo, "Advancing a Circular Economy for Solar Photovoltaics Exported for Reuse."

consumption, carbon emissions, and geopolitical dependencies.[49] This strategy also provides an opportunity to strengthen the manufacturing sector, ensuring the capacity to reutilize recycled materials when closed-loop systems are achieved, while maintaining the expertise needed for repair and maintenance activities. However, when PV modules are exported to meet the high demand for affordable off-grid energy in importing countries, regulatory changes are needed to minimize the negative impact on these nations. This includes extending the geographic scope of EPR, placing the financial responsibility for collection and recycling on the original manufacturers, regardless of the products' final destination. Such an approach would facilitate the transfer of financial resources from Europe to importing countries, like Nigeria or Ghana—a proposal these countries have also advocated for other products, such as textiles and plastics.

Recycle

Have you ever noticed a sparkling road surface? Often, this glitter comes from recycled glass particles, which might originate from old PV modules.[50]

Recycling is the last circular strategy at the end of the value chain and applies when products cannot be repaired or refurbished after reaching their operational end-of-life. Recycling involves breaking down used products into raw materials that can be reprocessed into new production cycles. Several methods are currently being researched and applied for recycling PV panels, each offering different approaches to material recovery.[51] [52] [53] However, despite these efforts to recycle materials as efficiently as possible, challenges remain, starting with the fact that a PV module is a composite product, meaning that all the different components, be it the glass, the solar cell, or the backsheet, are glued together to withstand weather conditions for around 30 years. For the recycling process, this leads to insufficient material purity and a high energy demand for separating the different materials. In addition, low legal requirements and cost barriers, which vary from region to region, make the recycling business even more difficult.

49 Brian Baldassarre, "Circular Economy for Resource Security in the European Union (EU): Case Study, Research Framework, and Future Directions," *Ecological Economics* 227 (January 2025): 108345, https://doi.org/10.1016/j.ecolecon.2024.108345.
50 Nyffenegger et al., "How Circular Is the European Photovoltaic Industry?"
51 The most common technologies include: 1. Mechanical recycling remains the currently most common recycling technology. PV panels are firstly broken down into smaller pieces using crushers or shredders, followed by sorting the materials based on their type (mostly as glass, metals, and plastics). 2. Thermal treatment: PV panels are heated to high temperatures to evaporate the plastic components, leaving behind glass and metals for further processing. 3. Elective etching: A chemical process that uses specific chemicals to dissolve and separate different materials within the panels, allowing for the precise extraction of valuable components like silicon and metals. 4. Hot knife cutting employs heated blades to slice through the layers of PV panels, making it easier to separate individual components for recycling (see Retrieve Project, "Home," 2024, https://www.retrieveproject.eu/). 5. Flash lamp annealing uses intense flashes of light to heat and separate materials by breaking down the bonds between different layers. 6. Solvent extraction uses solvents to dissolve and separate specific materials, particularly useful for extracting metals and other valuable components. 7. Water jet separation utilizes high-pressure water jets to cut and separate the materials in a clean and efficient manner, avoiding harmful chemicals. 8. Electrodynamic disintegration employs electrical discharges to create shock waves that break down the panels into their components, which are then separated and collected for recycling.
52 Retrieve Project "Home."
53 Karsten Wambach, "Technisches Recycling von PV Modulen Erfahrungen Und Neue Entwicklungen," 2022, https://www.swissolar.ch/de/angebot/fachveranstaltungen/vortraege-und-studien.

Currently, there are approximately 177 PV recyclers and recycling equipment manufacturers worldwide, of which some are not operating on an industrial scale. Thirty-eight of these facilities are in the U.S. where recycling is not yet mandatory. According to the U.S. National Renewable Energy Laboratory, less than 10% of the country's decommissioned panels are recycled.[54] In many cases, PV modules are simply collected for cable and frame removal before being sent to landfill, largely because it costs far less to dump them than to recycle them. The U.S. is expected to have the second largest cumulative PV panel waste by 2050, reaching up to 10 million tones,[55] which would soon exceed any landfill capacity under the current conditions. Consequently, researchers, practitioners, and policy makers are now racing to expand the recycling industry. In 2021, California introduced regulations classifying end-of-life PV modules as universal waste to reduce regulatory compliance and waste processing costs, providing a boost to domestic recycling facilities.[56] Furthermore, the U.S. government increased funding into research on solar recycling technologies (Bipartisan Infrastructure Law, 2022 and Department of Energy's Solar Energy Technologies Office [DOE SETO] action plan for PV system end-of-life management, 2023) and established a recycling cost-reduction target for 2030 to compete with the cost of landfill disposal.[57]

Similar challenges are currently applying to the Asia Pacific region, which is lacking policies that mandate the proper disposal of end-of-life PVs. Currently, there are

54 Chemical & Engineering News, "Solar Panels Face Recycling Challenge," 2022, https://cen.acs.org/environment/recycling/Solar-panels-face-recycling-challenge-photovoltaic-waste/100/i18.
55 Adriana Domínguez and Roland Geyer, "Photovoltaic Waste Assessment of Major Photovoltaic Installations in the United States of America," *Renewable Energy* 133 (April 1, 2019): 1188–1200, https://doi.org/10.1016/j.renene.2018.08.063.
56 Keiichi Komoto et al., "Task 12 PV Sustainability – Status of PV Module Recycling in Selected IEA PVPS Task 12 Countries," 2022, https://doi.org/10.2172/2308819.
57 International Energy Agency, "Advances in Photovoltaic Module Recycling."

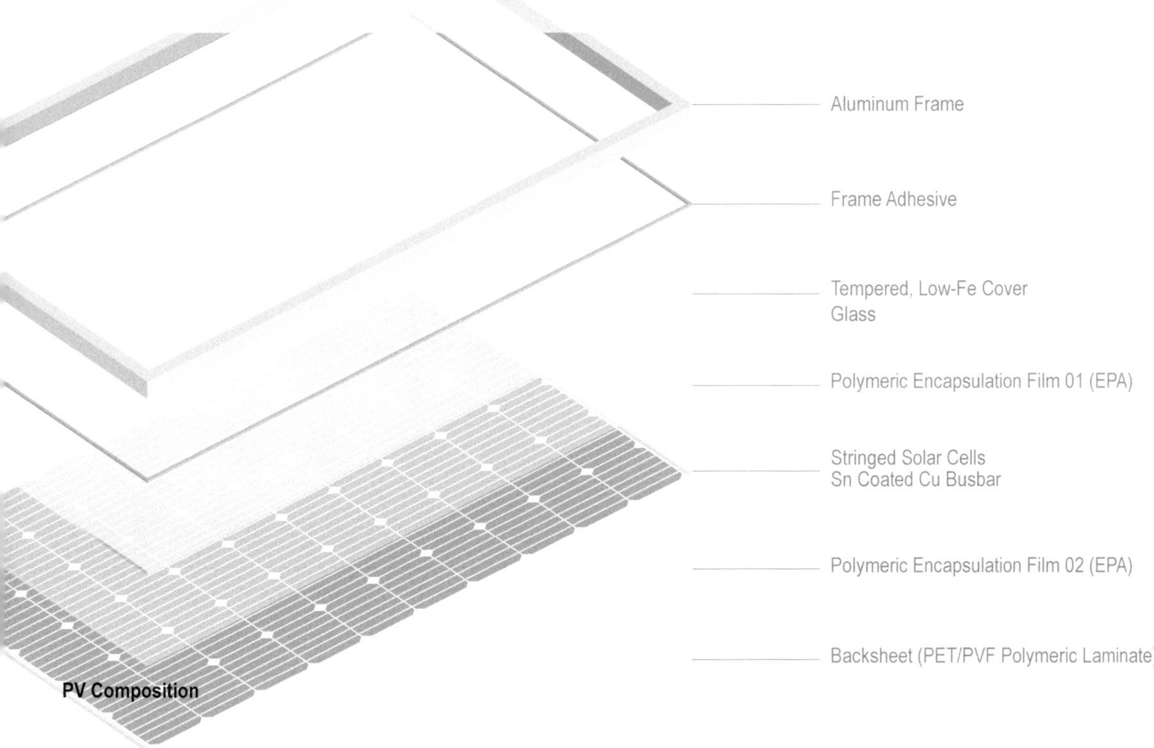

PV Composition

no recycling plants in Southeast Asia although many countries in this region, such as Vietnam or India, are experiencing an enormous increase in new installations. Thus, the primary challenges in managing solar waste revolve not only around technology, infrastructure, and regulation, but also economic viability. Costs associated with recycling a panel can be around $20–$30 while sending it to a landfill costs $1–$2. Without state support, no one will be prepared to bear these costs. Therefore, several countries in Asia Pacific have set initiatives to enhance PV recycling. In July 2022, for example, Japan obliged electricity producers to set up a reserve fund for PV systems, which will be returned to the companies at the end of the systems' life cycle to pre-finance their disposal.[58] China, which will be one of the most affected countries by PV waste in the near future, is aiming to set up regulations that clearly separate the recycling of solar panels, since current disposal is defined under the national solid waste law.[59] Singapore has made notable strides in PV sustainability with a spin-off from Nanyang Technological University that integrates both reuse and recycling practices for used PV modules. This initiative focuses on upgrading solar panels with high reusability potential, establishing a comprehensive recycling plant, and deploying a mobile recycling unit specifically designed to handle solar waste.

As mentioned earlier, in Europe, the recycling of PV modules is currently supported by the WEEE Directive, which prescribes the collection, recovery, and recycling targets for waste from electrical and electronic equipment, including PV panels. Since 2012, all EU members have implemented the regulation into national law, requiring PV panel manufacturers in the EU market to either operate their own take-back and recycling systems or join existing producer compliance schemes. According to the Directive, from 2019 onward, the minimum annual collection rate for Member States must be either 65% of the average weight of electrical and electronic equipment placed on the market over the preceding three years or 85% of the electronic waste generated within the Member State's territory.[60] For example, in 2021, around 30,000 tons of PV waste was collected from various Member States in 2021, with Germany, Italy, Portugal, and Belgium leading the way.[61]

The overall e-waste collection rate in the same year was at 46.2%.[62] However, industry stakeholders have expressed concerns about the WEEE directive's approach, which emphasizes recycling by weight rather than targeting specific high-value materials. This focus can foster a *fourth myth* around sustainability policies, creating the impression that meeting recycling quotas inherently leads to significant environmental impact. Yet, as one PV recycler highlighted, while the directive may address volume, it overlooks the quality and utility of recovered materials—an essential aspect to achieve truly circular outcomes: "The current

58 Asian Power, "How Is APAC Faring in Managing Solar PV Waste?," 2024, https://asian-power.com/exclusive/how-apac-faring-in-managing-solar-pv-waste.
59 Amjad Ali et al., "Policies and Regulations for Solar Photovoltaic End-of-Life Waste Management."
60 Komoto et al., "Task 12 PV Sustainability."
61 International Energy Agency, "Advances in Photovoltaic Module Recycling."
62 European Commission, "Waste Statistics – Electrical and Electronic Equipment," 2024, https://ec.europa.eu/eurostat/statistics-explained/index.php?title=Waste_statistics_-_electrical_and_electronic_equipment.

system poses challenges for PV modules, which are composed of mostly glass. If the legislation required that 85% of the silver in PV modules need to be recovered, then simply shredding would not be sufficient anymore." [63] Mechanical recycling is the benchmark, however, it does not currently allow recovery of silicon and silver. In addition, there remain challenges regarding the resulting purity of the recovered materials, which currently do not allow large-scale re-introduction in the production of new solar modules.[64] Further complicating Europe's recycling efforts is the cautious approach to scaling up recycling facilities, with a focus that has, until recently, been largely on developing new technologies rather than moving these solutions fully into industrial applications. Currently, industrial investors are hesitant to commit financial resources to large-scale operations without a guaranteed supply of materials to recycle. This reluctance is compounded by the fact that current volumes of end-of-life PV modules are insufficient to justify the business case for expanding recycling infrastructure. Again, the surveyed PV recycler explains: "Historically, we aimed to build a larger plant, but the business case was insufficient due to a lack of available old modules on the market. For example, we had a business case involving 7,500 tons, but much of the end-of-life PV modules ended up being reused rather than recycled. Consequently, investments in expanding recycling facilities have been delayed." As many old modules are currently being reused in secondary markets rather than entering the recycling stream, the anticipated feedstock for recycling facilities remains limited. One lever to counteract these challenges in Europe is to increase investments in research to optimize recycling (for example Retrieve 2024), also by an increased application of digital technologies.[65]

However, research alone will not fully address the challenges of missing recycling volumes, much of which appears to be disappearing into the Global South instead of being processed. These "missing materials" from end-of-life PVs represent a spectral mass—absent from the recycling stream but essential for the growth of a robust, efficient recycling industry. Without access to these missing materials, the progress of recycling technologies and the ability to attract industrial investments face significant hurdles. Current legislation lacks the right incentives, reinforcing the myth that ambitious recycling policies alone are enough to make a real impact. What is needed is a cohesive strategy where regulatory standards and industrial investment create a sustainable path forward, bridging the gap between policy promises and real-world recycling outcomes. By improving regulation, policymakers can help level the playing field for up- and downstream stakeholders committed to circular practices. This regulatory support is essential for ensuring that circularity not only becomes a market differentiator but also a fundamental aspect of the global PV industry's future growth.

63 Nyffenegger et al., "How Circular Is the European Photovoltaic Industry?"
64 International Energy Agency, "Advances in Photovoltaic Module Recycling."
65 Ässia Boukhatmi, Roger Nyffenegger, and Stefan N. Grösser, "Designing a Digital Platform to Foster Data-Enhanced Circular Practices in the European Solar Industry," *Journal of Cleaner Production* 418 (September 15, 2023): 137992.

Shaping a Sustainable Solar Future Beyond the Myths

The global solar industry stands as a modern myth in the narrative of sustainable energy—a source of clean power with the potential to reshape our environmental future. Yet, as we have seen, the industry's linear production model and reliance on resource extraction present significant challenges. Much like ancient myths that obscured complex truths beneath idealized visions, today's myth of solar energy as a perfect solution conceals a complex reality: sustainable solar growth requires more than just expansion; it demands a transformative, all-encompassing shift toward circularity.

Our analysis has shown that this shift must address key issues across the value chain. The current reliance on virgin materials and a linear production model perpetuates waste, resource depletion, and environmental harm, underscoring the need for policies and practices that rethink the full life cycle of PV modules. While regions like Africa are emerging as potential solar markets, they face barriers such as external funding dependence and insufficient local infrastructure. Meanwhile, Europe has made strides in policy-driven growth, but its sustainability directives often focus on volume over substance, creating the illusion of environmental impact without prioritizing high-value material recovery. In Asia-Pacific, production dominance—particularly China's—has driven down costs but also introduced ethical concerns, showing that a truly sustainable industry must balance growth with social responsibility. Finally, the Americas highlight both the promise and the obstacles of large-scale deployment, where recycling remains under-developed and industry growth faces regulatory and logistical hurdles.

Looking forward, the solar industry's shift to a circular economy demands a unified strategy to address several pressing challenges. First, "recycling by volume" policies must be replaced with material-specific policies that prioritize the recovery of valuable resources like silver and silicon, crucial for a sustainable supply chain. Second, reuse strategies must evolve to address the unintended consequences of module exports to regions with limited recycling infrastructure, ensuring that solar's benefits do not shift environmental burdens to less industrialized countries. Third, investment in scalable recycling technologies is essential to improve efficiency, reduce costs, and increase the quality of recovered materials, with, fourth, regulatory support guiding industry players toward these goals.

In this light, the solar industry of the future must be one where circularity is built into each stage of the value chain—from design and production to end-of-life management. Policies should not only incentivize local recycling but also enforce transparent reporting and tracking to ensure that solar modules are responsibly managed across borders. With the right investments and regulatory frameworks, the solar industry can transcend the myth of sustainability and become a leader in the circular economy, demonstrating that true environmental responsibility is achieved not through a single strategy but through a concerted, multifaceted approach that integrates reduce, reuse, and recycle practices at every level. In realizing this vision, the industry can fulfill its promise as a genuine agent of change

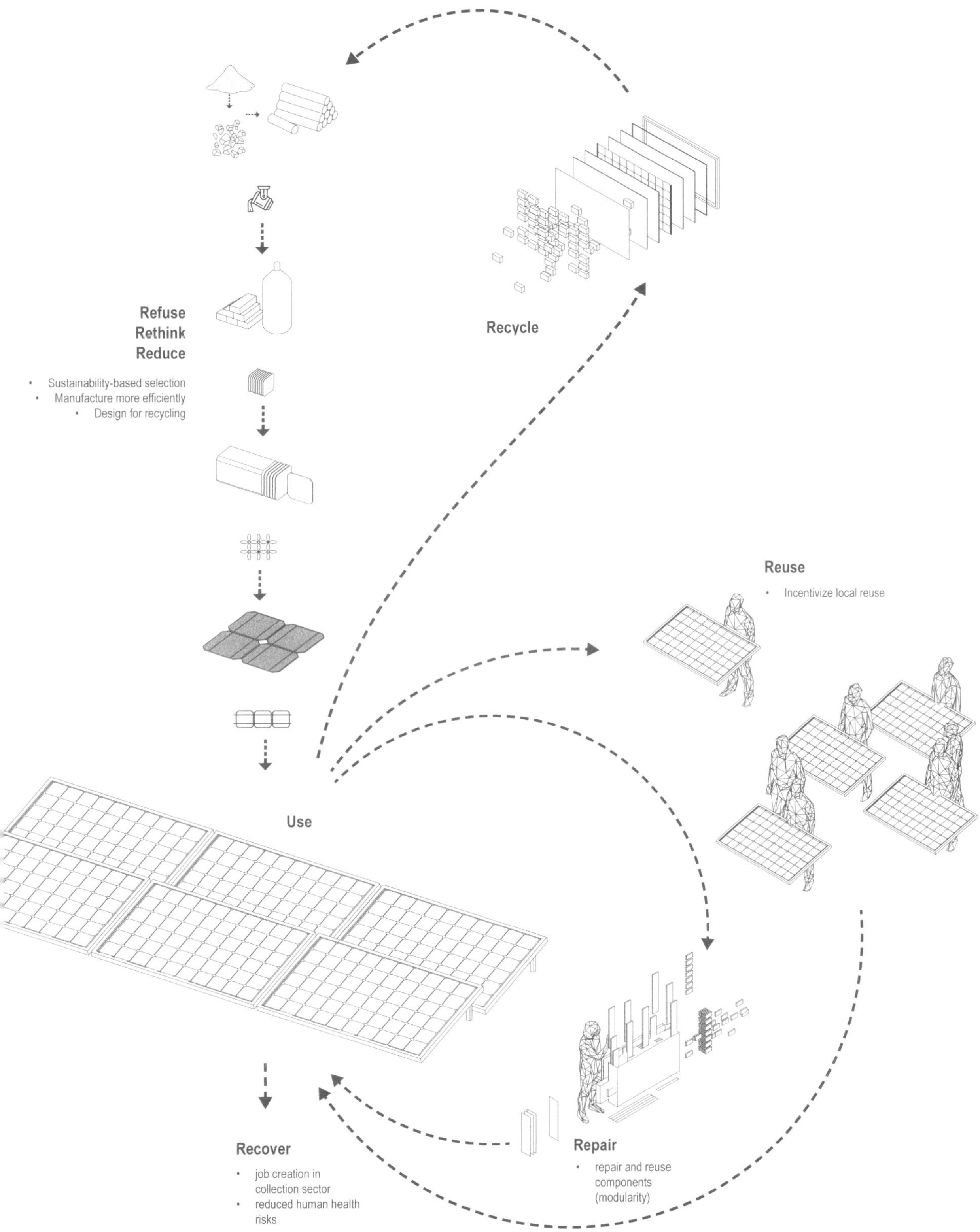

Circular Paradigm of PV Lifecycle

in the global transition to renewable energy. After all, myths endure because they contain a kernel of truth, revealing the greatness and potential of their subject; the sun, as the source of all life, embodies this promise, inviting us to responsibly harness its immense power for a sustainable future.

BIBLIOGRAPHY

Adenle, Ademola A. "Assessment of Solar Energy Technologies in Africa-Opportunities and Challenges in Meeting the 2030 Agenda and Sustainable Development Goals." *Energy Policy* 137 (February 1, 2020): 111180. https://doi.org/10.1016/j.enpol.2019.111180.

Aleksandra, Arcipowska, Blanco Perez Sara, Jakimów Małgorzata, Baldassarre Brian, Polverini Davide, and Cabrera Miguel. "Role of Solar PV in Net-zero Growth: An Analysis of International Manufacturers and Policies." *Progress in Photovoltaics: Research and Applications* 32, no. 9 (September 2024): 607–622. https://doi.org/10.1002/pip.3797.

Ali, Amjad, Sheraz Alam Malik, Md Shafiullah, Muhammad Zeeshan Malik, and Md Hasan Zahir. "Policies and Regulations for Solar Photovoltaic End-of-Life Waste Management: Insights from China and the USA." *Chemosphere* 340 (November 1, 2023): 139840. https://doi.org/10.1016/j.chemosphere.2023.139840.

Arya, Apoorva, and Arpit Bhutani. "Study on Items Shipped for Reuse and Extended Producer Responsibility Fees – A Case for Extending EU EPR Fees to Cover End-of-Life Activities of Products Shipped Outside the EU." Open Society Foundation and the LIFE Programme of the European Union. May 17, 2023. https://eeb.org/library/items-shipped-for-reuse-and-extended-producer-responsibility-fees-a-case-for-extending-eu-epr-fees-to-cover-end-of-life-activities-of-products-shipped-outside-the-eu/.

Asian Power. "How Is APAC Faring in Managing Solar PV Waste?" 2024. https://asian-power.com/exclusive/how-apac-faring-in-managing-solar-pv-waste.

Baldassarre, Brian. "Circular Economy for Resource Security in the European Union (EU): Case Study, Research Framework, and Future Directions." *Ecological Economics* 227 (January 2025): 108345. https://doi.org/10.1016/j.ecolecon.2024.108345.

Bauer, Sophia, Kasandra O'Malia, Shradhey Prasad, Gregor Clark, and Ingrid Behrsin. "A Race to the Top: Latin America 2023." *LATIN AMERICA*. 2023.

Boukhatmi, Ässia, Roger Nyffenegger, and Stefan N. Grösser. "Designing a Digital Platform to Foster Data-Enhanced Circular Practices in the European Solar Industry." *Journal of Cleaner Production* 418 (September 15, 2023): 137992. https://doi.org/10.1016/j.jclepro.2023.137992.

Burrows, Keith, and Vasilis Fthenakis. "Glass Needs for a Growing Photovoltaics Industry." *Solar Energy Materials and Solar Cells* 132 (January 1, 2015): 455–459. https://doi.org/10.1016/j.solmat.2014.09.028.

Chemical & Engineering News. "Solar Panels Face Recycling Challenge." *Chemical & Engineering News*. 2022. https://cen.acs.org/environment/recycling/Solar-panels-face-recycling-challenge-photovoltaic-waste/100/i18.

Cimadomo, Simone. "Advancing a Circular Economy for Solar Photovoltaics Exported for Reuse – Analysing the Institutional Feasibility of International Extended Producer Responsibility for EU-West Africa Transboundary Movements." Lund: Master thesis – IIIEE. Lund University, 2024.

Crawford, Alan, and Laura T. Murphy. "Uyghur Region Exposure Assessment for Solar Industry Sourcing." *Sheffield Hallam University Helena Kennedy Centre for International Justice*. November 2023.

Czajkowski, Adrian, et al. "Energies | Free Full-Text | Prediction of the Market of End-of-Life Photovoltaic Panels in the Context of Common EU Management System." 2022. https://www.mdpi.com/1996-1073/16/1/284.

Deng, Rong, Yuting Zhuo, and Yansong Shen. "Recent Progress in Silicon Photovoltaic Module Recycling Processes." *Resources, Conservation and Recycling* 187 (December 1, 2022): 106612. https://doi.org/10.1016/j.resconrec.2022.106612.

Domínguez, Adriana, and Roland Geyer. "Photovoltaic Waste Assessment of Major Photovoltaic Installations in the United States of America." *Renewable Energy* 133 (April 1, 2019): 1188–1200. https://doi.org/10.1016/j.renene.2018.08.063.

Ember. "Solar Power." 2022. https://ember-climate.org/topics/solar/.

European Commission. "Solar Energy." 2022. https://energy.ec.europa.eu/topics/renewable-energy/solar-energy_en.

European Commission. "Waste Statistics – Electrical and Electronic Equipment." 2024. https://ec.europa.eu/eurostat/statistics-explained/index.php?title=Waste_statistics_-_electrical_and_electronic_equipment.

European Parliament & Council. "Directive 2012/19/EU of the European Parliament and of the Council of 4 July 2012 on Waste Electrical and Electronic Equipment (WEEE)." Official Journal of the European Union, July 24, 2012.

European Union. "Directive 98/70/EC as Regards the Promotion of Energy from Renewable Sources." 2023. https://eur-lex.europa.eu/legal-content/EN/TXT/HTML/?uri=OJ%3AL_202302413.

Gill, Tom. "The 15 Largest Solar Farms in the World 2022." *The Eco Experts* (blog). 2022. https://www.theecoexperts.co.uk/solar-panels/biggest-solar-farms.

Hallam, Brett, Moonyong Kim, Yuchao Zhang, Li Wang, Alison Lennon, Pierre Verlinden, Pietro P. Altermatt, and Pablo R. Dias. "The Silver Learning Curve for Photovoltaics and Projected Silver Demand for Net-Zero Emissions by 2050." *Progress in Photovoltaics: Research and Applications* 31, no. 6 (2023): 598–606. https://doi.org/10.1002/pip.3661.

Heide, Arvid van der, Daniela Maria Godinho Ariolli, Guillermo Oviedo Hernandez, Serge Noels, and Jan Clyncke. “Re-Use of PV Modules: Progress in Standardisation and Learnings from a Real Case Study.” 2023. https://doi.org/10.4229/EUPVSEC2023/5DO.15.6.

International Energy Agency. “Advances in Photovoltaic Module Recycling – Literature Review and Update to Empirical Life Cycle Inventory Data and Patent Review.” IEA PVPS Task 12 Sustainability. 2024.

International Energy Agency. “Electricity 2025 – Analysis and Forecast to 2027.” February 2025.

International Energy Agency. *Special Report on Solar PV Global Supply Chains*. 2022. https://doi.org/10.1787/9e8b0121-en.

IRENA. “End-of-Life Management Solar Photovoltaic Panels.” 2016. https://www.irena.org/publications/2016/Jun/End-of-life-management-Solar-Photovoltaic-Panels.

IRENA. “IRENA – International Renewable Energy Agency.” 2024. https://www.irena.org/.

Komoto, Keiichi, Michael Held, Claire Agraffeil, Carmen Alonso-Garcia, Andrea Danelli, Jin-Seok Lee, Lv Fang, et al. “Task 12 PV Sustainability – Status of PV Module Recycling in Selected IEA PVPS Task 12 Countries.” 2022. https://doi.org/10.2172/2308819.

Lempkowicz, Bertrand, and Jade Liu. “RE-USE of PV Modules, Challenges and Opportunities of the Circular Economy – Study of Re-Used PV Modules.” Press Release – PV Cycle & imec. August 3, 2021. https://pvcycle.be/nl/wp-content/uploads/Press-Release-Reuse-08032021.pdf.

Nyffenegger, Roger, Ässia Boukhatmi, Tadas Radavičius, and Manuela Tvaronavičienė. “How Circular Is the European Photovoltaic Industry? Practical Insights on Current Circular Economy Barriers, Enablers, and Goals.” *Journal of Cleaner Production* 448 (April 5, 2024): 141376. https://doi.org/10.1016/j.jclepro.2024.141376.

Oviedo Hernandez, G., D. M. Godinho Ariolli, P. S. Enriquez Paez, and P. V. Chiantore. “Trends and Innovations in Photovoltaic Operations and Maintenance.” *Progress in Energy* 4, no. 4 (October 1, 2022): 042002. https://doi.org/10.1088/2516-1083/ac7c4f.

Oxford PV. 2024. https://www.oxfordpv.com/.

Potting, José, M.P. Hekkert, Ernst Worrell, and Aldert Hanemaaijer. *Circular Economy: Measuring Innovation in the Product Chain*. 2017.

pv magazine International. “Global Installed PV Capacity Tops 2 TW.” November 13, 2024. https://www.pv-magazine.com/2024/11/13/global-installed-pv-capacity-tops-2-tw/.

Retrieve Project. “Home.” 2024. https://www.retrieveproject.eu/.

Solar Energy Industries Association. “US Solar Market Insight.” 2024. https://www.woodmac.com/industry/power-and-renewables/us-solar-market-insight/thank-you/?__FormGuid=8e2fb1e9-c50c-43ef-9601-e9e96361421a&__FormLanguage=en&__FormSubmissionId=68e0222d-13e6-4be3-b19d-10f8e7dc1c27.

Solar Power Europe. “EU Market Outlook for Solar Power 2023–2027.” 2024. https://www.solarpowereurope.org/insights/outlooks/eu-market-outlook-for-solar-power-2023-2027#download.

Tawalbeh, Muhammad, Amani Al-Othman, Feras Kafiah, Emad Abdelsalam, Fares Almomani, and Malek Alkasrawi. “Environmental Impacts of Solar Photovoltaic Systems: A Critical Review of Recent Progress and Future Outlook.” *Science of The Total Environment* 759 (March 2021): 143528. https://doi.org/10.1016/j.scitotenv.2020.143528.

Thapa, Kaustubh, Walter J.V. Vermeulen, Pauline Deutz, and Olawale E Olayide. “Transboundary Movement of Waste Review: From Binary towards a Contextual Framing.” *Waste Management & Research* 41, no. 1 (January 1, 2023): 52–67. https://doi.org/10.1177/0734242X221105424.

United Nations. “The Paris Agreement | UNFCCC.” 2015. https://unfccc.int/process-and-meetings/the-paris-agreement.

“Uyghur Forced Labor Prevention Act | U.S. Customs and Border Protection.” Accessed February 19, 2025. https://www.cbp.gov/trade/forced-labor/UFLPA.

Wambach, Karsten. “Technisches Recycling von PV Modulen Erfahrungen Und Neue Entwicklunger.” 2022. https://www.swissolar.ch/de/angebot/fachveranstaltungen/vortraege-und-studien.

Yale School of Environment. “How China Became the World’s Leader on Renewable Energy.” 2024. https://e360.yale.edu/features/china-renewable-energy.

Yu, Pei-Hua. “What Does 1 GW of Installed Solar Energy Mean? How Many Houses Can It Power?” *Medium* (blog). 2021. https://peihuayu.medium.com/what-does-1-gw-of-installed-solar-energy-mean-how-many-houses-can-it-power-bb32ea6cc19.

IMAGE CITATIONS + CREDITS

Listed by Page Number

110 Map sources: Earthstar Geographics.

114 Map sources: Solar Potential data obtained from the Global Solar Atlas 2.0, a free, web-based application developed and operated by the company Solargis s.r.o. on behalf of the World Bank Group, utilizing Solargis data, with funding provided by the Energy Sector Management Assistance Program (ESMAP). Global Integrated Power Tracker, Global Energy Monitor, September 2024 release. World Bank Group, IMF.

117 Image references: "Analyzing the Performance of Identical PV Modules in a Semi-Arid Region over a 2-year Period – Scientific Figure on ResearchGate." Accessed May 16, 2025. https://www.researchgate.net/figure/PV-Cell-manufacturing-process-SINOVoltaicscom-2017_fig1_326380926.

118 Map sources: Leah Temper, Daniela del Bene, and Joan Martinez-Alier. "Mapping the Frontiers and Front Lines of Global Environmental Justice: the EJAtlas." *Journal of Political Ecology* 22 (2015): 255–278. https://journals.librarypublishing.arizona.edu/jpe/article/id/1932/; Global Forest Watch. 2014. *World Resources Institute*. Accessed May 16, 2025. www.globalforestwatch.org; Klaus J. Schulz and Joseph A. Briskey. 2005. "Major Mineral Deposits of the World: Open-File Report 2005-1294, U.S. Geological Survey, Reston, Virginia."

121 Image references: IEA. 2022. "Solar PV Global Supply Chains." Paris. https://www.iea.org/reports/solar-pv-global-supply-chains, Licence: CC BY 4.0.

122 Image references: IRENA. 2016. "End-of-Life Management Solar Photovoltaic Panels." https://www.irena.org/publications/2016/Jun/End-of-life-management-Solar-Photovoltaic-Panels.

131 Image references: Clean Energy Reviews. "Solar Panel Construction." Accessed May 16, 2025. https://www.cleanenergyreviews.info/blog/solar-panel-components-construction.

135 Image references: Stokvisch. 2022. "Strategies for Circular End-of-Life Management of Photovoltaic Panels on Amsterdam Rooftops."; Oviedo Hernandez et al. 2022. "Trends and Innovations in Photovoltaic Operations."; Farrell et al., 2020, "Technical Challenges and Opportunities in Realising a Circular Economy for Waste Photovoltaic Modules."; Salim et al., 2019, "Driver, Barriers and Enablers to End-of-Life Management of Solar Photovoltaic and Battery Energy Storage Systems: A Systematic Literature Analysis."

17°43'51"S
17°43'52"S
17°43'56"S
17°43'57"S
31°4'35"E
31°4'36"E
E-
loudness
light

WASTE

Reclaiming Digital Detritus + Forging a Sustainable Dawn in Zimbabwe

by Vusumuzi Maphosa

Introduction

The Fourth Industrial Revolution (4IR), a symphony of progress propelled by artificial intelligence (AI), Internet of Things, and cloud computing's boundless flow, yet dependent on critical minerals that enable this digital reign to sustainably expand. Sourcing and processing critical minerals through traditional ore mining often results in severe landscape degradation, deforestation, soil erosion, and water contamination, causing lasting damage to fragile ecosystems. Some critical minerals and rare earth metals are sourced from war-torn regions, posing risks of disruptions to the product supply chain. The relentless miniaturisation of electrical and electronic equipment (EEE), the growth of the digital economy, and declining prices fuel multiple device ownership[1] resulting in a tsunami of technological affluence overwhelming our capacity to manage its discarded remains. Mega satellite constellations in low Earth orbit improve broadband access and device ownership in underserved areas but increase digital waste, highlighting the environmental impact of digital systems. Addressing this complex issue requires comprehensive solutions to balance progress and sustainability.

1 Maphosa, Vusumuzi. "Rethinking Sustainability: A Bibliometric and Visualisation of E-Waste Management in Africa." *Journal of Higher Education Theory and Practice* 22, no. 1 (2022): 123–135.

Electronic waste encompasses discarded electrical and electronic equipment, regardless of functionality,[2] standing as a stark testament to our unquenchable thirst for technological advancement and the unsustainable waste it leaves in its path. In developing nations, particularly across Africa, formal e-waste management systems are critically underdeveloped, largely due to the absence of robust extended producer responsibility (EPR) initiatives. Zimbabwe exemplifies this challenge, where the lack of local electronics manufacturers and poorly regulated landfills has shifted e-waste recycling to the informal sector. Here, vulnerable groups, including women and children, dismantle electronics in hazardous conditions, releasing toxic substances that endanger both human health and the environment. Prolonged exposure to these pollutants leads to severe health issues, such as cancers, neurological disorders, and organ damage, with pregnant women and children facing heightened risks like premature births and developmental delays. Contaminated soil, water, and agricultural products further amplify the crisis, while municipalities worsen the situation by mixing e-waste with general waste. Urgent intervention is needed to protect vulnerable populations, enforce sustainable recycling practices, and advance a circular economy that prioritizes environmental safety and social equity.

Zimbabwe's Pomona dumpsite, once pristine wetland, now lies in ruin, transformed into an informal e-waste recycling hub where the desperate seek their meagre fortunes. Young migrants, driven by the cruel grip of poverty, dismantle the discarded electronics, extracting precious metals like copper and gold, but at a devastating cost. Toxic fumes and chemicals seep into the air, soil, and water, causing severe respiratory ailments and chronic illnesses for the nearby residents forced to bear this burden. Despite the hazardous conditions, this polluted and heterogeneous site remains a lifeline, highlighting the intricate link between environmental degradation and the struggle for survival.

In my paper with Mfowabo Maphosa we illuminated that the very production of electronic equipment involves the utilisation of over a thousand distinct components—both toxic and precious.[3] These elements range from the insidious presence of arsenic, chromium, cadmium, lead, and mercury, to the alluring bounty of gold, lanthanum, cerium, and neodymium. The manufacturing, usage, and ultimately disposal of digital systems contributes to the ballooning demand for limited critical and rare earth metals. Discarded electronics and outdated servers have led to the exponential growth of digital refuse,[4] devastating local landfills and landscapes with mountains of non-biodegradable waste leaching toxins into the soil and water. Where new electronic devices lie beyond the grasp of the many in developing countries, this public turns to second-hand electronic devices,

2 Vanessa Forti, Kees Baldé, Ruediger Kuehr, and Garam Bel, "Global E-Waste Monitor 2020," www.intu.itu, 2020, https://www.itu.int/en/ITU-D/Environment/Documents/Toolbox/GEM_2020_def.pdf.

3 Mfowabo Maphosa and Vusumuzi Maphosa, "A Bibliometric Analysis of the Effects of Electronic Waste on the Environment," *Global Journal of Environmental Science and Management* 8, no. 4 (2022): 592, https://doi.org/10.22034/GJESM.2022.04.10.

4 Vusumuzi Maphosa, "Sustainable E-Waste Management at Higher Education Institutions' Data Centres in Zimbabwe," *International Journal of Information Engineering and Electronic Business* 14, no. 5 (2022): 15–23, https://doi.org/10.5815/ijieeb.2022.05.02.

fuelling the relentless accumulation of e-waste in the receiving countries.[5] From my own experience, a significant portion of the imported EEE is refurbished and has a shorter lifespan, substantially contributing to the rapidly increasing e-waste discards in the country. This technological landscape offers both promise and risk. Challenges like inadequate recycling systems, rising costs, and geopolitical issues hinder progress. However, recycling e-waste can recover vital minerals, cut reliance on harmful mining, and provide a transformative solution to the growing e-waste crisis.

Global electronics supply chains and digital waste flows are shaped by persistent legacies of colonialism and inequality. The opacity of global systems enables industrialised nations to evade responsibility for the end-of-life (EoL) electronics they produce and use, undermining sustainability efforts and enabling greenwashing. "Green" metrics tend to focus on carbon emissions, rather than toxic legacies, informal labor, or long-term waste. This exemplifies the destructive nature of techno-colonialism, devastating vulnerable landscapes across the developing world. The global flows of e-waste exhibit a concerning neo-colonial dynamic, where prosperous nations offload the burdensome weight of e-waste management onto resource-constrained nations of the Global South.[6] This compounds the injustices of a system that rewards those at the top while offloading costs onto the vulnerable, perpetuating an inequitable cycle of technological refuse.

The unequal benefits of new technologies and seemingly green solutions highlight digital colonialism, where the Global South bears the environmental and health burdens of tech manufacturing, mining, and e-waste, while reaping minimal rewards. Ambiguous e-waste definitions in literature contribute to regulatory gaps, inefficient recycling, illegal trade, and improper disposal, worsening environmental and health risks and hindering effective management,[7] perpetuating inequitable burdens on low-income countries. This highlights the complexity of addressing the e-waste challenge and the need for cohesive global strategies: without shared, transnational understandings of what does and does not constitute e-waste, the distinction between reuse and disposal remains ambiguous, complicating enforcement and opening regulatory loopholes that may undermine the environmental objectives of regulation.[8] The Basel Action Network, the Organisation for Economic Co-operation and Development (OECD), the EU Directive on Electrical and Electronic Equipment, and the Solving the E-Waste Problem (StEP) Initiative each have a unique perspective on what constitutes e-waste, reflecting their specific focus areas and regulatory frameworks. The Basel Action Network emphasizes the environmental and health impacts of discarded electronics, while the OECD provides a broader definition

5 Mathias Nigatu Bimir, "Revisiting E-Waste Management Practices in Africa: Selected Countries' Cases," *Journal of the Air & Waste Management Association* 70, no. 7 (2020): 559–669, https://doi.org/10.1080/10962247.2020.1769769.
6 Maphosa and Maphosa, "Bibliometric Analysis," 591.
7 Forti et al., "Global E-Waste Monitor."
8 Devin N. Perkins, Marie-Noel Brune Drisse, Tapiwa Nxele, and Peter D. Sly, "E-Waste: A Global Hazard," *Annals of Global Health* 80, no. 4 (2014): 286–295, https://doi.org/10.1016/j.aogh.2014.10.001.

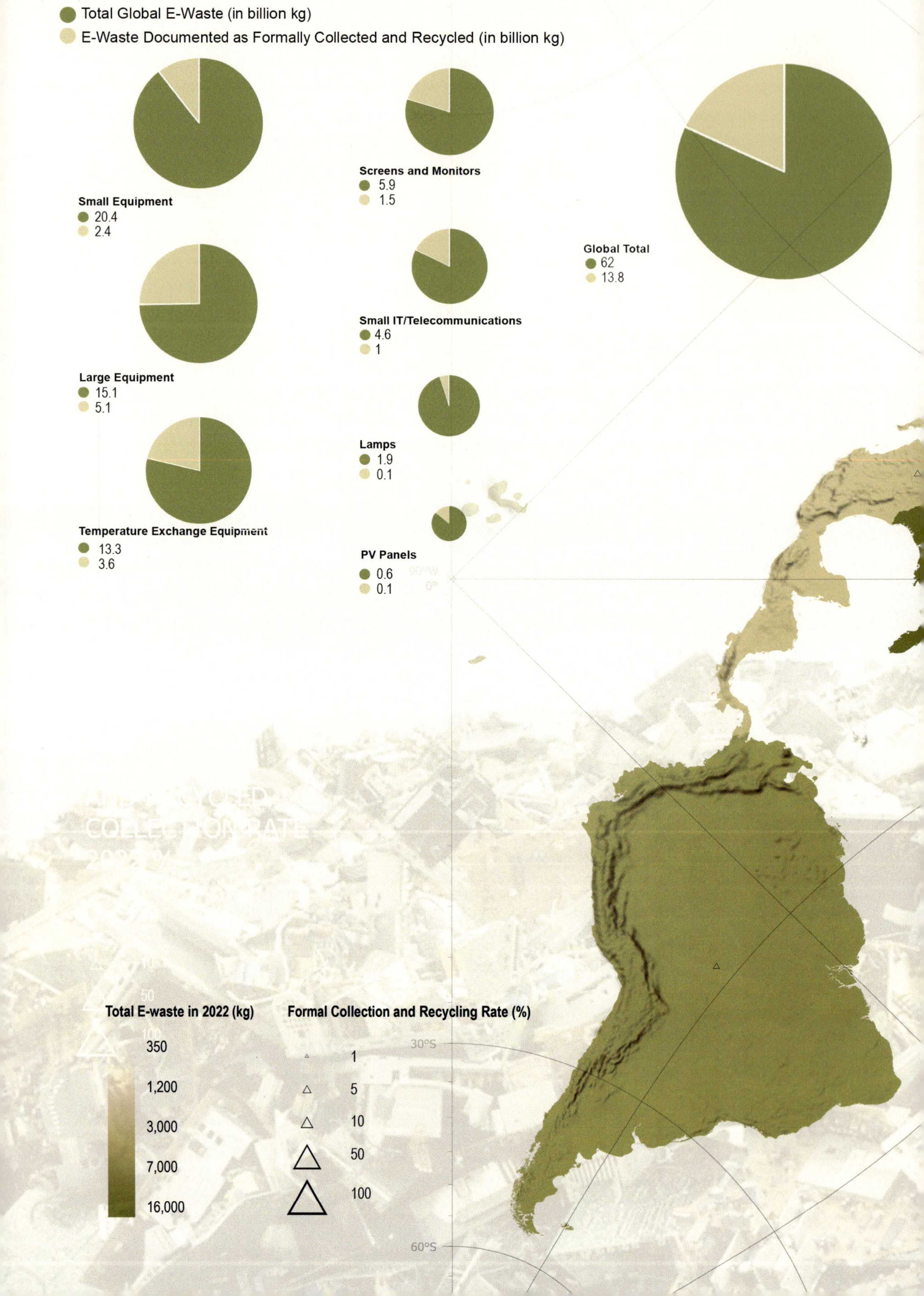
Total Global E-Waste (in billion kg)
E-Waste Documented as Formally Collected and Recycled (in billion kg)
Small Equipment
20.4
2.4
Screens and Monitors
5.9
1.5
Global Total
62
13.8
Large Equipment
15.1
5.1
Small IT/Telecommunications
4.6
1
Lamps
1.9
0.1
Temperature Exchange Equipment
13.3
3.6
PV Panels
0.6
0.1
90°W
0°
Total E-waste in 2022 (kg)
350
1,200
3,000
7,000
16,000
Formal Collection and Recycling Rate (%)
1
5
10
50
100
30°S
60°S

Global E-Waste Dynamics

that encompasses various types of electronic devices.[9] The EU Directive on EEE outlines specific criteria for hazardous waste management, and StEP focuses on sustainable solutions for e-waste disposal. Despite the noxious substances in the growing digital discards, this waste remains unclassified as a toxic scourge in developing nations like Zimbabwe.

Looking across Africa, the piles of e-waste relentlessly grow, presenting an overwhelming and staggering sight to behold. Research shows that less than 1% of Africa's e-waste is formally recycled,[10] paling in comparison to the global average of 30%. Across Africa, the essential infrastructure and legislation are sorely lacking,[11] rendering the entire e-waste value chain a daunting challenge as digital detritus mounts without recourse. Low-income countries like Zimbabwe, where recycling a computer costs just $2 compared to $20 in the U.S., often become dumping grounds for e-waste due to weak regulations and cheap labor, exposing communities to health and environmental risks and underscoring the need for global action to address this inequity.[12] E-waste recycling in Africa is largely informal and underdeveloped, posing significant health and environmental risks, but growing awareness and initiatives are beginning to address these challenges. Like many other African nations, Zimbabwe lacks the necessary e-waste recycling infrastructure and policies for effective management, from importation to disposal.[13] Zimbabwe's rapid adoption of digital technologies has outstripped its capacity to manage the resulting electronic waste. Traversing Zimbabwe's once pristine landscapes, the spectre of e-waste's burden is stark—sprawling dumpsites, scavengers sifting through mountains of e-waste, acid-leaching the soil, and noxious smoke—an urgent, devastating crisis demanding immediate action.

E-waste research is vital for Zimbabwe to address the environmental and public health risks of improper disposal, while also unlocking economic opportunities through sustainable management practices. To overcome the e-waste burden, community-based initiatives are needed to raise awareness, establish collection points, promote responsible management, and leverage community expertise for sustainable jobs. By quantifying impacts and identifying solutions, Zimbabwe can harness the benefits of a circular economy in the technology sector, making e-waste research a critical priority for the country's development.

---

9 Forti et al., "Global E-Waste Monitor."

10 Ruediger Kuehr, David Hirsch, and Terry Collins, "Global E-Waste Monitor 2024: Electronic Waste Rising Five Times Faster than Documented E-Waste Recycling," United Nations Institute for Training and Research, March 20, 2024, https://unitar.org/about/news-stories/press/global-e-waste-monitor-2024-electronic-waste-rising-five-times-faster-documented-e-waste-recycling.

11 Tamba S. Lebbie et al., "E-Waste in Africa: A Serious Threat to the Health of Children," *International Journal of Environmental Research and Public Health* 18, no. 16 (2021): 8488, https://doi.org/10.3390/ijerph18168488.

12 M. Sahadat Hossain, Sulala M. Al-Hamadani, and M. Toufiqur Rahman, "E-Waste: A Challenge for Sustainable Development," *Journal of Health and Pollution* 5, no. 9 (2015): 3–11.

13 Margaret Macherera, Vusumuzi Maphosa, and David Zezai, "Perceptions and Practices on E-Waste Management in Selected Institutions of Zimbabwe," *Water and Environmental Sustainability* 3, no. 2 (2023): 22–31, https://doi.org/10.52293/wes.3.2.2231.

Zimbabwe's Socio-Cultural Dynamics

Within this deluge, I see a glimmer of hope in emerging sustainable initiatives, which dovetail with Ubuntuism. I propose that Zimbabwe's e-waste crisis should be viewed through the socio-cultural context underpinned by Ubuntu, an indigenous ethical framework emphasising interconnected relationships between people, communities, and nature.[14] Socio-cultural factors include cultural perceptions of technology, gender roles, education levels, community trust in authorities, and practices of reuse and sharing, all of which shape how e-waste is generated, handled, and disposed of. Ubuntu challenges Western individualism, promoting values of solidarity, respect, and collaborative coexistence. Ubuntu philosophy considers that humans and nature have essential and practical value, viewing them as inter-related ends in themselves, each fulfilling unique purposes that contribute to the greater whole. Embracing Ubuntu's communal approach can foster decolonial conservation strategies focused on socio-ecological justice and equitable human-environment dynamics to address the e-waste crisis. This holistic approach can transform communities, helping them view end-of-life electronics as valuable resources rather than a burden. Communities can reclaim value from the technological refuse and reshape the narrative around this mounting environmental challenge, aligning with the collectivist ethos of Ubuntu guiding many Zimbabwean societies. This can shift public perceptions and catalyse more sustainable solutions rooted in the local fabric. Zimbabwe's collectivist cultural norms can be a powerful force in addressing the e-waste challenge by framing it as a community-based challenge, rather than viewing it solely as an individual issue.

E-Waste Sources

The one-way flow of digital waste from the prosperous North to the struggling South reflects neo-colonialism, unfairly burdening those least responsible for the accumulating digital debris.[15] The world produced an astonishing 62 million tons (Mt) of e-waste in 2022, an equivalent to the weight of about 41 million cars, and is projected to generate a staggering 82 Mt by 2030.[16] Only 17% of the 57.4 Mt of e-waste generated globally in 2019 was properly recycled, with over 47 Mt destined for dumping in landfills, primarily in the developing world,[17] leaving behind a toxic legacy that poisons the land and people.

Over the years, poor economic conditions, limited access to electronic devices, and inadequate infrastructure have hindered digital transformation efforts in low-income countries. To bridge the digital divide, many low-income countries

14 Mathew Bukhi Mabele, Judith E. Krauss, and Wilhelm Kiwango, "Going Back to the Roots: Ubuntu and Just Conservation in Southern Africa," *Conservation and Society* 20, no. 2 (2022): 92, https://doi.org/10.4103/cs.cs_33_21.

15 Vusumuzi Maphosa and Pfano Mashau, "The Conundrum: Transforming African E-Waste Landfills to Urban Mines," in *Advances and Challenges in Hazardous Waste Management*, ed. Selah M. Hosam, Amal I. Hassan, and Refaat F. Aglan (London: IntechOpen, 2023): 28–39, https://doi.org/10.5772/intechopen.1002419.

16 Kuehr, Hirsch, and Collins, "Global E-Waste Monitor 2024."

17 Forti et al., "Global E-Waste Monitor."

removed import customs duties on electronic equipment, which led to an increase in the importation of both new and used electrical and electronic equipment.[18] In my city, Bulawayo, as well as many others, small EEE retailers and informal traders have proliferated, supplying affordable second-hand phones, laptops, and accessories from Dubai, China, and high-income countries to meet local demand.[19] Over the years, I have seen the lifespan of electronics decline, accelerating e-waste production and raising concerns about the sustainability of our consumptive practices in managing the growing volume of discards. Short innovation cycles lead to rapid product obsolescence, as manufacturers design products with built-in limitations, intentionally encouraging frequent replacements and exacerbating the growing e-waste challenge and its environmental impact. Similarly, renewable energy technologies like wind turbines, batteries from electric vehicles, and solar panels are all destined for the landfill at the end of their lifespans.

In my country, unmanaged e-waste mounts due to insatiable demand for digital devices, driven by the new educational curricula, the fourth industrial revolution, AI, and the knowledge economy.[20] However, increased internet adoption has

18 Macherera, Maphosa, and Zezai, "Perceptions and Practices," 26.

19 Joseline Wadzanai Chitotombe, "Globalization of Information Communication Technology (ICT) and Consumerism in Developing Countries: Confronting the Challenges of E-Waste Disposal in Harare Urban, Zimbabwe," *International Journal of Environmental Sciences* 3, no. 6 (2014): 2172–2185, https://www.indianjournals.com/ijor.aspx?target=ijor:ijes&volume=3&issue=6&article=032.

20 Vusumuzi Maphosa and Mfowabo Maphosa, "E-Waste Management in Sub-Saharan Africa: A Systematic Literature Review," *Cogent Business & Management* 7, no. 1 (2020): 1–19, https://doi.org/10.1080/23311975.2020.1814503.

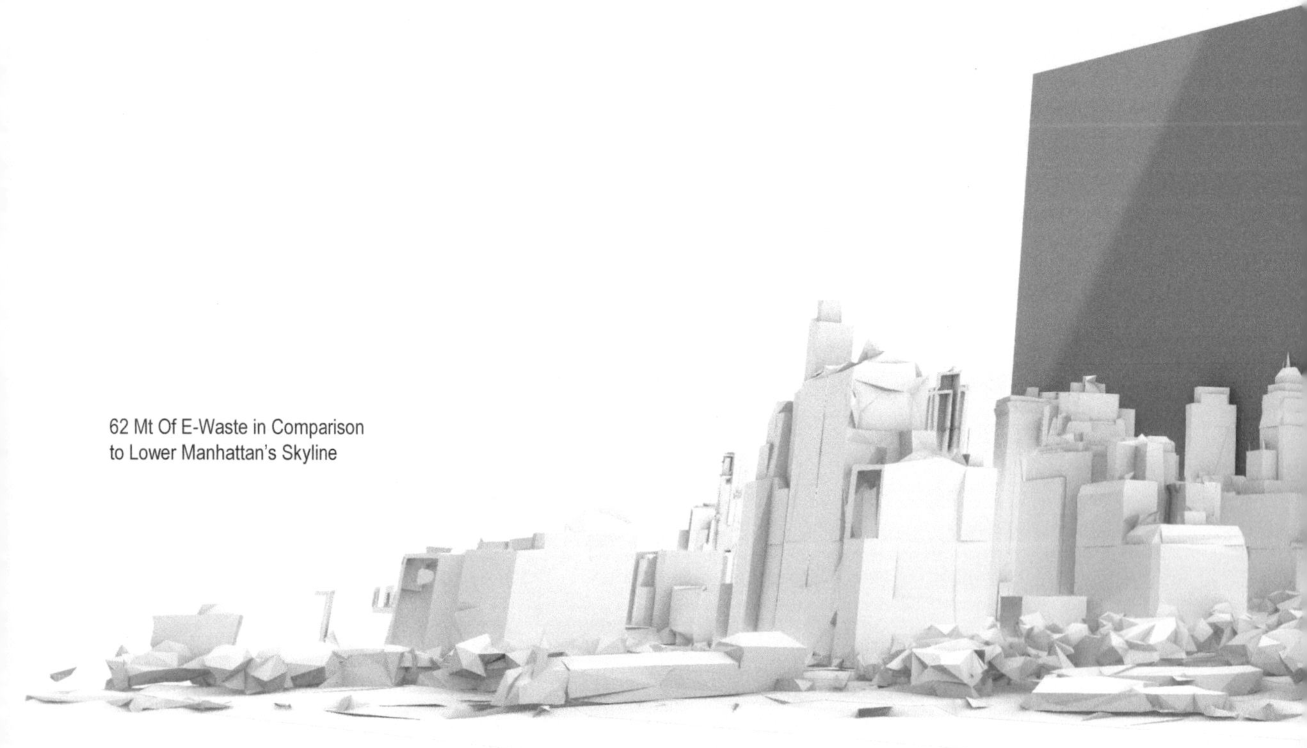

62 Mt Of E-Waste in Comparison
to Lower Manhattan's Skyline

led to a remarkable rise in e-waste generation. Back in 1995, there were just 500,000 internet users in the country. But by 2020, that number grew to around 8 million users,[21] a surging 1,500% increase that has transformed the technological topography of this nation. The fact that much used EEE is smuggled into the country further obscures the true scale of the problem.[22]

Zimbabwe's lack of effective e-waste management policies makes quantifying e-waste inflows difficult.[23] Approximately 500 containers filled with used EEE enter Nigeria monthly, while 600 arrive at the ports of Tema in Ghana.[24] A U.S. Environmental Protection Agency study found that exporting e-waste to India costs 10 times less than processing it domestically, creating huge incentives for legal and illegal exports.[25] The Basel Action Network (BAN) reported that 34% of e-waste from the EU and USA was exported to developing countries, with 93% of that going to Asian nations lacking proper recycling systems.[26] Over 80% of second-hand electronic equipment is illegally exported from the Global North to developing countries, where more than 75% of it is unusable and unrepairable,

---

21 Simon Kemp, "Digital in Zimbabwe: All the Statistics You Need in 2021," *DataReportal–Global Digital Insights*, February 12, 2021, https://datareportal.com/reports/digital-2021-zimbabwe.

22 Macherera, Maphosa, and Zezai, "Perceptions and Practices," 29.

23 Maphosa, "Sustainable E-Waste Management," 20.

24 Maphosa and Maphosa, "E-Waste Management," 11.

25 Pawan Kumar and Anita Singh, "New Technology and E-Waste: A Case of India," *International Journal of Current Research in Science and Technology* 1, no. 9 (2015): 19–29, https://crst.gfer.org/index.php/crst/article/view/43.

26 Basel Action Network, "*The 'Scam Recycling' Continues,*" 2018, http://wiki.ban.org/images/archive/b/b0/20170906214510!The_Scam_Recycling_Continues_2017_Phase_2.pdf.

undermining their digital progress.[27] This illegal trade exacerbates the e-waste crisis in these regions and raises ethical concerns related to environmental justice, accountability, and neo-colonialism. Addressing these issues requires more than regulation alone; it demands that the Global North provide technical expertise and funding to help the Global South build critical e-waste infrastructure and worker protections to safeguard public health.

E-Waste Management Systems

E-waste recycling provides jobs for millions, yet the essential infrastructure and legislation remain sorely lacking. Municipalities struggle to separate e-waste from the domestic refuse at source, leading to the contamination of pristine landscapes. I saw this for myself on a visit to the country's biggest dumpsite, the Pomona and Golden Quarry, where I observed the alarming, undifferentiated mixture of municipal, domestic, and industrial waste.[28] The Pomona dumpsite is a stark landscape of mixed waste, including discarded electronics, broken devices, tangled cables, shattered screens, plastic, organic, metal, hazardous materials, and industrial effluence, accumulating under open skies. Scavengers sift through sprawling dumpsites, searching for valuable materials like plastics and metals amid the digital detritus. Toxic fumes from burning materials permeate the air, and the soil beneath is contaminated by heavy metals, a hidden cost of technological advancement.

The narrative that solely blames the Global North for contaminating e-waste landfills and dumpsites in the Global South requires scrutiny. Cutting off imports of used EEE will not resolve the issue, as toxicants persist due to a lack of socially and environmentally responsible initiatives and waste management practices on the ground. Proposals to stop e-waste flows from developed to developing countries are unlikely to succeed, as the demand for used EEE is high due to low income levels and high unemployment. Meaningful change requires addressing systemic issues upstream in the design and production of electronics before they become waste, as well as enacting local policies to combat the problem.

The absence of take-back schemes in Zimbabwe, like the extended producer responsibility (EPR), leaves the management of the burgeoning e-waste crisis largely in the hands of the informal sector—a void that reflects the broader systemic challenges facing the nation. In Zimbabwe, much like in many developing countries, the recycling of e-waste often transpires within the shadowy realm of the informal economy,[29] where workers, bereft of protective gear, must undertake the perilous task of dismantling and extracting valuable components from the e-waste of our digital age. A precarious endeavour, where public health and the environment risks are too often sacrificed upon the altar

27 O. Osibanjo and I.C. Nnorom, "The Challenge of Electronic Waste (E-Waste) Management in Developing Countries," *Waste Management & Research* 25, no. 6 (2007): 489–501, https://doi.org/10.1177/0734242X07082028.
28 Maphosa, "Rethinking Sustainability," 125.
29 Gilbert Moyen Massa and Vasiliki-Maria Archodoulaki, "Electrical and Electronic Waste Management Problems in Africa: Deficits and Solution Approach," *Environments* 10, no. 3 (2023): 44, https://doi.org/10.3390/environments10030044.

of economic necessity. [30] Although international guidelines strongly advocate for the formalization of e-waste recycling, Zimbabwe faces significant barriers: weak policy frameworks, limited recycling infrastructure, insufficient funding, and low public awareness of the environmental consequences.

By embracing Ubuntu's principles of shared humanity and caring for mother earth, societies can transition from exploitative e-waste dumping to circular economies—where technology benefits all and no community bears the cost of irresponsible disposal. Grounded in Ubuntu's ethos of collective responsibility, green initiatives that prioritize repair, reuse, and community-based management can transform e-waste into a sustainable system. Households play a crucial role by donating old devices to refurbishment centers, rejecting illegal e-waste dumping, and advocating for eco-conscious practices—ensuring responsible e-waste management begins at household level. The rise of community-led electronics repair hubs, responsible disposal programs, and public awareness campaigns are key green initiatives driving this shift toward sustainability. Witnessing community-based initiatives contribute to improved e-waste management inspires hope for positive change aligning with the ethos of Ubuntu.

Informal Recyclers

The management of e-waste provides a valuable opportunity to examine the important but often overlooked work that supports our technological voyage. E-waste is increasingly viewed as a valuable resource, with its management becoming a priority for job creation in Zimbabwe's vastly informal economy.[31] The manual labor underpinning e-waste infrastructures offers insights into broader socio-technical transitions.

Informal e-waste workers toil amidst mountains of digital refuse, their lives shaped by disparities that drive them to undertake this perilous work, extracting value from society's technological castoffs. Their welfare and occupational health must be central to any solution. Surveying the dystopian e-waste dumpsites, I am overcome with sorrow—charred keyboards and toxic plumes of computer smoke wafting across a devastated landscape. My own devices come to mind, and I wonder about their ultimate fate. Where are the climate-conscious computing initiatives, the sustainable design and recycling practices to address this crisis? The human cost is overwhelming, and the scale of the problem demands action. We must rethink the entire electronics lifecycle, from eco-friendly manufacturing to responsible end-of-life management. Only then can we begin to heal this wounded environment and protect vulnerable communities.

One study highlighted that e-waste scavengers often lack the knowledge and skills needed to safely extract valuable materials from e-waste,[32] which

30 B. Mandevere and H. Wiechers, "E-Waste Related Heavy Metals in Harare, Zimbabwe, a Pending Environmental and Public Health Disaster," *International Journal of Research and Innovation in Social Science* 5, no. 12 (2021): 688–697.
31 Maphosa, "Rethinking Sustainability," 126.
32 Fred Gweme, Hasha Maringe, Luka Ngoyi, and Gertjan Van Stam, "E-Waste in Zimbabwe and Zambia" 1st

unfortunately heightens their exposure to toxic substances. The study reveals the grim reality of e-waste recycling—crude methods like burning components and acid leaching that put both workers and communities at risk of toxic exposures. I have witnessed informal e-waste workers haphazardly dismantle abandoned electronics with chisels and hammers, releasing noxious heavy metals and toxins devastating the local environment. The health impact on these vulnerable workers and nearby residents is devastating—neurological diseases, cancers, DNA alteration, and organ damage.[33]

In Zimbabwe, many unsecured landfills and open dumps are easily accessible to desperate e-waste scavengers. Informal e-waste workers toil ceaselessly, hauling carts of salvaged items from dumpsites to their humble workshops where the arduous work of sustaining our technological progress begins. The meticulous dismantling of detritus is marked by repetitive, physically-demanding actions—bending, twisting, leaning, and standing for long hours. Scholars indicate that the non-neutral postures and heavy lifting by informal workers significantly heighten the risk of musculoskeletal disorders.[34] This labour-intensive process underscores the formidable challenges faced by those tending our e-waste economy. Their resilience in the face of such adversity is a testament to the human spirit. However, sustainable solutions must prioritize worker safety, and not just livelihood support.[35] To confront these issues, I believe we must empower the embattled e-waste scavengers and formalize the very operations they undertake with such fervent purpose. Formalizing their operations can generate income safely, while enabling the government to collect taxes and enhance environmental protection.

Public Health Impact

Exposure to heavy metals can lead to serious conditions, including bone and blood disorders, kidney damage, reduced cognitive function, and neurological harm.[36] Brominated flame retardants can enter the human body through various routes, including ingestion, inhalation, dermal contact, and transplacental transfer from mother to foetus.[37] In e-waste recycling areas, the primary route of exposure for residents is likely dietary, followed closely by dust ingestion and

---

Institute of Lifelong Learning and Development Studies International Research Conference, Chinhoyi University of Technology, August 3, 2016, https://www.researchgate.net/publication/305807572_E-Waste_in_Zimbabwe_and_Zambia.

33 Heacock et al., 551

34 Augustine A. Acquah, Clive D'Souza, Bernard J. Martin, John Arko-Mensah, Duah Dwomoh, Afua Asabea Amoabeng Nti, and Lawrencia Kwarteng, "Musculoskeletal Disorder Symptoms among Workers at an Informal Electronic-Waste Recycling Site in Agbogbloshie, Ghana," *International Journal of Environmental Research and Public Health* 18, no. 4 (2021): 2055, https://doi.org/10.3390/ijerph18042055.

35 Kansuk Stephen, "Shaping Sustainable Paths for E-Waste Management in Ghana," *United Nations Development Programme (UNDP)* (blog), September 8, 2023, https://www.undp.org/ghana/blog/shaping-sustainable-paths-e-waste-management-ghana.

36 Hongwu Wang, Peng Huang, Ruibiao Zhang, Xueyong Feng, Qiulin Tang, Sixi Liu, and Feiqiu Wen, "Effect of Lead Exposure from Electronic Waste on Haemoglobin Synthesis in Children," *International Archives of Occupational and Environmental Health* 94, no. 5 (2021): 911–918, https://doi.org/10.1007/s00420-020-01619-1.

37 Emma A. R. Zuiderveen, J. Chris Slootweg, and Jacob de Boer, "Novel Brominated Flame Retardants – A Review of Their Occurrence in Indoor Air, Dust, Consumer Goods and Food," *Chemosphere* 255 (2020): 126816, https://doi.org/10.1016/j.chemosphere.2020.126816.

air inhalation. Moreover, studies indicate that e-waste workers are frequently exposed to a wide range of hazardous metals and suffer from numerous health complications, including burns, lacerations, respiratory problems, headaches, and musculoskeletal disorders.[38]

The exposure to persistent organic pollutants, dioxins, and furans can wreak devastation, damaging bone metabolism and impairing cognitive development. They imperil vital organs—the lungs, liver, kidneys, pancreas, and cardiovascular system[39]—as the toxic legacy of our technological progress poisons both body and mind. Lead, a prominent component of e-waste, has been associated with lower scores in neonatal behavioural and neurological development, as well as impaired thyroid function and various foetal defects.[40] These substances can persist in the environment and accumulate in living organisms, leading to a range of health issues, including developmental disorders, cancers, and reproductive system complications. Reports indicate that there have been deaths linked to the consumption of contaminated crops, meat, milk, fruits, and vegetables.[41]

The 2021 World Health Organisation (WHO) report reveals the gendered dimensions of the global waste crisis, with 12.9 million women employed in the informal waste industry risking exposure to toxic substances.[42] On my many drives through the country's dumpsites, I have seen pregnant women working in these hazardous conditions. Such exposure to toxic elements during pregnancy is linked to severe health outcomes: premature birth, stillbirth, and low birth weight and length.[43] Casting a shadow over the most vulnerable, it is not just the workers' lives but also unborn lives that are imperilled by the environmental burdens of this informal industry. Despite research confirming that toxic chemicals can easily pass from mothers to infants during pregnancy and breastfeeding, little has been done in terms of policies to minimise these adverse risks, or protections for women in the informal e-waste sector. This lack of action highlights a significant gap in addressing public health concerns related to e-waste exposure.

The sobering reports from WHO reveal that at least 18 million children and teenagers, some as young as five, are actively engaged in the informal waste sector.[44] I have witnessed children labouring in e-waste dumpsites first-hand. Disturbingly, children are often regarded as well suited to handle e-waste, as their

38 Vishal Singh, Javier Cortes-Ramirez, Leisa-Maree Toms, Thilakshika Sooriyagoda, and Shamshad Karatela, "Effects of Polybrominated Diphenyl Ethers on Hormonal and Reproductive Health in E-Waste-Exposed Population: A Systematic Review," *International Journal of Environmental Research and Public Health* 19, no. 13 (2022): 7820, https://doi.org/10.3390/ijerph19137820.

39 Lynda Andeobu, Santoso Wibowo, and Srimannarayana Grandhi, "Environmental and Health Consequences of E-Waste Dumping and Recycling Carried out by Selected Countries in Asia and Latin America," *Sustainability* 15, no. 13 (2023): 10405, https://doi.org/10.3390/su151310405.

40 Ceridwen Johnson, "Soaring E-Waste Affects the Health of Millions of Children, WHO Warns," World Health Organization, June 15, 2021, https://www.who.int/news/item/15-06-2021-soaring-e-waste-affects-the-health-of-millions-of-children-who-warns.

41 Anthony Boardman, Jeff Geng, and Bruno Lam, "The Social Cost of Informal Electronic Waste Processing in Southern China," *Administrative Sciences* 10, no. 1 (2020): 1-21, https://doi.org/10.3390/admsci10010007.

42 Johnson, "Soaring E-Waste."

43 Minghui Li, Xia Huo, Yukui Pan, Haoxing Cai, Yifeng Dai, and Xijin Xu, "Proteomic Evaluation of Human Umbilical Cord Tissue Exposed to Polybrominated Diphenyl Ethers in an E-Waste Recycling Area," *Environment International* 111 (2018): 362–371, https://doi.org/10.1016/j.envint.2017.09.016.

44 Johnson, "Soaring E-Waste."

smaller hands can access tight spaces. Because the organs that process and eliminate toxins are still developing, children exposed to a range of contaminants through skin contact, inhalation, and contaminated food or water are especially at risk, as their developing bodies are less capable of breaking down and eliminating hazardous substances.[45]

The young and vulnerable of the populace are imperilled by the technological refuse that has seeped into the very fabric of their environment.[46] The consequences can be dire, leading to the blighting of young minds, intellectual disabilities, and a host of other insidious health issues[47]—a dark legacy of our technological age, so often associated with progress and visions of the future.

Environmental Impact

The technological refuse of our age refuses to decompose, polluting the land, air, and water, subverting natural cycles, and contaminating the elements that sustain life, leading to blighted landscapes and environmental devastation. Crude processing techniques—like acid baths and open-air burning—release harmful substances into the environment, resulting in both immediate and long-term effects. In regions close to e-waste dumpsites, soils, grazing lands, and water bodies are often contaminated. This contamination can lead to the production of toxic crops, fruits, and aquatic organisms.[48] During visits, I have watched farmers and locals continuing to harvest potentially harmful produce, either unaware of risks or lacking safer alternatives.

Even so, researchers have been sounding the alarm on the significant risks posed by the release of toxic substances from deteriorating e-waste.[49] Recent studies indicate that leachate from e-waste is a major contributor to groundwater pollution, as hazardous substances seep into the soil and threaten drinking water supplies and surrounding ecosystems. In Bulawayo, research from 2017 confirmed that such contamination is widespread, as e-waste from landfills continues to infiltrate the groundwater.[50] Elevated levels of toxic elements like lead, arsenic, and cadmium compromise soil fertility, reduce agricultural productivity, and contaminate food crops, threatening food security and exposing

45 Acquah et al., "Musculoskeletal Disorder Symptoms."
46 Michelle Heacock, Carol Bain Kelly, Kwadwo Ansong Asante, Linda S. Birnbaum, Åke Lennart Bergman, Marie-Noel Bruné, and Irena Buka, "E-Waste and Harm to Vulnerable Populations: A Growing Global Problem," *Environmental Health Perspectives* 124, no. 5 (2015): 550–555, https://doi.org/10.1289/ehp.1509699.
47 Johnson, "Soaring E-Waste."
48 Gratien Twagirayezu, Olivier Irumva, Kui Huang, Hui Xia, Abias Uwimana, Jean Claude Nizeyimana, Habasi Patrick Manzi, Febronie Nambajemariya, and Auguste Cesar Itangishaka, "Environmental Effects of Electrical and Electronic-Waste on Water and Soil: A Review," *Polish Journal of Environmental Studies* 31, no. 3 (2022): 2507–2525, https://doi.org/10.15244/pjoes/144194.
49 Puneet Babbar, Swatantra Verma, and Gauhar Mehmood, "Groundwater Contamination from Non-Sanitary Landfill Sites–A Case Study on the Ghazipur Landfill Site, Delhi (India)," *International Journal of Applied Environmental Sciences* 12, no. 11 (2017): 1969–1991, https://doi.org/10.37622/ijaes/12.11.2017.1969-1991.; Li et al., "Proteomic Evaluation," 365.
50 Charles Teta and Tapiwa Hikwa, "Heavy Metal Contamination of Ground Water from an Unlined Landfill in Bulawayo, Zimbabwe," *Journal of Health and Pollution* 7, no. 15 (2017): 18–27, https://doi.org/10.5696/2156-9614-7.15.18.

both people and animals to long-term health hazards through dietary intake.[51] Waste accumulation, degradation, informal recycling, urban encroachment, heavy metal pollution, air contamination, and inadequate disposal practices all contribute to a growing crisis. Poor e-waste management in the areas I have visited has led to environmental devastation, health risks, and socioeconomic inequality. E-waste landfills devastate vast areas, transforming vibrant ecosystems into barren wastelands incapable of supporting any form of life. The insidious effects ripple outward, seeping into the very food chains upon which we all depend, heightening risks of exposure for humans, plants, and animals alike.

Fortunately, the recycling process is becoming safer and more efficient, allowing for the recovery of valuable components and materials. This not only minimises the risks associated with improper e-waste disposal but also supports a more circular economy by enabling the reuse of resources: the recovery of aluminium, copper, and iron from e-waste recycling in 2019 alone reduced CO_2 emissions by 15 megatons.[52]

Policies

The extended producer responsibility (EPR) model is widely regarded as a transformative strategy for tackling the e-waste crisis and promoting green initiatives. By holding manufacturers accountable for the full product lifecycle, the EPR incentivises sustainable design practices that prioritize recyclability and reuse, and minimise the environmental impact, driving a shift toward circularity. This is especially important as the expansion of sustainable energy technologies brings with it increased need for infrastructural middle-parts like inverters, controllers, capacitors, wires, and cooling systems. Though integral to the transition from fossil fuel dependence, these components will all contribute to electronic waste streams at the end of their functional life cycles. This not only promotes the recovery of critical minerals but also ensures that recycling and disposal are conducted through environmentally safe and economically sustainable methods.[53] EPR frameworks in developing regions could enhance global sustainability by engaging producers, but their implementation requires context-specific policies and stakeholder collaboration, ultimately contributing to the flourishing of a healthier and more vibrant planet.

Amidst the mounting tide of the digital detritus, only 13 African countries (24%) have established national laws to regulate the management of e-waste,[54]

---

51 Oladunni Bola Olafisoye, Tejumade Adefioye, and Otolorin Adelaja Osibote, "Heavy Metals Contamination of Water, Soil, and Plants around an Electronic Waste Dumpsite," *Polish Journal of Environmental Studies* 22, no. 5 (2013): 1431–1439, https://www.pjoes.com/pdf-89108-22967?filename=Heavy%20Metals.pdf.

52 Xu Pan, Christina W. Y. Wong, and Chunsheng Li, "Circular Economy Practices in the Waste Electrical and Electronic Equipment (WEEE) Industry: A Systematic Review and Future Research Agendas," *Journal of Cleaner Production* 365 (2022): 132671, https://doi.org/10.1016/j.jclepro.2022.132671.

53 P. Thakur and S. Kumar, "Evaluation of E-Waste Status, Management Strategies, and Legislations," *International Journal of Environmental Science and Technology* 19 (2022): 6957–6969, https://doi.org/10.1007/s13762-021-03383-2.

54 Charlotte Owen-Burge, "How Africa Is Leading the Way in Dealing with 'E-Waste,'" *Climate Champions*, March 28, 2022, https://climatechampions.unfccc.int/how-africa-is-leading-the-way-in-dealing-with-e-waste/.

a sobering statistic that reflects the systemic challenges to addressing this growing crisis across the continent. There is still a hope that implementing stricter regulations, recycling programs, manufacturer responsibility, and public awareness may improve e-waste management, encouraging greener business practices and easier consumer participation in recycling and waste reduction.[55] At the same time, despite an increase in the establishment of policies in Africa, research shows less than 1% of e-waste is formally recycled, indicating a troubling disconnect between policy and implementation.[56]

Current e-waste policies often rely on take-back schemes, such as the EPR model. Yet, in Africa, many key e-waste supply chain actors are unknown, hindering EPR enforcement. The EPR framework may not work in Zimbabwe due to the lack of transparency amongst key stakeholders and unavailability of local EEE producers to hold responsible. This lays bare the formidable challenge facing Zimbabwe as it grapples with the e-waste conundrum. As Zimbabwe develops its e-waste policy, the informal sector, whose labor and expertise are integral to any sustainable solution, will play a critical role. In many cases, these individuals are the first responders to e-waste challenges, often stepping in where formal systems fail. In my experience, these individuals are often the first responders to e-waste issues.

A truly effective policy would directly address barriers related to inadequate recycling infrastructure, limited funds, knowledge gaps, and weak collection and transportation systems. I have seen first-hand how the absence of effective regulations hamper the separation of e-waste from general solid waste at the source, complicating management and recycling efforts. Implementing stricter regulations on EEE imports would not only mitigate the risks associated with e-waste but also encourage the development of local recycling initiatives. Involving the informal sector in safer waste management can create opportunities for local businesses and empower communities to take ownership of their waste. This proactive approach could significantly enhance public health protections and environmental sustainability in Zimbabwe.

Economic Value

Manufacturing EEE relies on rare earth minerals like indium, lithium, and cobalt, as well as other precious metals. The extraction and use of these elements is intertwined with the digital detritus.[57] Global revenue from e-waste recycling is projected to reach $244 billion by 2032,[58] underscoring its significant economic potential. In 2019, e-waste recycling yielded $57 billion in recovered critical

55 Pan, Wong, and Li, "Circular Economy Practices," 132671.
56 Bouchra Bakhiyi, Sabrina Gravel, Diana Ceballos, Michael A. Flynn, and Joseph Zayed, "Has the Question of E-Waste Opened a Pandora's Box? An Overview of Unpredictable Issues and Challenges," *Environment International* 110 (2018): 173–192, https://doi.org/10.1016/j.envint.2017.10.021.
57 Perrine Chancerel, Max Marwede, Nils F. Nissen, and Klaus-Dieter Lang, "Estimating the Quantities of Critical Metals Embedded in ICT and Consumer Equipment," *Resources, Conservation and Recycling* 98 (2015): 9–18, https://doi.org/10.1016/j.resconrec.2015.03.003.
58 Allied Market Research, "E-Waste Management Market Is Projected to Hit $102.62 Billion by 2027," 2023, https://www.alliedmarketresearch.com/e-waste-management-market.

minerals and simultaneously reduced CO_2 emissions by 15 megatons through the recovery of aluminium, copper, and iron.[59] The Chinese government established over 1,000 e-waste recycling plants, targeting a recycling rate of 50% by 2025.[60] In countless homes across the globe, 5.3 billion mobile phones lie forgotten in drawers—nearly matching the world's population—while each year we add another billion into landfills, a staggering reminder of our neglect.[61] Recycling one million mobile phones can produce 24 kg of gold and 250 kg of copper,[62] a veritable treasure trove, if only we could harness its riches with wisdom and care. There are examples in which this care is apparent: the 5,000 medals awarded at the 2020 Tokyo Olympic Games were crafted from materials recovered from discarded mobile phones donated by residents.[63] Extracting precious metals conserves finite resources and supports environmental protection, cultivating the fertile ground for a true circular economy to take root and flourish.

Implementing more circular practices within urban mining, such as the recovery of precious metals and materials from e-waste, can create greener jobs with roles like recycling technicians, environmental engineers, and sustainability coordinators. These jobs not only provide employment but also empower active community engagement in the complex process of e-waste management. By recovering valuable materials such as gold, copper, and rare earth elements from cast off electronics, we both reduce the need for virgin resource extraction and breathe new life into discarded technologies. Ultimately, sustainable resource management supports the move toward a more circular economy, where materials are reused and recycled, contributing to both environmental and economic benefits. The latter benefits are worth noting: the density of critical minerals in e-waste is remarkable.[64] Studies show that the concentrations of valuable metals like copper and gold in e-waste can be 13–50 times higher than in primary mineral sources.[65] Recycling mobile phones can yield 130 kg of copper, 3.5 kg of silver, and 340 g of gold per ton of e-waste, far exceeding the 1.5 g of gold per ton from conventional ore.[66]

While the majority of African nations are yet to establish robust policies and take solid steps to manage the mounting e-waste crisis, South Africa, Egypt, and Ghana have emerged as beacons.[67] E-waste recycling generates approximately $105–$268 million in annual revenue and supports the livelihoods of around 200,000 people in Ghana.[68] Similarly, in Zimbabwe's informal economy, e-waste

---

59 Pan, Wong, and Li, "Circular Economy Practices," 132671.

60 World Economic Forum, *A New Circular Vision for Electronics: Time for a Global Reboot* (January 2019), https://www3.weforum.org/docs/WEF_A_New_Circular_Vision_for_Electronics.pdf.

61 Paige Bennett, "5 Billion Cell Phones Are Expected to Become E-Waste This Year," *World Economic Forum*, October 20, 2022, https://www.weforum.org/stories/2022/10/global-phone-technology-waste/.

62 World Economic Forum, *New Circular Vision*.

63 World Economic Forum, *New Circular Vision*.

64 MOL-E, "The New Generation of Scrapping: Urban Mining," July 27, 2024, https://www.mol-e.co/en/news/city-mining.

65 Maphosa and Mashau, "The Conundrum," 32.

66 Umicore, "When Your Phone Stops Ringing," November 19, 2020, https://www.umicore.com/en/newsroom/news/when-your-phone-stops-ringing-/.; Brian Owens, "Mining: Extreme Prospects," *Nature* 495, no. 7440 (2013): S4–S6, https://doi.org/10.1038/495s4a.

67 Bimir, "Revisiting E-Waste Management Practices," 562.

68 Martin Oteng-Ababio, Ebenezer Forkuo Amankwaa, and Mary Anti Chama, "The Local Contours of Scavenging for E-Waste and Higher-Valued Constituent Parts in Accra, Ghana," *Habitat International* 43 (2014): 163–71, https://doi.org/10.1016/j.habitatint.2014.03.003.

recycling is financially lucrative: workers earn 3.5 times more than those in general solid waste management,[69] and a study conducted in Harare found that some scavengers viewed the e-waste sector as a means of livelihood support amidst ongoing economic hardships in the country.[70]

By highlighting the economic opportunities of recycling like job creation and income generation, initiatives foster a shared sense of responsibility and ownership over waste management. This aligns well with the ethos of Ubuntu that underpins many Zimbabwean societies. Furthermore, elevating the role of local community leaders and elders in spearheading e-waste initiatives can lend cultural legitimacy and encourage broader community participation, positively contributing to a shared environment. Tapping into existing social structures and norms can create a greater sense of buy-in and long-term commitment. Offering socio-psychological incentives like community recognition, culturally integrated campaigns, educational workshops, and gamification makes e-waste recycling emotionally engaging and socially rewarding, empowering communities to become active environmental stewards through culturally relevant solutions. This holistic approach, sensitive to Zimbabwe's socio-cultural dynamics, has the potential to cultivate a more effective and responsible model for managing the e-waste crisis by effectively disseminating information about the benefits of sustainable e-waste management.

Circular economy

The prevailing linear economic model that dominates many developing countries struggles to maintain equilibrium. It leads to the inexorable and profound decline in biodiversity across devastated ecosystems and blighted landscapes, significantly contributing to the existential crisis of climate change[71]—a precarious imbalance that demands a fundamental rethinking of our relationship with the natural world. On my drives throughout the digital dumpsites of Zimbabwe, I have

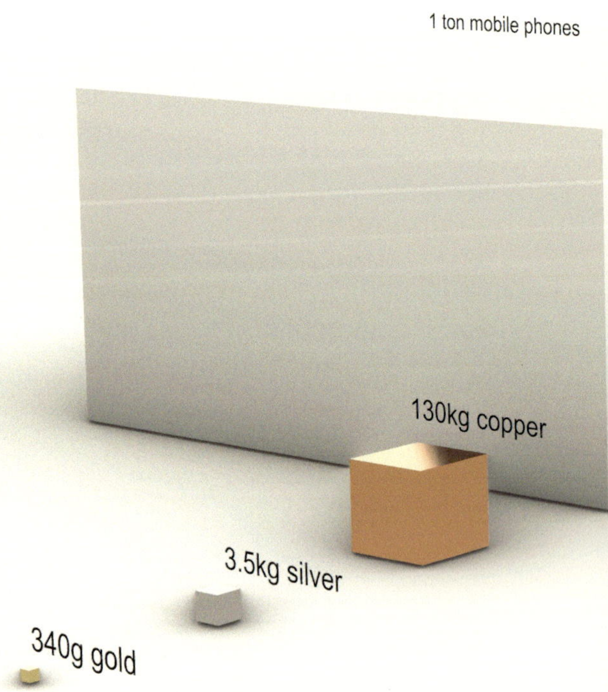

69 Maphosa and Maphosa, "E-Waste Management," 12.

70 Macherera, Maphosa, and Zezai, "Perceptions and Practices," 24; Chitotombe, "Globalization of Information Communication Technology," 2178.

71 Julian Kirchherr, Denise Reike, and Marko Hekkert, "Conceptualizing the Circular Economy: An Analysis of 114 Definitions," *Resources, Conservation and Recycling* 127 (2017): 221–232, https://doi.org/10.1016/j.resconrec.2017.09.005.

seen first-hand how the linear model fails to address the long-term consequences of waste and environmental degradation. The circular economy presents a compelling alternative: it is a sustainable approach inspired by natural systems, concentrating on maintaining ecological balance and the highest possible value of materials and eliminating harmful substances and waste.[72]

Industry experts emphasize that embracing circular economy principles can not only enhance productivity and efficiency but also alleviate the environmental and health burdens faced by communities living near landfills.[73] The circular economy reintegrates anthropogenic materials into the manufacturing process, reducing dependence on new resources, and minimizing the environmental impact.[74] This approach enables the recovery of valuable materials from e-waste for reuse in

72 World Economic Forum, "Helping Companies in China Recycle 50% of E-Waste by 2025," July 1, 2021, https://www.weforum.org/impact/helping-chinese-companies-reduce-recycle-e-waste/.
73 Mfowabo Maphosa and Vusumuzi Maphosa, "Using the Circular Economy to Mitigate the Global Electronic Waste Challenge: A Systematic Literature Review Approach," *European Journal of Social Impact and Circular Economy* 5, no. 3 (2024): 28–54, https://doi.org/10.13135/2704-9906/10799.
74 Sami Kara, Michael Hauschild, John Sutherland, and Tim McAloone, "Closed-Loop Systems to Circular Economy: A Pathway to Environmental Sustainability?" *CIRP Annals* 71, no. 2 (2022): 505–528, https://doi.org/10.1016/j.cirp.2022.05.008.

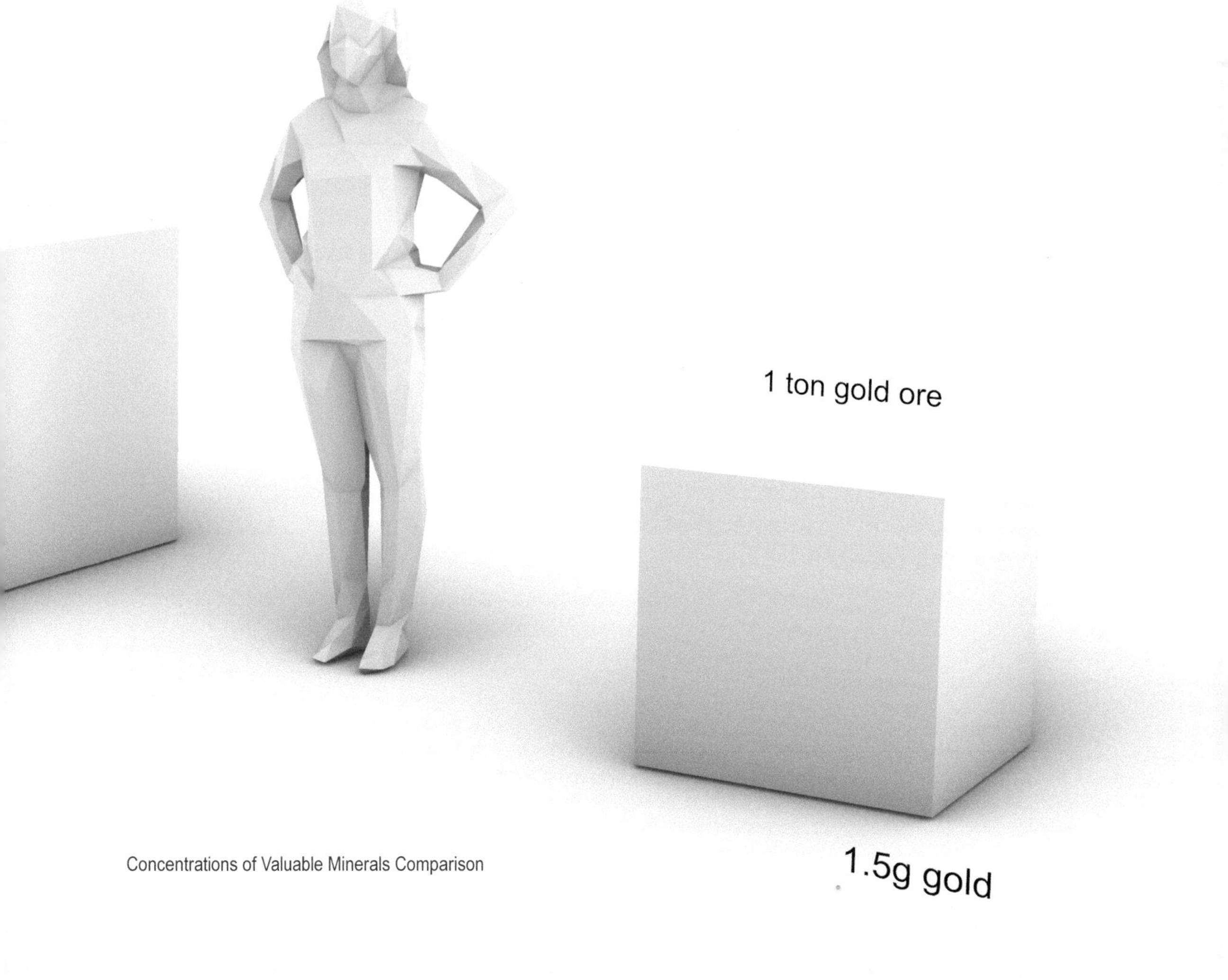

Concentrations of Valuable Minerals Comparison

new products, while also reducing pollution,[75] improving product design, and extending the lifespan of existing equipment through repair, refurbishment, and remanufacturing.[76]`

One of the goals of adopting the circular economy is to slow down our consumptive desires and rejuvenate resource restorative cycles. This approach minimises the extraction of new materials and significantly decreases waste production.[77] Globally, approximately 300 tons of gold can be recovered annually from e-waste if appropriate recycling methods are practiced.[78] Central to the circular economy are the "7 R's": redesign, reduce, reuse, refuse, rethink, recover, and recycle.[79] The principles of reduce, reuse, refuse, and rethink resonate with Ubuntuism, emphasizing community, respect for resources, and the interconnectedness of all living beings in promoting sustainable practices. For instance, redesigning products can make them more sustainable and easier to recycle. Refusing unnecessary products and rethinking our consumption habits are also vital steps individuals can take toward a more sustainable future. Concepts like product-as-a-service are promising models where, instead of owning a product, consumers pay for access to its function or service over time, promoting sustainability by encouraging longer product lifecycles and reducing waste.[80]

While circular economy principles hold transformative potential, there are numerous challenges to implementing them in Zimbabwe, requiring nuanced, context-specific solutions to truly unlock the benefits of sustainable practices. Insufficient policies and funding, a lack of awareness, and inadequate collection and transportation systems are just a few of the hurdles we face. In my experience, the contrast between Zimbabwe and more developed nations is stark: many developing countries lack established e-waste collection points and the critical transportation systems needed to manage the growing stream of e-waste responsibly. This gap in infrastructure forces people to dispose of e-waste alongside regular waste, a practice that not only undermines sustainable development but also jeopardizes both environmental and public health.[81]

Conclusion

The rapid march of technological progress, fueled by the Fourth Industrial Revolution, has led to an explosion of electronic devices, each with a tragically short lifespan. As demand for critical minerals surges and obsolete gadgets pile up, e-waste has become the world's fastest-growing solid waste stream,

75 Pan, Wong, and Li, "Circular Economy Practices." 8.
76 Kara et al., "Closed-Loop Systems," 513.
77 Kirchherr, Reike, and Hekkert, "Conceptualizing the Circular Economy," 225.
78 Víctor Oestreicher, Carolina S. García, Rodrigo Pontiggia, Melina B. Rossi, Paula C. Angelomé, and Galo J. A. A. Soler-Illia, "E-Waste Upcycling for the Synthesis of Plasmonic Responsive Gold Nanoparticles," *Waste Management* 117 (2020): 9–17, https://doi.org/10.1016/j.wasman.2020.07.037.
79 Piero Morseletto, "Targets for a Circular Economy," *Resources, Conservation and Recycling* 153 (2020): 104553, https://doi.org/10.1016/j.resconrec.2019.104553.
80 World Economic Forum, "Helping Companies in China."
81 Massa and Archodoulaki, "Electrical and Electronic Waste Management," 44.

poisoning landscapes and communities. Yet within this crisis lies an opportunity: discarded electronics contain far higher concentrations of precious metals than mined ore, offering a chance to turn toxic trash into a sustainable resource through responsible recycling. But the global system persists in its broken state, with wealthy nations offloading defunct electronics to developing countries while professing to close the digital divide. A practice that perpetuates colonial exploitation and environmental injustice. This toxic trade thrives on policy gaps and weak enforcement, as developing nations lack the infrastructure, funding, and technical know-how to manage the e-waste deluge. But solutions exist—stronger regulations, investment in recycling systems, and international cooperation can empower these countries to reclaim value from waste while mitigating health and ecological risks.

For instance, Zimbabwe can leverage the spirit of Ubuntu, rooted in community solidarity, to mobilize grassroots action and integrate informal waste pickers into a formal circular economy. By pairing economic incentives with cultural values, nations can transform e-waste from a crisis into an opportunity, fostering green jobs and reducing reliance on destructive mining. The shift from a linear "take-make-waste" model to a circular economy is not just an environmental imperative—it is a chance to redefine progress. Recovering metals from discarded electronics through urban mining could slash the need for destructive extraction while protecting ecosystems and stabilizing supply chains. But success hinges on global equity: wealthy nations must stop offshoring their waste and instead support sustainable systems worldwide. By combining policy innovation, community empowerment, and ethical recycling, we can dismantle the toxic legacy of e-waste—turning a looming catastrophe into a catalyst for a fairer, greener future. Now is the time to act, as further delay risks irreversible consequences.

BIBLIOGRAPHY

Acquah, Augustine A., Clive D'Souza, Bernard J. Martin, John Arko-Mensah, Duah Dwomoh, Afua Asabea Amoabeng Nti, and Lawrencia Kwarteng. "Musculoskeletal Disorder Symptoms among Workers at an Informal Electronic-Waste Recycling Site in Agbogbloshie, Ghana." *International Journal of Environmental Research and Public Health* 18, no. 4 (February 19, 2021): 1–21. https://doi.org/10.3390/ijerph18042055.

Akcil, Ata, Ceren Erust, Chandra Sekhar Gahan, Mehmet Ozgun, Merve Sahin, and Aysenur Tuncuk. "Precious Metal Recovery from Waste Printed Circuit Boards Using Cyanide and Non-Cyanide Lixiviants – a Review." *Waste Management* 45 (November 2015): 258–271. https://doi.org/10.1016/j.wasman.2015.01.017.

Allied Market Research. "E-Waste Management Market Is Projected to Hit $102.62 Billion by 2027." 2023. https://www.alliedmarketresearch.com/e-waste-management-market.

Andeobu, Lynda, Santoso Wibowo, and Srimannarayana Grandhi. "Environmental and Health Consequences of E-Waste Dumping and Recycling Carried out by Selected Countries in Asia and Latin America." *Sustainability* 15, no. 13 (January 1, 2023): 10405. https://doi.org/10.3390/su151310405.

Babbar, Puneet, Swatantra Verma, and Gauhar Mehmood. "Groundwater Contamination from Non-Sanitary Landfill Sites – a Case Study on the Ghazipur Landfill Site, Delhi (India)." *International Journal of Applied Environmental Sciences* 12, no. 11 (November 30, 2017): 1969–1991. https://doi.org/10.37622/ijaes/12.11.2017.1969–1991.

Bakhiyi, Bouchra, Sabrina Gravel, Diana Ceballos, Michael A. Flynn, and Joseph Zayed. "Has the Question of E-Waste Opened a Pandora's Box? An Overview of Unpredictable Issues and Challenges." *Environment International* 110 (January 2018): 173–192. https://doi.org/10.1016/j.envint.2017.10.021.

Basel Network Action. "The 'Scam Recycling' Continues." 2018. http://wiki.ban.org/images/archive/b/b0/20170906214510!The_Scam_Recycling_Continues_2017_Phase_2.pdf

Bennett, Paige. "5 Billion Cell Phones Are Expected to Become E-Waste This Year." World Economic Forum. October 20, 2022. https://www.weforum.org/stories/2022/10/global-phone-technology-waste/.

Bimir, Mathias Nigatu. "Revisiting E-Waste Management Practices in Africa: Selected Countries' Cases." *Journal of the Air & Waste Management Association* 70, no. 7 (May 13, 2020): 559–669. https://doi.org/10.1080/10962247.2020.1769769.

Bleischwitz, Raimund, Miying Yang, Beijia Huang, Xiaozhen Xu, Jie Zhou, Will McDowall, Philip Andrews-Speed, Zhe Liu, and Geng Yong. "The Circular Economy in China: Achievements, Challenges and Potential Implications for Decarbonisation." *Resources, Conservation and Recycling* 183 (2022): 106350. https://doi.org/10.1016/j.resconrec.2022.106350.

Boardman, Anthony, Jeff Geng, and Bruno Lam. "The Social Cost of Informal Electronic Waste Processing in Southern China." *Administrative Sciences* 10, no. 1 (February 5, 2020): 1–21. https://doi.org/10.3390/admsci10010007.

Cervantes Puma, Genesis Camila, Adriana Salles, and Luís Bragança. "Nexus between Urban Circular Economies and Sustainable Development Goals: A Systematic Literature Review." *Sustainability* 16, no. 6 (January 1, 2024): 2500. https://doi.org/10.3390/su16062500.

Chakuya, Jeremiah, Norman Munkuli, Courage Mutema, and Edson Gandiwa. "An Assessment of the Impact of Illegal Artisanal Gold Mining on the Environment in Parts of Chewore Safari Area, Northern Zimbabwe." *Environmental Research Communications* 5, no. 7 (July 1, 2023): 075005. https://doi.org/10.1088/2515-7620/ace23b.

Chancerel, Perrine, Max Marwede, Nils F. Nissen, and Klaus-Dieter Lang. "Estimating the Quantities of Critical Metals Embedded in ICT and Consumer Equipment." *Resources, Conservation and Recycling* 98 (May 2015): 9–18. https://doi.org/10.1016/j.resconrec.2015.03.003.

Chitotombe, Joseline Wadzanai. "Globalization of Information Communication Technology (ICT) and Consumerism in Developing Countries: Confronting the Challenges of E-Waste Disposal in Harare Urban, Zimbabwe." *International Journal of Environmental Sciences* 3, no. 6 (2014): 2172–2185. https://www.indianjournals.com/ijor.aspx?target=ijor:ijes&volume=3&issue=6&article=032.

Forti, Vanessa, Cornelis Peter Baldé, Ruediger Kuehr, and Garam Bel. "Global E-Waste Monitor 2020." www.itu.int. 2020. https://www.itu.int/en/ITU-D/Environment/Documents/Toolbox/GEM_2020_def.pdf

Gweme, Fred, Hasha Maringe, Luka Ngoyi, and Gertjan Van Stam. "E-Waste in Zimbabwe and Zambia." 1st Institute of Lifelong Learning and Development Studies International Research Conference, Chinhoyi University of Technology. August 3, 2016. https://www.researchgate.net/publication/305807572_E-Waste_in_Zimbabwe_and_Zambia.

Heacock, Michelle, Carol Bain Kelly, Kwadwo Ansong Asante, Linda S. Birnbaum, Åke Lennart Bergman, Marie-Noel Bruné, and Irena Buka. "E-Waste and Harm to Vulnerable Populations: A Growing Global Problem." *Environmental Health Perspectives* 124, no. 5 (September 29, 2015): 550–555. https://doi.org/10.1289/ehp.1509699.

Hossain, M. Sahadat, Sulala M. Al-Hamadani, and M. Toufiqur Rahman. "E-Waste: A Challenge for Sustainable Development". *Journal of Health and Pollution* 5, no. 9 (2015): 3–11. https://doi.org/10.5696/2156-9614-5-9.3

Johnson, Ceridwen. "Soaring E-Waste Affects the Health of Millions of Children, WHO Warns." World Health Organization. 2021. https://www.who.int/news/item/15-06-2021-soaring-e-waste-affects-the-health-of-millions-of-children-who-warns.

Kara, Sami, Michael Hauschild, John Sutherland, and Tim McAloone. "Closed-Loop Systems to Circular Economy: A Pathway to Environmental Sustainability?" *CIRP Annals* 71, no. 2 (July 2022): 505–528. https://doi.org/10.1016/j.cirp.2022.05.008.

Kemp, Simon. "Digital in Zimbabwe: All the Statistics You Need in 2021." DataReportal – Global Digital Insights. February 12, 2021. https://datareportal.com/reports/digital-2021-zimbabwe.

Kirchherr, Julian, Denise Reike, and Marko Hekkert. "Conceptualizing the Circular Economy: An Analysis of 114 Definitions." *Resources, Conservation and Recycling* 127, no. 1 (December 2017): 221–232. https://doi.org/10.1016/j.resconrec.2017.09.005.

Kuehr, Ruediger, David Hirsch, and Terry Collins. "Global E-Waste Monitor 2024: Electronic Waste Rising Five Times Faster than Documented E-Waste Recycling." United Nations Institute for Training and Research. March 20, 2024. https://unitar.org/about/news-stories/press/global-e-waste-monitor-2024-electronic-waste-rising-five-times-faster-documented-e-waste-recycling.

Lebbie, Tamba S., Omosehin D. Moyebi, Kwadwo Ansong Asante, Julius Fobil, Marie Noel Brune-Drisse, William A. Suk, Peter D. Sly, Julia Gorman, and David O. Carpenter. "E-Waste in Africa: A Serious Threat to the Health of Children." *International Journal of Environmental Research and Public Health* 18, no. 16 (January 1, 2021): 8488. https://doi.org/10.3390/ijerph18168488.

Li, Minghui, Xia Huo, Yukui Pan, Haoxing Cai, Yifeng Dai, and Xijin Xu. "Proteomic Evaluation of Human Umbilical Cord Tissue Exposed to Polybrominated Diphenyl Ethers in an E-Waste Recycling Area." *Environment International* 111 (February 1, 2018): 362–371. https://doi.org/10.1016/j.envint.2017.09.016.

Mabele, Mathew Bukhi, Judith E. Krauss, and Wilhelm Kiwango. "Going Back to the Roots: Ubuntu and Just Conservation in Southern Africa." *Conservation and Society* 20, no. 2 (2022): 92. https://doi.org/10.4103/cs.cs_33_21.

Macherera, Margaret, Vusumuzi Maphosa, and David Zezai. "Perceptions and Practices on E-Waste Management in Selected Institutions of Zimbabwe." *Water and Environmental Sustainability* 3, no. 2 (June 10, 2023). 22–31. https://doi.org/10.52293/wes.3.2.2231.

Mandevere, B., and H. Wiechers. "E-Waste Related Heavy Metals in Harare, Zimbabwe, a Pending Environmental and Public Health Disaster." *International Journal of Research and Innovation in Social Science* V, no. XII (2021): 688–697.

Maphosa, Mfowabo, and Vusumuzi Maphosa. "A Bibliometric Analysis of the Effects of Electronic Waste on the Environment." *Global Journal of Environmental Science and Management* 8, no. 4 (October 1, 2022): 589–606. https://doi.org/10.22034/GJESM.2022.04.10.

Maphosa, Mfowabo, and Vusumuzi Maphosa. "Using the Circular Economy to Mitigate the Global Electronic Waste Challenge: A Systematic Literature Review Approach." *European Journal of Social Impact and Circular Economy* 5, no. 3 (2024): 28–54. https://doi.org/10.13135/2704-9906/10799.

Maphosa, Vusumuzi. "Rethinking Sustainability: A Bibliometric and Visualisation of E-Waste Management in Africa." *Journal of Higher Education Theory and Practice* 22, no. 1 (2022): 123–135. https://articlearchives.co/index.php/JHETP/article/view/3290/3265.

Maphosa, Vusumuzi. "Sustainable E-Waste Management at Higher Education Institutions' Data Centres in Zimbabwe." *International Journal of Information Engineering and Electronic Business* 14, no. 5 (October 8, 2022): 15–23. https://doi.org/10.5815/ijieeb.2022.05.02.

Maphosa, Vusumuzi. "The Rise of Artificial Intelligence and Emerging Ethical and Social Concerns." *AI, Computer Science and Robotics Technology* 3, no. 1 (December 8, 2024): 1–20. https://doi.org/10.5772/acrt.20240020.

Maphosa, Vusumuzi, and Mfowabo Maphosa. "E-Waste Management in Sub-Saharan Africa: A Systematic Literature Review." Edited by Albert W. K. Tan. *Cogent Business & Management* 7, no. 1 (January 1, 2020). 1–19. https://doi.org/10.1080/23311975.2020.1814503.

Maphosa, Vusumuzi, and Pfano Mashau. "The Conundrum: Transforming African E-Waste Landfills to Urban Mines." In *Advances and Challenges in Hazardous Waste Management*, ed. Selah M. Hosam, Amal I. Hassan, and Refaat F. Aglan. London: IntechOpen, 2023: 28–39. https://doi.org/10.5772/intechopen.1002419.

MOL-E. "The New Generation of Scrapping: Urban Mining." Mol-e.co. July 27, 2024. https://www.mol-e.co/en/news/city-mining.

Morseletto, Piero. "Targets for a Circular Economy." *Resources, Conservation and Recycling* 153 (February 2020): 104553. https://doi.org/10.1016/j.resconrec.2019.104553.

Moyen Massa, Gilbert, and Vasiliki-Maria Archodoulaki. "Electrical and Electronic Waste Management Problems in Africa: Deficits and Solution Approach." *Environments* 10, no. 3 (March 5, 2023): 44. https://doi.org/10.3390/environments10030044.

Oestreicher, Víctor, Carolina S. García, Rodrigo Pontiggia, Melina B. Rossi, Paula C. Angelomé, and Galo J. A. A. Soler-Illia. "E-Waste Upcycling for the Synthesis of Plasmonic Responsive Gold Nanoparticles." *Waste Management* 117 (November 2020): 9–17. https://doi.org/10.1016/j.wasman.2020.07.037.

Olafisoye, Oladunni Bola, Tejumade Adefioye, and Otolorin Adelaja Osibote. "Heavy Metals Contamination of Water, Soil, and Plants around an Electronic Waste Dumpsite." *Pol. J. Environ. Stud.* 22, no. 5 (2013): 1431–1439. https://www.pjoes.com/pdf-89108-22967?filename=Heavy%20Metals.pdf.

Osibanjo, O., and I. C. Nnorom. "The Challenge of Electronic Waste (E-Waste) Management in Developing Countries." *Waste Management & Research* 25, no. 6 (December 2007): 489–501. https://doi.org/10.1177/0734242x07082028.

Oteng-Ababio, Martin, Ebenezer Forkuo Amankwaa, and Mary Anti Chama. "The Local Contours of Scavenging for E-Waste and Higher-Valued Constituent Parts in Accra, Ghana." *Habitat International* 43 (July 2014): 163–171. https://doi.org/10.1016/j.habitatint.2014.03.003.

Owens, Brian. "Mining: Extreme Prospects." *Nature* 495, no. 7440 (March 2013): S4–S6. https://doi.org/10.1038/495s4a.

Owen-Burge, Charlotte. "How Africa Is Leading the Way in Dealing with 'E-Waste.'" Climate Champions. March 28, 2022. https://climatechampions.unfccc.int/how-africa-is-leading-the-way-in-dealing-with-e-waste/.

Pawan, Kumar, and Anita Singh. "New Technology and E-waste: A Case of India". *International Journal of Current Research in Science and Technology* 1, no. 9 (2015): 19–29. https://crst.gfer.org/index.php/crst/article/view/43.

Pan, Xu, Christina W. Y. Wong, and Chunsheng Li. "Circular Economy Practices in the Waste Electrical and Electronic Equipment (WEEE) Industry: A Systematic Review and Future Research Agendas." *Journal of Cleaner Production* 365 (September 2022): 132671. https://doi.org/10.1016/j.jclepro.2022.132671.

Perkins, Devin N., Marie-Noel Brune Drisse, Tapiwa Nxele, and Peter D. Sly. "E-waste: A Global Hazard." *Annals of Global Health* 80, no. 4 (2014): 286–295. https://doi.org/10.1016/j.aogh.2014.10.001.

Singh, Vishal, Javier Cortes-Ramirez, Leisa-Maree Toms, Thilakshika Sooriyagoda, and Shamshad Karatela. "Effects of Polybrominated Diphenyl Ethers on Hormonal and Reproductive Health in E-Waste-Exposed Population: A Systematic Review." *International Journal of Environmental Research and Public Health* 19, no. 13 (June 25, 2022): 7820. https://doi.org/10.3390/ijerph19137820.

Stephen, Kansuk. "Shaping Sustainable Paths for E-Waste Management in Ghana United Nations Development Programme." United Nations Development Programme. September 8, 2023. https://www.undp.org/ghana/blog/shaping-sustainable-paths-e-waste-management-ghana.

Teta, Charles, and Tapiwa Hikwa. "Heavy Metal Contamination of Ground Water from an Unlined Landfill in Bulawayo, Zimbabwe." *Journal of Health and Pollution* 7, no. 15 (September 2017): 18–27. https://doi.org/10.5696/2156-9614-7.15.18.

Thakur, P., and S. Kumar. "Evaluation of E-Waste Status, Management Strategies, and Legislations." *International Journal of Environmental Science and Technology* 19 (July 3, 2022): 16957–69669. https://doi.org/10.1007/s13762-021-03383-2.

Twagirayezu, Gratien, Olivier Irumva, Kui Huang, Hui Xia, Abias Uwimana, Jean Claude Nizeyimana, Habasi Patrick Manzi, Febronie Nambajemariya, and Auguste Cesar Itangishaka. "Environmental Effects of Electrical and Electronic-Waste on Water and Soil: A Review." *Polish Journal of Environmental Studies* 31, no. 3 (April 12, 2022): 2507–2525. https://doi.org/10.15244/pjoes/144194.

Umicore. "When Your Phone Stops Ringing." November 19, 2020. https://www.umicore.com/en/newsroom/news/when-your-phone-stops-ringing-/.

UNIDO. "Circular Economy Development of Recycling Industries." 2018. https://www.unido.org/our-focus-cross-cutting-services-circular-economy/circular-economy-development-recycling-industries-old.

United Nations. "Transforming Our World: The 2030 Agenda for Sustainable Development." 2015. https://sustainabledevelopment.un.org/post2015/transformingourworld/publication.

Wang, Hongwu, Peng Huang, Ruibiao Zhang, Xueyong Feng, Qiulin Tang, Sixi Liu, and Feiqiu Wen. "Effect of Lead Exposure from Electronic Waste on Haemoglobin Synthesis in Children." *International Archives of Occupational and Environmental Health* 94, no. 5 (January 20, 2021): 911–918. https://doi.org/10.1007/s00420-020-01619-1.

World Economic Forum. "A New Circular Vision for Electronics Time for a Global Reboot in Support of the United Nations E-Waste Coalition Platform for Accelerating the Circular Economy Platform." January 2019. https://www3.weforum.org/docs/WEF_A_New_Circular_Vision_for_Electronics.pdf.

World Economic Forum. "Helping Companies in China Recycle 50% of E-Waste by 2025." July 1, 2021. https://www.weforum.org/impact/helping-chinese-companies-reduce-recycle-e-waste/.

Zuiderveen, Emma A. R., J. Chris Slootweg, and Jacob de Boer. "Novel Brominated Flame Retardants – A Review of Their Occurrence in Indoor Air, Dust, Consumer Goods and Food." *Chemosphere* 255 (September 2020): 126816. https://doi.org/10.1016/j.chemosphere.2020.126816.

IMAGE CITATIONS + CREDITS

Listed by Page Number

140 Map sources: Esri, Maxas, Earthstar Geographics, and the GIS User Community.

142 Collage: Matthew Seibert.

146 Map sources: Cornelis P Baldé, Ruediger Kuehr, Tales Yamamoto, Rosie McDonald, Elena D'Angelo, Shahana Althaf, Garam Bel, Otmar Deubzer, Elena Fernandez-Cubillo, Vanessa Forti, Vanessa Gray, Sunil Herat, Shunichi Honda, Giulia Iattoni, Deepali S. Khetriwal, Vittoria Luda di Cortemiglia, Yuliya Lobuntsova, Innocent Nnorom, Noémie Pralat, and Michelle Wagner. "Global E-Waste Monitor 2024." Geneva/Bonn: International Telecommunication Union (ITU) and United Nations Institute for Training and Research (UNITAR), 2024. https://ewastemonitor.info/wp-content/uploads/2024/03/GEM_2024_18-03_web_page_per_page_web.pdf.

150 Image references: Cornelis P Baldé, Ruediger Kuehr, Tales Yamamoto, Rosie McDonald, Elena D'Angelo, Shahana Althaf, Garam Bel, Otmar Deubzer, Elena Fernandez-Cubillo, Vanessa Forti, Vanessa Gray, Sunil Herat, Shunichi Honda, Giulia Iattoni, Deepali S. Khetriwal, Vittoria Luda di Cortemiglia, Yuliya Lobuntsova, Innocent Nnorom, Noémie Pralat, and Michelle Wagner. "Global E-Waste Monitor 2024." Geneva/Bonn: International Telecommunication Union (ITU) and United Nations Institute for Training and Research (UNITAR), 2024. https://ewastemonitor.info/wp-content/uploads/2024/03/GEM_2024_18-03_web_page_per_page_web.pdf.

160 Image references: Brian Owens. "Mining: Extreme Prospects." *Nature* 495, no. 7440 (March 2013): S4–S6. https://doi.org/10.1038/495s4a; Umicore. "When Your Phone Stops Ringing." Umicore.com. Accessed: November 19, 2020. https://www.umicore.com/en/newsroom/news/when-your-phone-stops-ringing-/.

2009
2022
HYDROGEN
y River
Site C Dam

Landscape and Literacy along Canada's Peace River

by Douglas Robb

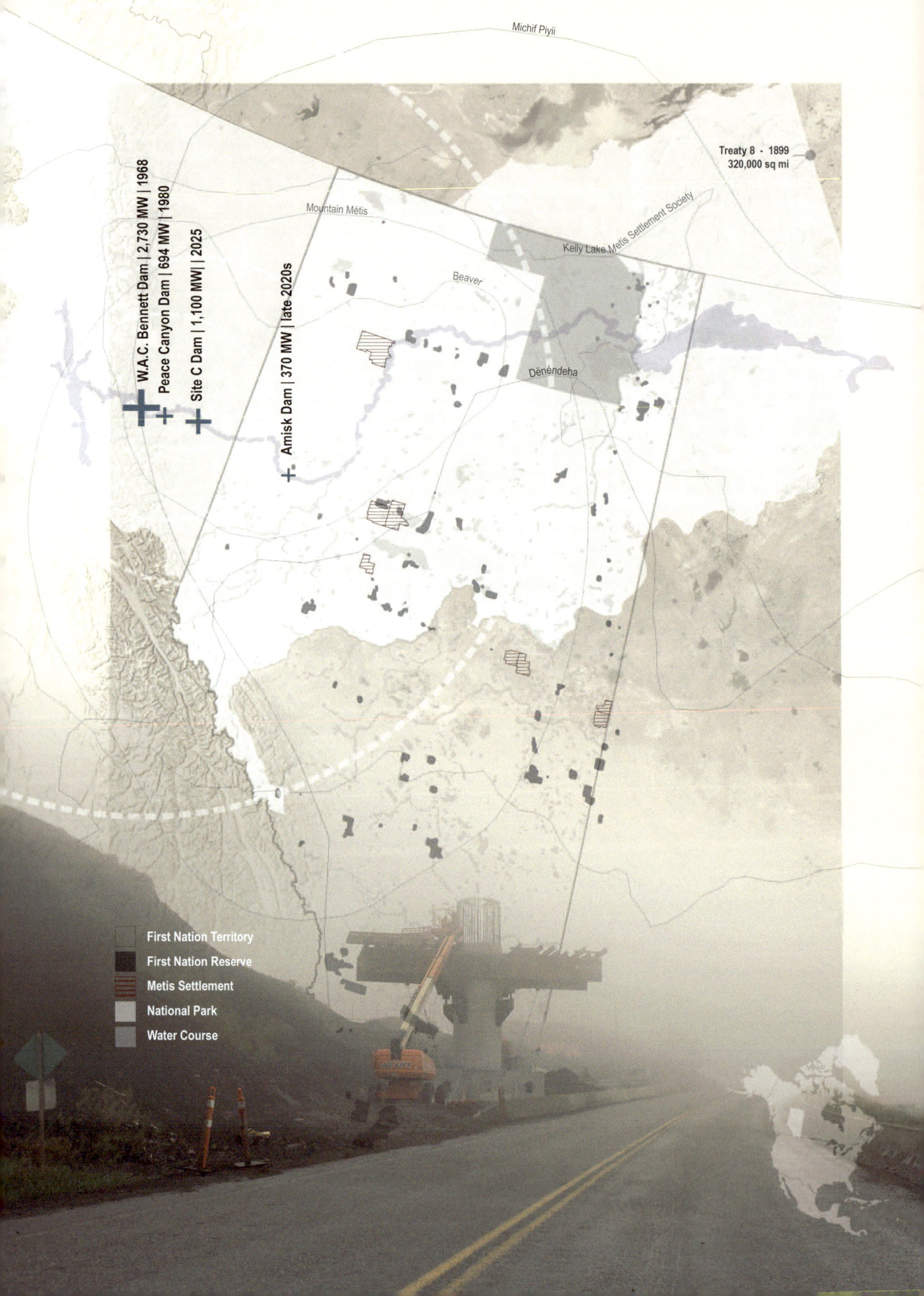

Michif Piyii
Treaty 8 - 1899
320,000 sq mi
Mountain Métis
Kelly Lake Metis Settlement Society
Beaver
Dënéndeha
W.A.C. Bennett Dam | 2,730 MW | 1968
Peace Canyon Dam | 694 MW | 1980
Site C Dam | 1,100 MW| | 2025
Amisk Dam | 370 MW | late 2020s
First Nation Territory
First Nation Reserve
Metis Settlement
National Park
Water Course

A Blurred Landscape

I pulled open my hotel room curtains. Outside, the early morning sky was obscured by thick yellow smoke. I could barely see the faintest outline of nearby trees and buildings through the dense haze.

Damn, I thought, *I hope I didn't come all this way for nothing.*

I had spent the past three days driving 1,100 km (nearly 700 mi) from my home in Vancouver to the small town of Hudson's Hope, located on the Peace River in northeastern British Columbia, Canada. The purpose of my trip was to conduct fieldwork on the social and environmental impacts of hydroelectricity projects along the river. I planned to start my fieldwork at the W.A.C. Bennett Dam located near Hudson's Hope before heading downstream to the Site C Dam. Site C is one of Canada's newest—and most controversial—hydroelectric projects.[1] When construction is completed in 2025, it will be one of the country's largest concrete structures, cutting straight across the Peace River valley. This valley is a beautiful and biodiverse landscape home to unique geological formations and some of Canada's northernmost agricultural land. The Peace River (or Saaghii Naachii as it is known in the local Dane-zaa language) is also a traditional and ancestral homeland for Indigenous peoples across the region, including the Dane-zaa, Cree, Saulteaux, Dene Thá, and Métis peoples. Once Site C is complete, the landscapes of the Peace River between Hudson's Hope and the dam will be forever drowned beneath its reservoir.

1 Sarah Cox, "The Most Expensive Dam in Canadian History: Cost of B.C.'s Site C Dam Balloons to $16 Billion," *The Narwhal*, February 26, 2021. https://thenarwhal.ca/bc-site-c-dam-16-billion-horgan/.

My goal was to conduct ethnographic research and visual documentation of the river valley before it was lost. But the weather had other plans. Summer 2021 was shaping up to be one of the worst wildfire seasons on record in British Columbia, and the thick smoke from hundreds of fires threatened to undermine my project. I had managed to persuade my partner to join me with the promise of a scenic road trip through beautiful mountainous landscapes, but each morning we awoke to red eyes and irritated sinuses from the oppressive smoke. When I wasn't driving, I obsessively refreshed the wildfire forecast on my weather app. The outlook was grim: no rain predicted for days. As we drove further north, the sky continued to darken and I grew increasingly worried. At any moment, the tinder-dry forests around us could ignite, blocking the few roads north and forcing us to turn back—or worse. I hoped that by the time we reached Hudson's Hope the smoke would clear, but no luck. On our first day in the field, I could barely see the landscape I had come to document.

We began at the W.A.C. Bennett Dam, located a few kilometers west of Hudson's Hope. As we drove across the crest of the dam toward a designated viewpoint, it was hard not to feel frustrated. I had previously visited the site in February 2020, just weeks before the COVID-19 pandemic placed my fieldwork on indefinite hold. During that trip, the cold winter air allowed for clear views of the dam and the vast Williston Reservoir it impounded. This reservoir—the largest "lake" in British Columbia—is a testament to the scale and scope of landscape transformations that have reshaped the Peace Region over the past half-century. Completed in 1968, the W.A.C. Bennett Dam supplies over one-third of British Columbia's power demand with renewable hydroelectricity, but at a terrible price. The dam and its reservoir have significantly altered the Peace River's hydrology, ecology, and even local weather patterns. Ongoing shoreline erosion poses serious geotechnical issues, and airborne dust caused by the reservoir's fluctuating water levels creates significant health risks for nearby communities. This is especially true for the Tsay Keh Dene First Nation, the Kwadacha (Fort Ware) First Nation, and other Indigenous communities that were forcibly relocated from the Peace Valley when the Williston Reservoir was flooded; today, they bear a disproportional share of the dam's social and environmental impacts.[2] I recounted this grim history to my partner as we stood at the viewpoint. Together, we turned our gaze outward to survey the transformed landscape, but all we could see was a wall of thick wildfire smoke.

Visibility didn't improve as we drove downstream toward the Site C Dam. With my partner behind the wheel, I was poised, camera in hand, to photograph the vast earthworks and forest clear-cuts shaping Site C's future reservoir. But the smoke persisted. Other than a few half-constructed bridge piers and shoreline fortifications, it was impossible to survey the full extent of the project. Annoyed, we pulled onto the shoulder of the highway so we could reassess our plans. As we sat by the side of the road, dozens of tanker trunks zoomed past, sending up a spray of gravel that skittered across the hood of our car. I knew from speaking

2 Dennis Izony and Hadi Dowlatabadi, "Opinion: B.C. Hydro and Duty of Care," *The Vancouver Sun*, September 29, 2016. https://vancouversun.com/opinion/opinion-bc-hydro-and-duty-of-care.

with local residents that these trucks were likely hauling fuel or water to one of the numerous hydraulic fracturing sites peppered throughout the region.

In addition to being one of British Columbia's foremost sources of hydroelectricity, the Peace Region is also an emerging fossil fuel landscape driven by the exponential growth in hydraulic fracturing over the past decade. Like hydropower, hydraulic fracturing—or fracking, as it is commonly known—has reconfigured the Peace River's landscapes. Vast networks of service roads, dams, pipelines, and seismic lines have carved the region's subalpine forests into a fragmented patchwork, resulting in significant environmental harms. Humans have also suffered in this degraded landscape, particularly local Indigenous peoples who are no longer able to engage in traditional land-based cultural and spiritual practices—rights they are formally guaranteed though treaties with the Government of Canada. I checked the status of the wildfires on my app: a new fire had ignited near the south shore of the Williston Reservoir. Overhead, the sky seemed to be darkening. With a heavy sigh, we pulled back onto the highway and continued driving.

Days later, en route back to Vancouver, I was convinced that my fieldwork had been an utter failure. My pessimism was confirmed when I uploaded my photos and found that I had taken hundreds of blurry, smoke-filled pictures. *So much for documenting the river*, I grumbled. I stashed the images in a folder and tried to forget about them as I continued writing my dissertation. It was not until much later, when asked to describe how energy extraction had impacted the Peace region, that I realized my photos might hold a possible answer.

In the 1960s, the construction of the W.A.C. Bennett Dam flooded the headwaters of the Peace and degraded the river's physical and cultural landscapes. Since the early 2000s, fossil fuel extraction has fragmented regional ecosystems into a threadbare patchwork. Today, the Site C Dam is poised to drown yet another segment of the Peace River, this time in the name of renewable energy. Each of these episodes tells a different story of colonial occupation, technological progress, political agendas, and environmental change. The cumulative impacts of these inter-related stories have rendered the Peace River's landscapes nearly unrecognizable. Likewise, my blurry photos reflected the landscape's increasing illegibility; in my case, from the smoke of hundreds of wildfires exacerbated by anthropogenic climate change. And yet, even as I try to make sense of these stories, a new energy resource is poised to rewrite the landscapes of Peace Region once again: hydrogen.

Reading Hydrogen Landscapes

As I was trying, unsuccessfully, to document the smoky landscapes of the Peace River, a parallel energy story was unfolding on Canada's Atlantic coast. In August 2022, the German Chancellor Olaf Scholz concluded a high-profile visit to Canada with an unlikely stop in the small seaside community of Stephenville, located on the island of Newfoundland. Once a small resource town dependent

Infrastructural Networks around Site C Dam

on agriculture and the pulp and paper industry, Stephenville has recently become an important site for Canada's emerging hydrogen sector. Earlier that year, an investment consortium headed by World Energy GH2 announced plans to construct a 3 GW wind-to-hydrogen facility along Stephenville's coastline, capable of producing up to 250,000 tons of hydrogen fuel per year.[3]

News of the proposed development captured the attention of many European political leaders, including the German Chancellor. Speaking before a room of journalists and trade delegates, Scholz expounded on the importance of Canadian hydrogen in the global race toward net-zero decarbonization.[4] "Hydrogen will play a major role in our future energy supply" he announced, before stating that hydrogen fuel produced in Atlantic Canada "presents a huge opportunity" for Germany's low-carbon transition.[5] Canadian Prime Minister Justin Trudeau responded enthusiastically to the Chancellor's message, affirming that Canada "can be the reliable supplier of the clean energy a net-zero world needs."[6] But is this claim necessarily true? Will hydrogen become the renewable fuel of the future, and is Canada poised to become a global hydrogen superpower?

Scholz's visit culminated with the signing of the Canada–Germany Hydrogen Alliance, which affirms Canada's commitment to supply Germany with *green* hydrogen produced from water.[7] However, at the time of writing, there are no operational green hydrogen facilities anywhere in the world. Instead, Canada only produces *grey* and *blue* hydrogen made from fossil fuels—specifically natural gas—sourced from the western provinces of Alberta and British Columbia, including the Peace Region.

What do these colors mean, and why does the distinction between different colors of hydrogen matter? As Scholz's trip suggests, hydrogen is quickly emerging an important fuel for the global low-carbon transition, as well as a central pillar in Canada's net-zero decarbonization strategy.[8] But not all hydrogen is created equal. There are many different chemical and industrial processes used to

3 World Energy GH2, "About Project Nujio'qonik," accessed December 1, 2022, https://worldenergygh2.com/about/.

4 "Global Warming of 1.5°C An IPCC Special Report on the Impacts of Global Warming of 1.5°C Above Pre-Industrial Levels and Related Global Greenhouse Gas Emission Pathways, in the Context of Strengthening the Global Response to the Threat of Climate Change, Sustainable Development, and Efforts to Eradicate Poverty," Intergovernmental Panel of Climate Change, accessed December 1, 2022. https://www.ipcc.ch/sr15/download/#full.

5 The Canadian Press, "Trudeau, Scholz Push Hydrogen Plans in Newfoundland," *The Globe and Mail* (video), August 24, 2022: 1:46. https://www.theglobeandmail.com/canada/video-trudeau-scholz-push-hydrogen-plans-in-newfoundland.

6 The Canadian Press, "Trudeau, Scholz Push Hydrogen Plans in Newfoundland," The Globe and Mail (video), August 24, 2022: 1:31. https://www.theglobeandmail.com/canada/video-trudeau-scholz-push-hydrogen-plans-in-newfoundland/..

7 Natural Resources Canada,"Joint Declaration of Intent Between the Government of Canada and the Government of the Federal Republic of Germany on Establishing a Canada-Germany Hydrogen Alliance," accessed December 1, 2022. https://www.nrcan.gc.ca/climate-change-adapting-impacts-and-reducing-emissions/canadas-green-future/the-hydrogen-strategy/joint-declaration-intent-between-the-government-canada-and-the-government-the-federal/24607

8 Natural Resources Canada,"Hydrogen Strategy for Canada: Seizing the Opportunities for Hydrogen—A Call to Action," Accessed December 2, 2022. https://www.nrcan.gc.ca/sites/nrcan/files/environment/hydrogen/NRCan_Hydrogen-Strategy-Canada-na-en-v3.pdf.

produce hydrogen fuel, each with their own capital and labor requirements, regulatory regimes, extractive logics, and greenhouse gas emissions. In energy industry jargon, these distinct processes are color-coded: brown, grey, blue, green, black, and even pink and turquoise. Hydrogen's colorful shorthand is useful to fast-track technical discussions, but it can also make it difficult for non-experts to differentiate between the costs and benefits of proposed energy projects. This presents a barrier to informed discussion and limits opportunities for public participation in nuanced debates over infrastructure approval, siting, and construction. This gap in our collective vocabulary can also make it difficult to describe emerging hydrogen landscapes, or, more dangerously, this gap can be exploited by individuals and corporations seeking to advance particular energy futures.

Hydrogen is a colorless and odorless gas, but its semantic opacity presents a political problem. We can't debate what we can't perceive, and developing a shared vocabulary to discuss the costs and benefits of hydrogen production will be essential to negotiate a just transition to a post-fossil fuel world. This is especially true in Canada, where, as my blurry photos of the Peace Region illustrate, the race to develop new energy resources can obscure ongoing processes of colonial violence and settler entitlement to land.[9] Without a shared vocabulary to describe our evolving energy landscapes, we risk perpetuating the injustices of the past in our rush to implement the post-carbon future.

Landscape (Il)literacy

One way to counteract this violence is by foregrounding concepts of landscape in conversations surrounding the low-carbon transition. The idea of *landscape literacy* is useful to guide this work. First articulated by Anne Whiston Spirn, landscape literacy refers to a set of cultural practices that "entails both understanding the world and transforming it."[10] Spirn uses the example of the postwar public education movement in the United States to demonstrate how verbal and written literacy were valued as a precondition for a citizen's full participation in democratic society. Likewise, landscape literacy argues for the importance of shared, place-based knowledge in collective processes of spatial decision-making.

Most people rely on words like "natural," "urban," or "wild" to describe a landscape. However, in the context of the Anthropocene, these familiar categories begin to blur, as my photos of the Peace River attest. As a result, it can be difficult to observe and understand emerging energy landscapes that appear simultaneously natural, technological, and cultural. Consider the landscapes of the Peace River: for some, the Williston Reservoir is an environmental

9 In making this argument, I do not intend to discredit the work of governments, academics, and environmental organizations in advancing a legitimate decarbonization agenda, such as the Government of Canada's Net-Zero Emissions Accountability Act and ongoing efforts by the federal Net-Zero Advisory Board.

10 Anne Whiston Spirn, "Landscape Literacy and Design for Ecological Democracy: The Nature of Mill Creek, West Philadelphia," in Henrik Ernstson and Sverker Sörlin (eds.), *Grounding Urban Natures: Histories and Futures of Urban Ecologies* (Cambridge: MIT Press, 2019): 128.

The Landscapes of the Peace River
W.A.C. Bennett Dam | 2,730 MW | 1968
Peace Canyon Dam | 694 MW | 1980

+ 1515 ft (164 ft higher than current aveage level)
Flood Reserve Level
Area: 23,054.93 acre
3,714.35 acre
Length: 51.6 mile
Width: 2–3 times the original width
Site C Dam | 1,199 MW | 2025

tragedy that drowned a fragile ecosystem and displaced settler and Indigenous communities; for others, Williston Lake is a beloved recreational site for boating and fishing.

Landscapes hold simultaneous and often contradictory stories. As Jane Wolff suggests, we often lack the words to name what we see: "most people don't have an adequate vocabulary to characterize the essential qualities of places where they live or discuss the possibilities that lie ahead. The lack of nuanced, commonly understood language limits public discussion, and that makes effective agency difficult for technical experts, policymakers, politicians, and citizens—for anyone with a stake in the future."[11] In the context of the climate emergency—and specifically with respect to emerging and unfamiliar landscapes of hydrogen fuel production—designers bear a responsibility to make these complex places legible so that we may collectively negotiate our shared energy future.

A Commodity of Many Colors

Before we can understand how hydrogen is poised to transform the landscapes of the Peace River, let us begin with a brief primer on hydrogen's many colors. Hydrogen is the most abundant substance in the universe, comprising nearly all organic compounds. It is also the lightest chemical element, made up of one proton and one neutron. However, hydrogen is rarely found in its gaseous form on Earth. Instead, hydrogen atoms must be split from other substances, such as water (H_2O), or methane (CH_4). The resulting gas can then be mixed with other substances to produce a synthetic form of natural gas (syngas) that can be utilized by conventional internal combustion engines.[12] Alternatively, pure hydrogen (H_2) can be burned as a fuel in specially designed hydrogen engines (HICEs); it can also be used to make hydrogen fuel cells—a kind of battery that generates electricity by combining hydrogen with air.[13] Whereas syngas and HICEs produce varying amounts of greenhouse gas emissions, the chemical reaction contained within hydrogen fuel cells produces only water and heat as by-products.

Like oil and gas, hydrogen has a high energy density, or the amount of energy stored per unit of mass.[14] This makes it a valuable energy carrier, and offers a distinct advantage over intermittent sources of renewable energy such as wind and solar. Replacing fossil fuels with hydrogen presents a significant opportunity for so-called "hard to decarbonize" sectors of the economy, including steel and cement production, aviation, shipping, and long-haul trucking.[15] Incorporating

---

11 Jane Wolff, *Bay Lexicon* (Montreal and Kingston: McGill-Queen's University Press, 2021), 4.

12 International Energy Agency, "Hydrogen: Energy System Overview," accessed 1 December, 2022, https://www.iea.org/reports/hydrogen.

13 Nurettin Demirdoven and John Deutch, "Hybrid Cars Now, Fuel Cell Cars Later," *Science* 305, no. 5686 (2004): 974.

14 Marc A. Rosen and Seama Koohi-Fayegh, "The Prospects for Hydrogen as an Energy Carrier: an Overview of Hydrogen Energy and Hydrogen Energy Systems," *Energy, Ecology and Environment* 1, no. 1 (2016): 11.

15 Noting, of course, that "hard-to-decarbonize" is as much a political category as it is a technical one. As Holly Jean Buck states in *Ending Fossil Fuels: Why Net-Zero is Not Enough* (2021), "[w]hile it is true that there are biophysical aspects that make decarbonization of various activities more difficult than others, at present there is

hydrogen into these sectors could significantly reduce their overall emissions, provided the right "color" of hydrogen fuel is utilized.

Over 96% of the hydrogen fuel produced in the world today is generated from fossil fuels.[16] Brown and black hydrogen refer to fuel produced from coal, and constitute a relatively small proportion of total global production.[17] The vast majority of hydrogen produced today—approximately 75%—is grey hydrogen, which is made from natural gas that has undergone an industrial process known as steam methane reformation (SMR).[18] In this process, methane derived from natural gas is subject to various chemical reactions to separate its hydrogen and carbon atoms. SMR is a relatively well-established technology, but it requires significant energy inputs and produces considerable carbon dioxide emissions.[19]

Enter blue hydrogen. Like the black, brown, and grey varieties, blue hydrogen is also produced from fossil fuels, but with the addition of carbon dioxide removal (CDR) technologies designed to prevent greenhouse gas emissions from escaping into the atmosphere. These technologies typically include chemical or mechanical "scrubbers" that capture carbon dioxide directly from the SMR exhaust so it can be stored in carbon sinks, such as underground salt caverns or depleted oil fields.[20] Alternatively, the captured carbon can be used as a feedstock for other industrial processes; this is why many proposed hydrogen facilities are located in "hydrogen hubs" alongside related industries and manufacturing centers.[21]

Blue hydrogen is a tempting commodity for oil and gas-producing countries like Canada because it appears to offer a technically viable strategy for net-zero decarbonization while also maintaining—and perhaps even increasing—current levels of fossil fuel extraction. According to Jackie Forrest, executive director of the ARC Energy Research Institute, "[if] we turn our natural gas into hydrogen, then it is now part of the long-term future for energy… There is no need to decline our production because we have found a pathway that has growing demand."[22] In theory, blue hydrogen can enable countries like Canada to retain existing configurations of capital investment and social organization geared toward fossil fuel production while simultaneously reducing greenhouse gas emissions. There is also a business case to be made for rapid investment in blue

no standard definition of 'difficulty.' In the real world, these distinctions of 'difficulty' will be political and economic, not just technical" (45).

16 World Energy Council, "New Hydrogen Economy: Hope or Hype?," 2019, accessed December 2, 2022. https://www.worldenergy.org/assets/downloads/WEInnovation-Insights-Brief-New-Hydrogen-Economy-Hype-or-Hope.pdf.

17 International Energy Agency, "Global Hydrogen Review 2021," accessed 2 December 2022. https://iea.blob.core.windows.net/assets/5bd46d7b-906a-4429-abda-e9c507a62341/GlobalHydrogenReview2021.pdf.

18 International Energy Agency, "Global Hydrogen Review 2021," accessed 2 December 2022, https://iea.blob.core.windows.net/assets/5bd46d7b-906a-4429-abda-e9c507a62341/GlobalHydrogenReview2021.pdf.

19 US Department of Energy Office of Fossil Energy and Carbon Management, "Hydrogen Strategy: Enabling a Low-Carbon Economy," accessed 2 December, 2022. https://www.energy.gov/sites/prod/files/2020/07/f76/USDOE_FE_Hydrogen_Strategy_July2020.pdf.

20 International Energy Agency, "Global Hydrogen Review 2021."

21 Alicia Eastman, *Scaling Green Hydrogen for a Global Market*, interview with Andy Stone, podcast audio, September 27, 2022. https://kleinmanenergy.upenn.edu/podcast/scaling-green-hydrogen-for-a-global-market/.

22 Kevin Orland, "Home of the Oil Sands Eyes Cleaner Future as Hydrogen Superpower," *Bloomberg News*, August 13, 2020. https://www.bloomberg.com/news/articles/2020-08-13/home-of-the-oil-sands-eyes-cleaner-future-as-hydrogen-superpower?leadSource=uverify%20wall.

hydrogen infrastructure. The International Energy Agency forecasts that global hydrogen demand will increase six times by the year 2050, amounting to 10% of total energy consumption.[23] Financial barriers to blue hydrogen production are also likely to diminish over time as greater investment is directed toward CDR development, thereby lowering overall costs.

Critics, however, are skeptical that blue hydrogen will be a miracle fuel for the low-carbon transition. First, there are a number of environmental risks associated with the use of hydraulic fracturing to extract natural gas, which is the primary feedstock required to make blue hydrogen. The cumulative environmental impacts of hydraulic fracturing are well-documented, ranging from surface and groundwater contamination,[24] methane and hydrogen sulphide emissions,[25] and induced seismicity,[26] amongst others. Hydraulic fracturing is also a spatially extensive operation, resulting in widespread landscape fragmentation and biodiversity decline. Once extraction is complete, fugitive emissions from abandoned and improperly sealed wells also present considerable environmental risks.[27] Methane—the primary constituent of natural gas and the main feedstock for blue hydrogen—is a powerful greenhouse gas: in a unit-to-unit comparison, methane is one hundred times more potent than carbon dioxide as an atmospheric warming agent.[28] Approximately 25% of the net global warming that has occurred in recent decades is estimated to be due to methane.[29]

There are also a number of technical uncertainties regarding the CDR technologies required to scrub carbon dioxide from the SMR process. Internal reporting from Shell's Quest Hydrogen facility in Alberta—at present, one of only two commercial-scale blue hydrogen plants in the world—suggests that the total percentage of carbon captured falls short of government and industry benchmarks.[30] While there are many different kinds of CDR technologies, the ones typically used in hydrogen fuel production tend to be very energy intensive, resulting in their own considerable carbon emissions. Some of this demand can be met by reusing waste energy produced by the SMR process itself, but it is not enough to lower blue hydrogen's overall carbon intensity. One recent analysis

23 International Energy Agency,"Global Hydrogen Review 2021."

24 Joshua Wisen, Romain Chesnaux, Gilles Wendling, John Werring, Florent Barbecot, and Paul Baudron, "Assessing the Potential of Cross-Contamination from Oil and Gas Hydraulic Fracturing: A Case Study in NorthEastern British Columbia, Canada," *Journal of Environmental Management* 246 (2019): 275.

25 Atherton, "Mobile Measurement of Methane Emissions," 12406.

26 Gail M. Atkinson, David W. Eaton, Hadi Ghofrani, Dan Walker, Burns Cheadle, Ryan Schultz, Robert Shcherbakov, et al. "Hydraulic Fracturing and Seismicity in the Western Canada Sedimentary Basin." *Seismological Research Letters* 87, no. 3 (2016): 631.

27 Atherton, "Mobile Measurement of Methane Emissions," 12405.

28 "Climate Change 2013: The Physical Science Basis. Contribution of Working Group I to the Fifth Assessment Report of IPCC the Intergovernmental Panel on Climate Change," Intergovernmental Panel on Climate Change, 2013. Accessed November 20, 2022. https://www.ipcc.ch/report/ar5/wg1/

29 "Climate Change 2013: The Physical Science Basis. Contribution of Working Group I to the Fifth Assessment Report of IPCC the Intergovernmental Panel on Climate Change," Intergovernmental Panel on Climate Change, 2013. Accessed November 20, 2022. https://www.ipcc.ch/report/ar5/wg1/.

30 The percentage of carbon that must be captured in order for grey hydrogen to be considered blue varies depending on jurisdiction; there is currently no international standard. Europe's CertifHy project identifies a carbon intensity threshold of 4.37 kg CO_2 per kg of hydrogen including upstream emissions; other regulatory bodies appear to be coalescing around a minimum 90% carbon capture efficacy threshold.

suggests that blue hydrogen's greenhouse gas footprint "is more than twenty percent greater than burning natural gas or coal for heat."[31]

Green hydrogen is altogether different. Unlike black, brown, grey, and blue hydrogen, which are produced from fossil fuels, green hydrogen is made by splitting water molecules into hydrogen and oxygen atoms using renewable energy such as wind, solar, geothermal, or hydropower. This process is called electrolysis, and requires approximately 9 tons of water to produce 1 ton of green hydrogen.[32] Water can come from either freshwater sources, or seawater can be desalinated at an additional cost.[33] The electricity used to power the electrolysis reaction is the most expensive part of the overall production process; therefore, for green hydrogen to be economically viable, production facilities require access to plentiful renewable energy.[34]

As a result, many proposed green hydrogen hubs are located near stable sources of baseload renewable electricity, such as in deserts where solar energy is abundant, or along windy coastlines where wind energy can be harnessed.[35] Although global hydrogen production is currently dominated by grey and blue varieties, green hydrogen may soon overtake them both. In August 2022, the European Union approved up to €5.2 billion in funding and subsidies for the development of large-scale electrolysers and renewable electricity infrastructure to power new green hydrogen hubs.[36] These European initiatives mirror similar policies in North America, such as the U.S. Infrastructure Investment and Jobs Act and the Inflation Reduction Act, both of which allocate significant financial resources to scale up green hydrogen production, utilization, and export.[37]

Grounding Hydrogen

Tracing the flow of hydrogen across western Canada is a complicated task. Most of Canada's hydrogen fuel is currently produced in the province of Alberta, which has been making grey hydrogen from natural gas for the past half-century. The Government of Alberta's 2021 *Hydrogen Roadmap* situates the province's hydrogen industry within a century-long historical arc. "In 1923," the report begins, "a 130-kilometer natural gas pipeline was built from Viking to Edmonton, Alberta, allowing a switch from coal to natural gas for heating, lighting, and cooking. Nearly a century later, Alberta is leading another energy shift with clean

31 Robert W. Howarth and Mark Z. Jacobson, "How Green is Blue Hydrogen?," *Energy Science & Engineering* 9, no. 10 (2021): 1676.
32 International Energy Agency,"Global Hydrogen Review 2021."
33 International Energy Agency, "Global Hydrogen Review 2021."
34 Alicia Eastman, "Scaling Green Hydrogen for a Global Market."
35 The latter describes the proposed green hydrogen development in Stephenville. Nujio'qonik, the project's name, is derived from an Indigenous placename meaning "where the sand blows" in the local Mi'kmaq language.
36 "State Aid: Commission approves up to €5.2 billion of public support by thirteen Member States for the second Important Project of Common European Interest in the hydrogen value chain," European Commission, accessed December 1, 2022. https://ec.europa.eu/commission/presscorner/detail/en/ip_22_5676.
37 These Acts lay out specific carbon intensity thresholds that must be met for hydrogen production to receive subsidies, but they notably do not stipulate which hydrogen production processes are to be used.

hydrogen."[38] This story is a familiar cultural narrative in western Canada, where concepts of progress and modernity are closely linked to material histories of resource extraction and technological innovation in the oil and gas sector.[39]

Much of Alberta, and parts of the adjoining provinces of British Columbia and Saskatchewan, lie on top of a vast geological formation known as the Western Sedimentary Basin. This formation holds the majority of Canada's oil and gas reserves, which were known and utilized by Cree and Dene peoples prior to the arrival of European colonization.[40] It was not until 1875 that John Macoun, a scientist for the Geologic Survey of Canada, first arrived in northern Alberta and took interest in the "tar conglomerate" now known as the Athabasca Oil Sands.[41] Macoun's geological observations sparked a longstanding political economic project to capitalize on the region's energy resource wealth. Today, the violent legacies of this colonial project are inscribed in the province's physical landscapes in the form of pipelines, pumpjacks, orphaned wells, tailings ponds, and vast surface mines.

As Canada looks to expand its hydrogen fuel sector, fossil fuel extraction will almost certainly rise in tandem. As of 2021, annual hydrogen production in Alberta totalled approximately 2.4 million tons, the vast majority of which is consumed by various industrial processes in province's oil and gas sector.[42] Much of this work is concentrated northeast of the provincial capital, Edmonton, in an area known as the Alberta Industrial Heartland (AIH). The AIH is one of Canada's largest petrochemical processing hubs.[43] It is also home to Shell's Quest blue hydrogen facility, and new CDR technologies, and is the endpoint of thousands of kilometers of natural gas pipelines linking the AIH to sites of fossil fuel extraction scattered throughout the region. According to Alberta's *Hydrogen Roadmap*, the AIH's expanded landscape is forecast to grow in size as blue hydrogen production increases: "[e]xporting 10 million tons of clean hydrogen per year from Alberta to international markets by 2050 could demand about 4 billion cubic feet per day in natural gas" above current supply.[44] This staggering extractive footprint is plainly visible in satellite images of the Peace Region, where the tell-tale pattern of thousands of hydraulic fracturing drill pads punctuate the landscape.

Hydraulic fracturing has been used in Canada since the mid-1950s, but its use has grown exponentially since the early 2010s.[45] Over the past decade, thousands of wells have been drilled along the Peace River, consuming

38 "Alberta Hydrogen Roadmap," Government of Alberta, 8. Note the use of the term "clean hydrogen" instead of "blue hydrogen."

39 Jordan B. Kinder, "From Dirty Oil to Ethical Oil: Petroturfing and the Cultural Politics of Canadian Oil After Social Media," *Journal of Environmental Media* 1, no. 2 (2020): 167.

40 As the geographer Michael Simpson (2019: 3) describes, "bitumen, a sticky substance that oozes out of the banks of the Athabasca River... was known and used long before it was ever targeted for extraction. Cree and Dene peoples, who used bitumen to patch their canoes, first introduced European traders to the material as far back as 1715"; Michael Simpson, "Resource Desiring Machines: The Production of Settler Colonial Space, Violence, and the Making of a Resource in the Athabasca Tar Sands," *Political Geography* 74 (2019): 3.

41 Michael Simpson, "Resource Desiring Machines": 5.

42 "Alberta Hydrogen Roadmap," Government of Alberta, Ministry of Energy, 12, accessed December 1, 2022. https://open.alberta.ca/publications/alberta-hydrogen-roadmap.

43 Alberta Industrial Heartland Association, "Investor Overview," accessed December 1, 2022.

44 Government of Alberta, "Alberta Hydrogen Roadmap," 49.

45 Council of Canadian Academies, "Environmental Impact of Shale Gas Extraction in Canada," 18.

tremendous volumes of water and contributing to significant landscape fragmentation and biodiversity decline.[46] Increased natural gas production also requires additional sources of energy—ideally renewable energy which can be used to reduce the carbon intensity associated with gas extraction, processing, and transportation. Critics argue that need for renewable electricity is the true *raison d'être* for the controversial Site C Dam. By providing renewable baseload power, electricity produced at Site C can reduce the carbon intensity of natural gas production. The resulting "renewable natural gas"—a common misnomer[47]—can then be burned as a source of energy or greenwashed as "clean" hydrogen fuel. As a result, investments in hydrogen fuel production will likely also result in increased fossil fuel production across the Peace Region.[48]

The social and environmental impacts of increased natural gas extraction are often elided in discourses surrounding "clean" hydrogen fuel. These semantics complicate the broader challenge of making hydrogen landscapes legible. Reducing, and hence depoliticizing, the potential impacts of hydrogen production to a series of colors limits our ability to understand hydrogen's cumulative impacts, and narrows our collective ability to imagine alternative energy futures. Thinking back to my smoke-filled fieldwork, I can admit that none of these landscapes' relationships were obvious to me at the time. Standing on the riverbank in Hudson's Hope, hundreds of miles away from Alberta's hydrogen hubs, it was difficult to comprehend how the Peace River valley—which would soon be plunged beneath a hydroelectric reservoir—was also part of a complex hydrogen landscape. I initially blamed the wildfires for my inability to read the landscape, but I realize now that the smoke was merely a symptom of our fossil-fuel-dependant energy system. Looking back on my fieldwork photos, I now recognize how small details—the tanker trucks, the transmission lines, the half-constructed bridge piers—each provide a glimpse into a larger landscape narrative connecting hydropower, hydraulic fracturing, and hydrogen.

Building Hydrogen Landscape Literacy

Landscape stories matter in conversations surrounding energy futures. When faced with the challenge of confronting the climate emergency, "narratives, rather than climate information per se, play the decisive role in motivating or demotivating climate action."[49] Building shared literacy around energy landscapes is important work for landscape educators and practitioners. By leveraging our ability to analyze and visualize complicated places, landscape architects can narrate contested sites of the post-fossil fuel transition in terms that are

46 Ecotrust Canada, "Atlas of Cumulative Landscape Disturbance in the Traditional Territory of Blueberry River First Nations," 82.

47 Sarah Cox, "Alberta is on a Blitz to Promote B.C.'s LNG Industry," *The Narwhal*, February 7, 2023. https://thenarwhal.ca/alberta-bc-lng/.

48 Douglas Robb, Philippe Le Billon, and Karen Bakker, "Landscapes of Recarbonization: Carbon Neutrality, Settler Colonialism, and Cumulative Environmental Effects in the Peace River Region, Canada." *Annals of the American Association of Geographers* (2024): 1–18.

49 Jochen Hinkel, Diana Mangalagiu, Alexander Bisaro, and J. David Tàbara, "Transformative Narratives for Climate Action," *Climatic Change* 160, no. 4 (2020): 495.

accessible to everyone, thereby filling a gap in our collective vocabulary. This vision for the future of the profession sees design as a form of public discourse, facilitating our transition toward more just and inclusive energy futures.

As my experiences in the Peace Region attest, this work is often slow and improvisational—qualities that are often at odds with the urgent need for rapid decarbonization. However, in the rush to build the post-fossil-fuel future, I worry that we risk losing sight of the social, cultural, and ecological landscapes that we create and destroy in the process. As geographer Kirby Calvert writes, "emphasis on the sociotechnical dimensions of energy transition has not been balanced by emphasis on the socioecological dimensions of energy transition."[50] I similarly wonder if the speed of decarbonization required to avoid the most damaging effects of climate change outpaces our collective ability to recognize, name, and navigate the new human-environment relationships of our evolving energy system.

Grounding hydrogen in specific landscapes forces us to confront these complex and challenging questions. How can an average person make sense of the political and technological dynamics associated with hydrogen production? How should one weigh the economic opportunities of hydrogen with the social and environmental costs of continued fossil fuel extraction? What vocabulary does a community need in order to begin this conversation and ensure that everyone can participate in the discussion?

The sematic opacity surrounding hydrogen can perhaps be attributed, in part, to its relative novelty as an energy source; it is still rare to have encountered a hydrogen-powered vehicle, let alone a hydrogen production facility. As a result, most people are likely unable to visualize the people and landscapes implicated in the making of hydrogen fuel (as opposed to Canada's Athabasca Oil Sands,

50 Kirby Calvert, Kirsten Greer, and Margot Maddison-MacFadyen, "Theorizing Energy Landscapes for Energy Transition Management: Insights from a Socioecological History of Energy Transitions in Bermuda," *Geoforum* 102 (2019): 192.

which have been the subject of longstanding activist and media attention).[51] As I discovered through my fieldwork, it can be difficult to find the right words to describe the linkages between local environmental impacts and faraway sites of energy production. This, to me, seems like essential work that landscape architects are perhaps uniquely qualified to address.

In the absence of a shared energy vocabulary, unscrupulous actors may seek to conflate the differences that exist between hydrogen's different colors. As Randy Boissonnault, Canada's Minister of Tourism and Associate Minister of Finance, recently stated, "whether it's green, whether it's blue, whether it's turquoise, we're throwing the color wheel out, because what matters is having the right mix of renewables."[52] Achieving the right combination of energy sources is indeed an important goal, however the framing of this statement is both inaccurate and misleading in potentially harmful ways. According to Robert Howarth, "the idea behind blue hydrogen is a new one… and it is totally the creation of the oil and gas industry, promoted through the Hydrogen Council since 2018."[53] Indeed, documents prepared by the Hydrogen Council—an industrial development organization whose membership includes many oil and gas multinationals—rarely make reference to the distinctions between hydrogen's many colors; instead, their reports uses catch-all terms such as "clean hydrogen" or "renewable hydrogen."[54] This is not to say that blue hydrogen is uniquely "bad." Green hydrogen projects can also elicit social and environmental conflict in landscapes where renewable

51 Patrick McCurdy, "Bearing Witness and the Logic of Celebrity in the Struggle over Canada's Oil/Tar Sands," in Benedetta Brevini and Graham Murdock (eds.), *Carbon Capitalism and Communication* (Cham: Palgrave Macmillan, 2017): 131.
52 Taylor C. Noakes, "Why Justin Trudeau's 'Blue Hydrogen' Dream Is Not Really Green," *The Toronto Star*, November 13, 2022. https://www.thestar.com/opinion/contributors/2022/11/13/why-justin-trudeaus-blue-hydrogen-dream-as-an-energy-alternative-for-canada-is-not-really-green.html.
53 Taylor C. Noakes, "Why Justin Trudeau's 'Blue Hydrogen' Dream Is Not Really Green," The Toronto Star, November 13, 2022. https://www.thestar.com/opinion/contributors/2022/11/13/why-justin-trudeaus-blue-hydrogen-dream-as-an-energy-alternative-for-canada-is-not-really-green.html.
54 Hydrogen Council, "Global Hydrogen Flows," accessed December 1, 2022. https://hydrogencouncil.com/wp-content/uploads/2022/10/Global-Hydrogen-Flows.pdf.

energy infrastructures collide with existing land uses and fragile ecologies. While green hydrogen may appear to resolve some of the environmental challenges associated with blue hydrogen, it can also conceal harmful landscape impacts through moral appeals to the benefits of renewable energy.

Back in 2021, as I anxiously drove through the smoke-filled landscapes of the Peace River, I thought my project of documenting processes of landscape transformation was an utter failure. Upon reflection, I now recognize that this trip—and each subsequent visit—was essential in the slow work of becoming literate in this landscape. I was learning its stories; specifically, hydro stories—hydropower, hydrocarbons, hydrogen. Learning to recognize these stories required me to look past my technical knowledge of energy systems and my ever-expanding vocabulary of industry jargon. Building landscape literacy meant relying on fragmented stories and spatial clues at a time when wildfire smoke made it impossible to see clearly—a condition which has worsened with each passing year.

The Peace River provided a through-line to navigate these stories, connecting diverse geographies, bodies, and changing ecologies along the way. Learning to see the river as a series of inter-related stories opened up new ways for me to think about hydrogen and its role in shaping the landscape. This was no easy task; hydrogen is abundant, invisible, and constantly changing. However, by framing hydrogen through the lens of the river—that is, seeing hydrogen as less of a *thing* and more like a series of place-based *relationships*—then we can create new ways to tell landscape stories, and, collectively, to imagine new landscape futures.

BIBLIOGRAPHY

Åhman, Max, Lars J. Nilsson, and Bengt Johansson. "Global Climate Policy and Deep Decarbonization of Energy-Intensive Industries." *Climate Policy* 17, no. 5 (2017): 634–649.

Alberta Industrial Heartland Association. "Investor Overview." Accessed December 1, 2022. https://industrialheartland.com/opportunity/#investor.

Atherton, Emmaline, David Risk, Chelsea Fougere, Martin Lavoie, Alex Marshall, John Werring, James P. Williams, and Christina Minions. "Mobile Measurement of Methane Emissions From Natural Gas Developments in Northeastern British Columbia, Canada." *Atmospheric Chemistry and Physics* 17, no. 20 (2017): 12405–12420.

Atkinson, Gail M., David W. Eaton, Hadi Ghofrani, Dan Walker, Burns Cheadle, Ryan Schultz, Robert Shcherbakov, et al. "Hydraulic Fracturing and Seismicity in the Western Canada Sedimentary Basin." *Seismological Research Letters* 87, no. 3 (2016): 631–647.

Behn, Caleb, and Karen Bakker. "Rendering Technical, Rendering Sacred: The Politics of Hydroelectric Development on British Columbia's Saaghii Naachii/Peace River." *Global Environmental Politics* 19, no. 3 (2019): 98–119.

Bridge, Gavin. "The Map is Not the Territory: A Sympathetic Critique of Energy Research's Spatial Turn." *Energy Research & Social Science* 36 (2018): 11–20.

Bridge, Gavin, and Ludger Gailing. "New Energy Spaces: Towards a Geographical Political Economy of Energy Transition." *Environment and Planning A: Economy and Space* 52, no. 6 (2020): 1037–1050.

Buck, Holly Jean. *Ending Fossil Fuels: Why Net Zero is Not Enough*. London: Verso, 2021.

Calvert, Kirby, Kirsten Greer, and Margot Maddison-MacFadyen. "Theorizing Energy Landscapes for Energy Transition Management: Insights from a Socioecological History of Energy Transitions in Bermuda." *Geoforum* 102 (2019): 191–201.

CertifHy. "Certification Schemes." Accessed December 1, 2022. https://www.certifhy.eu/

Council of Canadian Academies. *Environmental Impact of Shale Gas Extraction in Canada*. Ottawa: Expert Panel on Harnessing Science and Technology to Understand the Environmental Impacts of Shale Gas Extraction. 2014.

The Canadian Press. "Trudeau, Scholz Push Hydrogen Plans in Newfoundland." *The Globe and Mail* (video). August 24, 2022. https://www.theglobeandmail.com/canada/video-trudeau-scholz-push-hydrogen-plans-in-newfoundland/.

Demirdoven, Nurettin, and John Deutch. "Hybrid Cars Now, Fuel Cell Cars Later." *Science* 305, no. 5686 (2004): 974–976.

Eastman, Alicia. *Scaling Green Hydrogen for a Global Market*. Interview with Andy Stone. Podcast audio. September 27, 2022. https://kleinmanenergy.upenn.edu/podcast/scaling-green-hydrogen-for-a-global-market/.

European Commission. "State Aid: Commission Approves up to €5.2 Billion of Public Support by Thirteen Member States for the Second Important Project of Common European Interest in the Hydrogen Value Chain." Accessed December 1, 2022. https://ec.europa.eu/commission/presscorner/detail/en/ip_22_5676.

Finkel, Madelon L. "The Impact of Oil Sands on the Environment and Health." *Current Opinion in Environmental Science & Health* 3 (2018): 52–55.

Hinkel, Jochen, Diana Mangalagiu, Alexander Bisaro, and J. David Tàbara. "Transformative Narratives for Climate Action." *Climatic Change* 160, no. 4 (2020): 495–506.

Howarth, Robert W., and Mark Z. Jacobson. "How Green is Blue Hydrogen?" *Energy Science & Engineering* 9, no. 10 (2021): 1676–1687.

Hydrogen Council. "Global Hydrogen Flows." Prepared for the Hydrogen Council by McKinsey & Company. Accessed December 1, 2022. https://hydrogencouncil.com/wp-content/uploads/2022/10/Global-Hydrogen-Flows.pdf.

Intergovernmental Panel on Climate Change. "Climate Change 2013: The Physical Science Basis. Contribution of Working Group I to the Fifth Assessment Report of IPCC the Intergovernmental Panel on Climate Change." 2013. Accessed November 20, 2022. https://www.ipcc.ch/report/ar5/wg1/.

Intergovernmental Panel on Climate Change. "Global Warming of 1.5°C: An IPCC Special Report on the Impacts of Global Warming of 1.5°C Above Pre-Industrial Levels and Related Global Greenhouse Gas Emission Pathways, in the Context of Strengthening the Global Response to the Threat of Climate Change, Sustainable Development, and Efforts to Eradicate Poverty." 2018. Accessed December 1, 2022. https://www.ipcc.ch/sr15/download/#full.

International Energy Agency. "Global Hydrogen Review 2021." Accessed December 2, 2022. https://iea.blob.core.windows.net/assets/5bd46d7b-906a-4429-abda-e9c507a62341/GlobalHydrogenReview2021.pdf.

International Energy Agency. "Hydrogen: Energy System Overview." Accessed December 1, 2022. https://www.iea.org/reports/hydrogen.

Jefferson, Michael. "A Crude Future? COVID-19s Challenges for Oil Demand, Supply and Prices." *Energy Research & Social Science* 68 (2020): 101669.

Kinder, Jordan B. "From Dirty Oil to Ethical Oil: Petroturfing and the Cultural Politics of Canadian Oil After Social Media." *Journal of Environmental Media* 1, no. 2 (2020): 167–183.

Macdonald, Eliana. *Atlas of Cumulative Landscape Disturbance in the Traditional Territory of Blueberry River First Nations*. Vancouver: Ecotrust Canada, 2016.

Mac Dowell, Niall, Paul S. Fennell, Nilay Shah, and Geoffrey C. Maitland. "The Role of CO_2 Capture and Utilization in Mitigating Climate Change." *Nature Climate Change* 7, no. 4 (2017): 243–249.

McCurdy, Patrick. "Bearing Witness and the Logic of Celebrity in the Struggle over Canada's Oil/Tar Sands." In Brevini, Benedetta, and Graham Murdock (eds.), *Carbon Capitalism and Communication*, 131–145. Cham: Palgrave Macmillan, 2017.

Ministry of Energy. "Alberta Hydrogen Roadmap." *Government of Alberta* (November 2021). Accessed December 1, 2022. https://open.alberta.ca/publications/alberta-hydrogen-roadmap.

Ministry of Energy. "Getting Alberta Back to Work: Natural Gas Vision and Strategy." *Government of Alberta* (October 2020). Accessed December 1, 2022. https://open.alberta.ca/publications/getting-alberta-back-to-work-natural-gas-vision-and-strategy.

Natural Resources Canada. "Energy Factbook 2021–2022." *Government of Canada* (December 2021). Accessed December 1, 2022. https://www.nrcan.gc.ca/sites/nrcan/files/energy/energy_fact/2021-2022/PDF/2021_Energy-factbook_december23_EN_accessible.pdf.

Natural Resources Canada. "Joint Declaration of Intent Between the Government of Canada and the Government of the Federal Republic of Germany on Establishing a Canada-Germany Hydrogen Alliance." *Government of Canada* (August 23, 2022). Accessed December 1, 2022. https://www.canada.ca/en/natural-resources-canada/news/2022/08/canada-and-germany-sign-agreement-to-enhance-german-energy-security-with-clean-canadian-hydrogen.html.

Natural Resources Canada. "Hydrogen Strategy for Canada: Seizing the Opportunities for Hydrogen—A Call to Action." *Government of Canada* (December 2022). Accessed December 2, 2022. https://www.nrcan.gc.ca/sites/nrcan/files/environment/hydrogen/NRCan_Hydrogen-Strategy-Canada-na-en-v3.pdf.

Noakes, Taylor C. "Why Justin Trudeau's 'Blue Hydrogen' Dream is not Really Green." *The Toronto Star*, November 13, 2022.https://www.thestar.com/opinion/contributors/2022/11/13/why-justin-trudeaus-blue-hydrogen-dream-as-an-energy-alternative-for-canada-is-not-really-green.html.

Orland, Kevin. "Home of the Oil Sands Eyes Cleaner Future as Hydrogen Superpower." *Bloomberg News*, August 13, 2020. https://www.bloomberg.com/news/articles/2020-08-13/home-of-the-oil-sands-eyes-cleaner-future-as-hydrogen-superpower?leadSource=uverify%20wall.

Palmer-Wilson, Kevin Gordon. "How Do We Power Decarbonization? Land and Other Resources in Canada's West." PhD diss. Victoria: University of Victoria, 2020.Continuing to Support Fossil Fuels." *Columbia Journal of Environmental Law* 45 (2020).

Ridington, Robin, and Jillian Ridington. *Where Happiness Dwells: A History of the Dane-Zaa First Nations*. Vancouver: UBC Press, 2013.

Robb, Douglas, Harrison Cole, Jennifer Baka, and Karen Bakker. "Visualizing Water-Energy Nexus Landscapes." *Wiley Interdisciplinary Reviews: Water* 8, no. 6 (2021): e1548.

Robb, Douglas, Philippe Le Billon, and Karen Bakker. "Landscapes of Recarbonization: Carbon Neutrality, Settler Colonialism, and Cumulative Environmental Effects in the Peace River Region, Canada." *Annals of the American Association of Geographers* (2024): 1–18.

Rosen, Marc A., and Seama Koohi-Fayegh. "The Prospects for Hydrogen as an Energy Carrier: An Overview of Hydrogen Energy and Hydrogen Energy Systems." *Energy, Ecology and Environment* 1, no. 1 (2016): 10–29.

Simpson, Michael. "Resource Desiring Machines: The Production of Settler Colonial Space, Violence, and the Making of a Resource in the Athabasca Tar Sands." *Political Geography* 74 (2019): 1–12.

Spirn, Anne Whiston. "Landscape Literacy and Design for Ecological Democracy: The Nature of Mill Creek, West Philadelphia." In Ernstson, Henrik, and Sverker Sörlin (eds.), *Grounding Urban Natures: Histories and Futures of Urban Ecologies*, 109–136. Cambridge: MIT Press, 2019.

Department of Energy Office of Fossil Energy and Carbon Management. "Hydrogen Strategy: Enabling a Low-Carbon Economy." Accessed 2 December, 2022. https://www.energy.gov/sites/prod/files/2020/07/f76/USDOE_FE_Hydrogen_Strategy_July2020.pdf.

Westman, Clinton N., and Tara L. Joly. "Oil Sands Extraction in Alberta, Canada: A Review of Impacts and Processes Concerning Indigenous Peoples." *Human Ecology* 47, no. 2 (2019): 233–243.

Wisen, Joshua, Romain Chesnaux, Gilles Wendling, John Werring, Florent Barbecot, and Paul Baudron. "Assessing the Potential of Cross-Contamination From Oil and Gas Hydraulic Fracturing: A Case Study in Northeastern British Columbia, Canada." *Journal of Environmental Management* 246 (2019): 275–282.

Wolff, Jane. *Bay Lexicon*. Montreal and Kingston: McGill-Queen's University Press, 2021.

World Energy Council. "New Hydrogen Economy: Hope or Hype?" Accessed December 2, 2022. https://www.worldenergy.org/assets/downloads/WEInnovation-Insights-Brief-New-Hydrogen-Economy-Hype-or-Hope.pdf.

World Energy GH2. "About Project Nujio'qonik." Accessed December 1, 2022. https://worldenergygh2.com/about/.

World Energy GH2. "Project Nujio'qonik GH2 Environmental Assessment Registration." Accessed December 1, 2022. https://www.gov.nl.ca/ecc/files/2202-Registration-Document.pdf.

Yahey v. British Columbia. BC Supreme Court 1287, paragraph 1894. Accessed December 1, 2022. https://www.bccourts.ca/jdb-txt/sc/21/12/2021BCSC1287.htm.

IMAGE CITATIONS + CREDITS

Listed by Page Number

168 Map sources: Maxar.

170 Photo: Douglas Robb.

Map sources: Canada, Natural Resources. "Aboriginal Lands of Canada Legislative Boundaries – Open Government Portal." Accessed May 16, 2025. https://open.canada.ca/data/en/dataset/522b07b9-78e2-4819-b736-ad9208eb1067.

174 Photo: Douglas Robb.

Map sources: "North American Cooperation on Energy Information (NACEI). Mapping Resources on Energy Infrastructure and Potential. Collaborative data by U.S. Department of Energy, Natural Resources Canada, and Ministry of Energy of Mexico." https://open.canada.ca/data/en/dataset/aae6619f-f9f3-435d-bc32-42decd58b674. License viewable here: https://open.canada.ca/en/open-government-licence-canada. Accessed June 9, 2025.

178 Photo: Douglas Robb.

Map sources: Maxar.

186 Photo: Douglas Robb.

188 Photo: Douglas Robb.

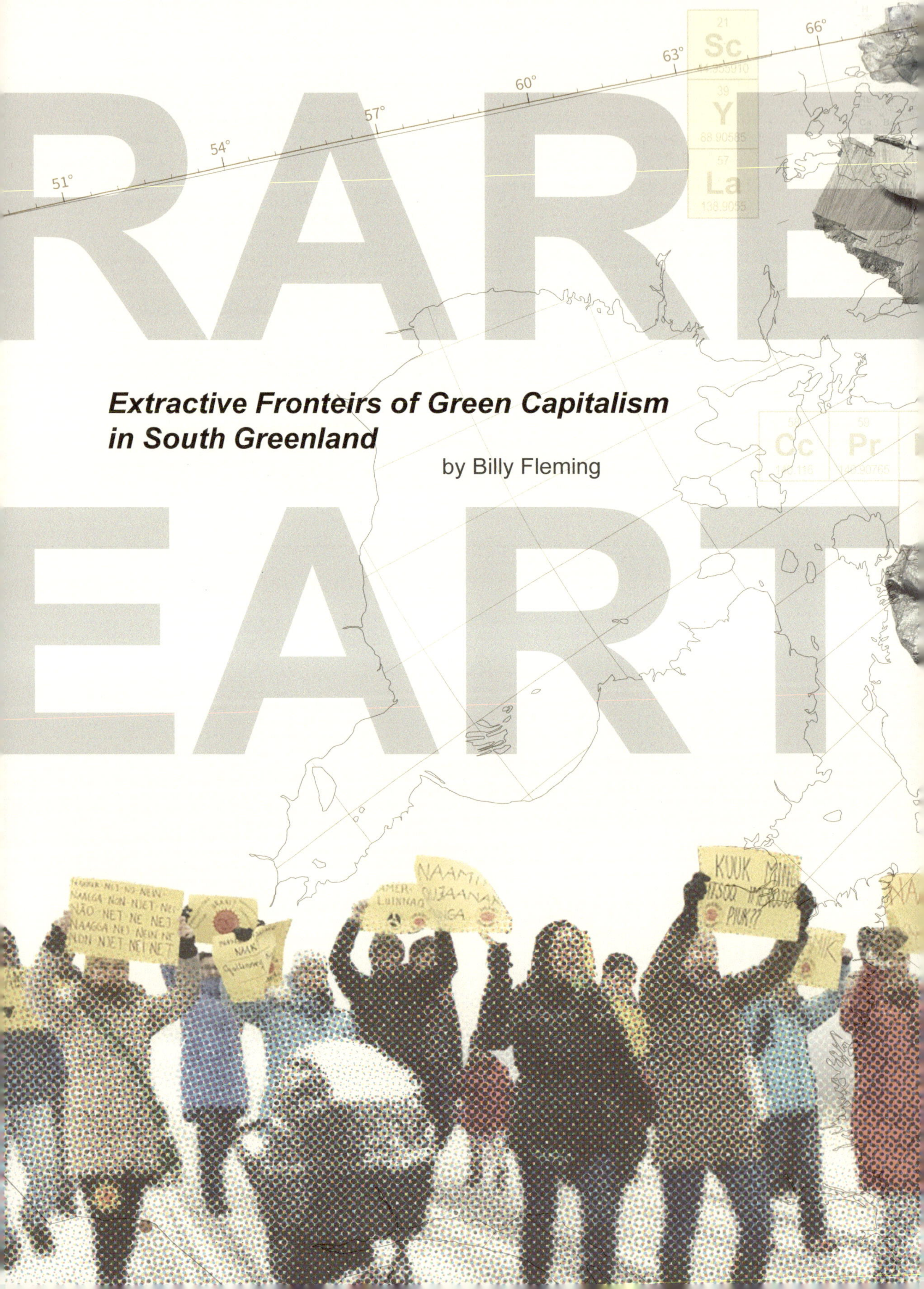

Extractive Fronteirs of Green Capitalism in South Greenland

by Billy Fleming

72°
Rare Earth Element Assessment Tract
ICE BASE MELT STATUS (probable)
Frozen
Uncertain
Thawed
Sm
Eu
Gd
Tb
Dy
Ho
Er
Tm
Lu
15
SARFARTOQ
MOTZFELDT SØ
QAQARSSUK
NIAQORNAKASSAK
UMIAMMAKKU NUNAA
NUUK
TIKIUSAAQ
QAQQARSUAQ
NARSARSUAQ
KVANEFIELD
NARSAQ
QAQORTOQ
KRINGLERNE
URAN

25 September 2022
10:24 am GMT
Elevation: 24,000′
Temperature: -40C (in the air)
Latitude/Longitude: 63.286590°, -42.801193° (somewhere over the Ice Sheet in East Greenland)
Modes of Transport: Shuttle Van, Propeller Plane, School Bus, Boat Ferry, Feet

I slowly wake and turn my head to glimpse the chop and churn of the North Atlantic Ocean—its deep blues, stippled with the white foam and flotsam of its heaving waves, extends beyond the horizon in all directions.

My window is saturated in their hues—water and sky; crest and cloud, blending together in a cerulean swirl. On board the plane itself, two sounds dominate—the whir of the De Havilland Canada DHC-8-202 Dash 8's propeller engines, droning endlessly through my noise-cancelling headphones; and the booming snores of a few tired architecture students in the back of the aircraft, recovering from a long layover in Reykjavik and an early morning departure.

The plane has room for 90 passengers, but there are only 19 people on board—10 Yale architecture students, my collaborator and Greenland field guide Zane Griffin Talley Cooper,[1] 2 American tourists, one Danish construction worker, one local Greenlander, three crew, and myself. Our flight is scheduled to last 3.5 hours or so, but Hurricane Fiona is battering the North Atlantic. The pilot comes on the intercom to let us know that Fiona's strong headwinds are going to add an hour or so to our trip.

1 For more on Zane Griffin Talley Cooper's fieldwork in South Greenland and the Arctic, see his dissertation: Zane Griffin Talley Cooper, "Bit/Coin, Rare/Earth: Data, Energy, and Extraction Across the Arctic." Phd diss. University of Pennsylvania, 2014.

When our pilot finishes his spiel, I unbuckle and leave my seat to wander the sparsely populated plane. I make my way forward to the galley to ask our flight attendant for a coffee—there are so many empty seats and so many sleeping passengers on board that beverage service is only provided upon request. As I melt back into my row of empty seats and sip a cup of burnt, watered-down coffee, I peer outside and notice something odd: a break in the endless expanse of blue sky and sea, punctured by a blinding white light to our right. As the white mass slowly comes into focus, I realize that we're about to traverse the Greenlandic Ice Sheet—a massive glacier covering some 660,000 square miles, roughly 80% of the nation's total land area.[2]

Giddy, I burst out of my seat and begin waking my students, gesturing toward the window and watching as their faces shift from groggy annoyance to complete rapture. For a brief period, it seemed as if we might never make it here—the remnants of Fiona battered our stopover service in Reykjavik, threatening to maroon us in Iceland for the duration of our studio trip. Zane and I even went so far as to plan a few spontaneous days touring the Iceland countryside and energy infrastructure. But as the weather broke this morning and our plane surged into the sky, those contingency plans were abandoned and the place we'd spent our semester dreaming of was suddenly, literally, beneath our feet.

For the next hour, I pace up and down the aisle, talking through what we're seeing with a cohort of architecture students unfamiliar with Arctic landscapes. I describe the outlet glaciers, moraines, tundra, and occasional mountain peaks bursting through the ice sheet; we discuss the brilliant blues (meltwater) and browns (glacial rock flour) spilling out of the ice flow and into the sea, and how rapidly rising temperatures are transforming Arctic nations like Greenland into new resource frontiers for multinational mining and oil and gas operators.[3] At some point, the pilot comes on the intercom to tell us to take our seats and buckle in—the winds above the ice sheet are often rough, and we're heading into a bout of turbulence that shakes and slides the plane as if it were walking along the ice sheet without crampons.

---

2 For more on the role that Greenland's ice sheet plays in developing planetary models of oceanic and atmospheric systems, including the advent of climate science as a field of study, see: D. Dahl-Jensen et al., "Past Temperatures Directly from the Greenland Ice Sheet," *Science* 282, no. 5387 (1998): 268–271.; Phillip Huybrechts and Jan de Wolde, "The Dynamic Response of the Greenland and Antarctic Ice Sheets to Multiple-Century Climactic Warming," *Journal of Climate* 12, no. 8 (1999): 2169–2188.; Heiko Goelzer et al., "Recent Progress in Greenlandic Ice Sheet Modelling," *Glaciology and Climate Change* 3 (2017): 291–302. For more on the cultural role that ice plays in Greenland, see: Amanda Boetzkes and Jeff Diamanti, "At the Moraine," *e-flux: Accumulations*, September 2020, https://www.e-flux.com/architecture/accumulation/341644/at-the-moraine/.; Susan Carruth, "Developing Renewable Energy in Discontiguous Greenland: An Infrastructural Urbanism of 'Material Practices,'" *Journal of Landscape Architecture* 11, no. 1 (2016): 66–79.; Lenore Grenoble, Hilary McMahan, and Alliaq Kleist Petrussen, "An Ontology of Landscape and Seascape in Greenland: An Encoding of Land in Kalaallisut," *International Journal of American Linguistics* 85, no. 1 (2019): 1–43.

3 For more on the expansion of Arctic mining operations, see: Zoe Garbis et al., "Governing the Green Economy in the Arctic," *Nature Climate Change* 176 (2023): article 33, https://doi.org/10.1007/s10584-023-03506-3.; Dorothee Cambou and Greg Poelzer, *Enhancing Energy Justice in the Arctic: An Appraisal of the Participation of Arctic Indigenous Peoples in the Transition to Renewable Energy* (London: Routledge, 2021); and the recent press release from the Executive Director of the Arctic Economic Council, Mads Qvist Frederiksen, "If We Want an Energy Transition, We Must Have More Mining," Arctic Economic Council, accessed May 21, 2025, https://arcticeconomiccouncil.com/news/if-we-want-an-energy-transition-we-must-have-more-mining/.

We hardly notice. Instead, we're collectively raptured by the alien landscape beneath and the prospect of what we will find after we touch down. Soon, our pilot announces we're preparing to descend into Narsarsuaq—a former military base (like many other settlements in Greenland) with a small airfield, hotel, and ferry service to our final location, Narsaq.[4]

The Narsarsuaq Airport is infamous, known as one of the most dangerous places to fly in or out of in the world. It's short runway (just a touch over 6,000') with perpetual high winds and fog, and sits in a hemmed-in location among the fjords and mountains of South Greenland, where the microclimate can force days or even weeks of flight cancellations in or out of the country (remember that, it will be important later in this dispatch). It requires a 90-degree turn immediately after takeoff (or before landing) and serves more small charter flights than commercial aircraft. Air Greenland—the primary carrier to and from the nation—only operates eight winged aircraft, and most seat fewer than 50 passengers.[5] As the Danish construction worker on our flight lets us know, the Narsarsuaq Airport is slated for decommissioning in 2025—a new airport is under construction in Qaqortoq, a much larger settlement a few fjords away (though I'm told its weather conditions are even worse, and locals expect regular flight cancellations to continue after the shift). He's here to help manage that project, and is meeting his crew in Narsarsuaq en route to Qaqortoq, where they'll all live for most of the next three years.[6]

We fall out of the sky, onto the tarmac, and quickly deplane. As students gather their backpacks, Zane and I make our way to the Disko Line Ferry desk—our ride from Narsarsuaq to Narsaq. Like much of the rest of the Arctic, there are no roads between settlements in Greenland. It's a nation of 56,653 people, nearly all of whom are Inuit. The largest city is its capital, Nuuk, with 17,635 people. Most other settlements here range from as little as two (we would pass several such places on our various boat rides through South Greenland) to 123 in Narsarsuaq to 1,346 in Narsaq. Travel is only possible via aircraft, boat, snowmobile or ATV, dogsled, or foot. The seasonal and diurnal heave of thawing permafrost and soil makes building and maintaining roadways between settlements and through the mountains and ice sheet impossible.

-----

4 For more on the militarization of South Greenland, see Anthony Dzik, "Interplays of Site and Situation along Tunalliarfik Fjord, South Greenland," *Journal of Settlements & Spatial Planning* 5, no. 2 (2014): 67–89.
5 For more on the dangers of air travel in South Greenland, see: Hannu Heikkinen, Lill Rastad Bjorst, and Albina Pashkevich, "Challenging Tourism Landscapes of Southwest Greenland: Identifying Social and Cultural Capital for Sustainable Tourism Development," *Arctic Anthropology* 57, no. 2 (2020): 212–228.
6 For more on the historical and ongoing efforts to expand Greenland's aviation industry, see: Bert De Jonghe, *Inventing Greenland: Designing an Arctic Nation* (New York: Actar, 2022).

25 September 2022
12:24 pm WGST
Elevation: 121′
Temperature: 14C
Lat/Long: 61.159311°, -45.440902° (Narsarsuaq Airport)
Modes of Transport: School Bus, Boat Ferry, Feet

We're nearly an hour late landing in Narsarsuaq, and for a moment I begin to panic—will the ferry still be there to take us to Narsaq? What will we do if we literally missed the boat?

Zane quickly reminds me that we are in South Greenland—we are quite literally the only passengers on this particular ferry, it would never leave without us. Sure enough, we meet our boat captain and shuttle bus driver in the baggage claim area and quickly get underway. It's a short school bus ride from the airport to the docks, where we descend down a series of slippery ramps into our vessel—a 12 passenger ship called "Nilaq" or Blue Ice, skippered by a husband and wife who live in a settlement an hour or so Southeast of Narsaq.

The ride takes about an hour—and it's one of the most spectacular hours I've ever experienced. The water is smooth and glassy between the fjords, and the scenery is mesmerizing. A brilliant palette of lichen, moss, and other Arctic plants are smeared across the mountains, cliffs, rocks, and farmlands that frame our ride. A queue of cerulean icebergs float down the channel, some close enough that we can nearly touch them from the bow of the ship.

As our ship nears the righthand turn into Narsaq, we pass beneath a set of high-voltage wires that stretch—and hum—across the full length of the channel. This is not unique to Narsaq. Each settlement in Greenland has its own, purpose-built hydroelectric plant powering its people. Many of these plants are relatively new—towns in Greenland ran on diesel generators for most of the 20th century, and many are still maintained as backups in the event of a prolonged outage. But the energy grid itself remains unchanged—every settlement is built on a microgrid, with its own unique power source. There is no Greenlandic or regional power pool.[7]

These new hydroelectric facilities are designed and built by Icelandic engineers, then owned and operated by Greenlanders forced to live in isolation for months or, in some cases, years. We are slated to visit Narsaq's plant in Qorlortorsuaq in a few days, once we've settled in a bit. The low buzz of the wires is a gateway of sorts—a signal that we're entering one of South Greenland's largest settlements.

Our approach takes us through flocks of seabirds and a slalom course of icebergs. Our captain weaves through them all before docking beneath a whaling mural at a pier that connects to the local petrol station and brewery (yes, Greenlandic breweries are amazing). As we grab our bags and disembark, Zane leads us on a short walk to our accommodations: the pastel yellow Hotel Narsaq. It's a ten-minute or so walk up a small hill that takes us past a field of debris and trash, small seasonal boats, the local slaughterhouse (Neqi), and a few brightly colored homes. Greenland is many things, but it's clear in our first hour on the ground that it isn't the dreary, inhospitable ice sheet we're often told. Its architecture, infrastructure, and landscapes are as brilliantly colored as any in the world. We arrive at the hotel, situated at the base of Mount Qaqqarsuaq (elevation 2,250'), where its proprietor, Augusta, quickly checks everyone into their rooms. We drop our bags and reconvene outside for our first round of fieldwork.

-----

7 For more on Greenland's hydroelectricity industry, see: Janina Priebe, "Glacial Energy Futures? The History of Unbuilt Hydropower in Greenland from the 1950s to the 1970s," *Water History* 16 (2024): 271–290.; and Rasmus Gjedsø Bertelsen, Jens Christian Justinussen, and Coco Smits, "Energy as a Developmental Strategy: Creating Knowledge-Based Energy Sectors in Iceland, the Faroe Islands, and Greenland," in *Handbook of the Politics of the Arctic* (London: Elgar, 2015), 3–25.

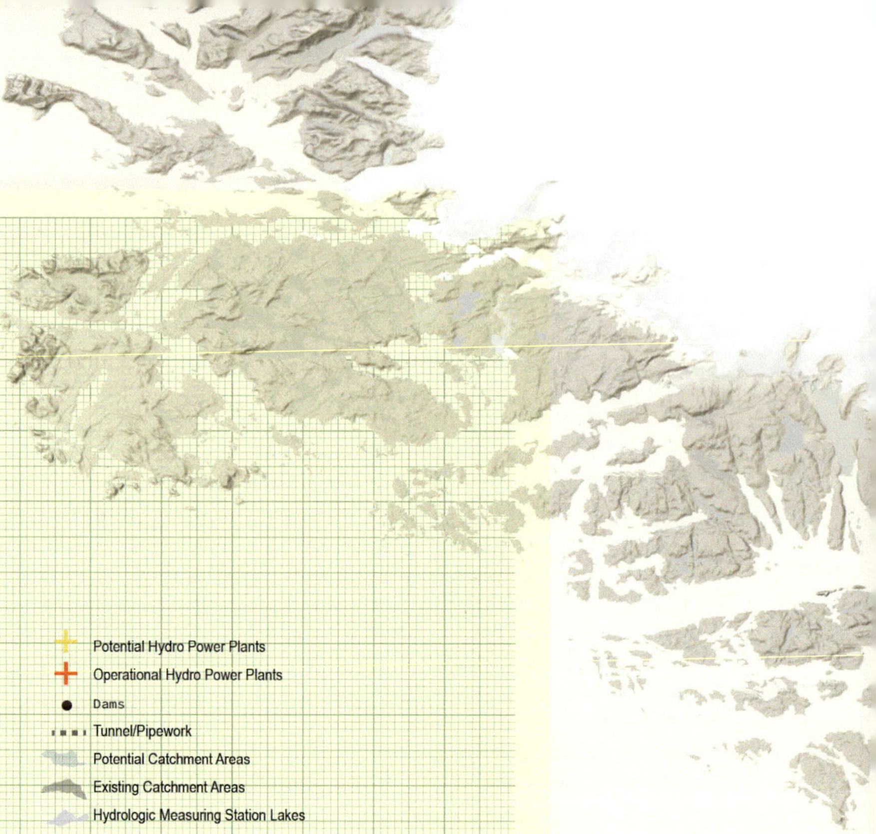

25 September 2022
1:56 pm WGST
Elevation: 979′
Temperature: 15C
Latitude/Longitude: 60.919180°, -46.029924°
(Water Intake for Narsaq)
Modes of Transport: Feet

0 5 10 15 20 Miles

The goal for the rest of today is straightforward—we're going to walk the physical infrastructure of Narsaq's local water system. We're doing this for a few reasons. One is to emphasize the precarity of life in rural Arctic communities. Narsaq has a simple intake system and a small backup reservoir, both of which rely on spring and rainwater to be replenished. But the increased precipitation, melting ice sheet, and thawing permafrost have quickly overwhelmed their basic infrastructure systems, washing our trails and dirt roads and, now, threatening to damage the town's water supply.

Our route is an easy hike that includes stops at the settlement's backup reservoir, southeast of the docks; a trek along the base of Mount Qaqqarsuaq, to see the streams and springs feeding into the lower elevation backup storage and intake system within the town; and a longer hike into the fjords to see the primary water intake system, fed by the ice sheet's outflow creeks and the alpine lake, Taseq, at the base of the glacier. Zane has made this hike several times, and remarks repeatedly on the extreme nature of this one (Hurricane Fiona has transformed creeks one could step over into impassable, roaring rivers near town).

As we leave Hotel Narsaq, our route is east along the gravel path that snakes between some of Narsaq's brightly colored Danish colonial homes until we reach the top of a small knoll. From there, we have panoramic views of the settlement, its water infrastructure, and the south face of Mount Qaqqarsuaq. It's a tangle of Euclidean forms and earth tones (roadways, waterlines, power lines, and highly engineered shorelines) in a sea of undulating, brilliantly colored hills and mountains and waterways. Students collect rocks, take soil and water pH samples, and begin to field test the Geiger counters we brought with us. Some just sit and stare. Some sketch. Some dip their toes in the water.

Narsaq is home to the Kvanefjeld mine[8]—a strategic uranium deposit for U.S. and European nuclear energy and weapons programs for at least the last half-century. Like most other mines of this kind, the open-pit operations here spewed radioactive waste in the water, air, and soil, contributing to the exceptionally high cancer rates in South Greenland. It was decommissioned in the early 1990s as a national network of Inuit activists known as Urani? Naamik! (Uranium? No Thanks!) waged a successful, years-long campaign to win a uranium mining ban in the country. Their work began as residents of Narsaq, Qaqortoq, and other small villages in the region began to notice an increase in cancer diagnoses among their families and game herds. It featured protests and direct actions at critical logistical nodes (for example, ports, airfields, and mining community meetings), where members of Urani? Naamik! could confront mining industry executives and sympathetic Members of Parliament (MPs). Their work culminated in an electoral win that swept an entire slate of new MPs into office on the promise that they enact a ban on all uranium mining in Greenland. That victory lasted nearly 25 years—but the toxic and radioactive slag that preceded it would persist for millions more.

8 For more on the historical and ongoing efforts to develop the Kvanefjeld plateau into an open pit mine, see: Jose Ricardo Carvajal Gamboa, "Rare Earth Mining and Sovereign Statehood in Greenland: 'Sustainable Engagement' with Extractivism in Kvanefjeld." Thesis. University of British Columbia, 2022.; Damien Krebs, "Greenland's Unique Kvanefjeld Multi-Element Rare Earth Project," *Technology Review* 61, no. 2 (2000): 154–164.; and Henning Sorensen et al., *The Uranium Deposit at Kvanefjeld, the Ilimaussaq Intrusion, South Greenland* (Copenhagen: The Geological Survey of Greenland, 1974).

Still, as we traverse the village, this extant era of industrial-scale mining is all around us. We can see it in the abandoned earth-moving technology and in the sifters and other abandoned machinery at the base of the mine; we can see it in the radioactive core samples strewn across the largest agricultural plot in the area; and we can see it in the mounds of discarded ore and waste material rimming the bay below Kvanefjeld. At various points, I have to remind students not to take one of the highly radioactive core samples or piles of crushed stone home with them.

We're here now in part because so many of those anti-extractivist victories are being undone. Though oil and gas operators have been prospecting (without much success) in Greenland for decades, the recent shift toward renewable energy and digital infrastructure has placed geological deposits known as rare earth elements (REE) and critical minerals (CM) at the center of the mining industry. Several companies have come here in recent years to conduct exploratory analyses of the deposits in and around decommissioned mines like this one. In the course of their work, mining executives learned that the Kvanefjeld site is the primary home to the second densest collection of these REE/CMs in the world—needed to manufacture things like lithium-ion batters, supermagnets, solar photovoltaics, and all manner of turbines at the core of the energy transition. More specifically, the Kvanefjeld contains some of the world's most abundant and accessible deposits of fluorospar, neodymium, praseodymium, dysprosium, and terbium.[9]

The mine has become the object of focus for an Australian mining company known as Greenland Minerals. After failing to secure an exploitation license to begin operations there in 2019, the company began the process of challenging Greenland's resource sovereignty in the Danish court system—Greenland remains a protectorate of Denmark, and thus lacks true control over its land and resources. That case could be decided any time and so one of our goals here is to measure and map the radioactive hotspots around the mine, the town of Narsaq and the places in between. We want to know if Greenland Minerals is correct—that there is not any excess ambient radiation from previous mining regimes in the area—or if, as local activists suspect, there is an exceptionally high level of ambient radiation spewing from Kvanefjeld. If the latter is the case, crucially, there is simply no way to access the REE/CM deposits below without spreading even more toxic material in and around the region.[10]

We continue on our walk, cruising past the base of Mt. Qaqqarsuaq and through a series of artificial stream banks and erosion control measures in town. Students stop to explore and measure a set of retention ponds, bioswales, and other forms of green infrastructure in the crop and grazing fields on the way back out of town. Soon, we're at the tailwaters of the stream that drains most of the watershed around Narsaq. Zane tells us this stream is usually 2–3' wide and easy to traverse—last time he was here, he simply stepped over it. But today it is more

9 For more on the recent controversies surrounding the mine, see: Patrick Greenfield and Phoebe Watson, "Fearing Toxic Waste, Greenland Ended Uranium Mining. Now, They Could Be Forced to Restart—or Pay $11 Billion," *The Guardian*, March 5, 2025, https://www.theguardian.com/environment/2025/mar/05/greenland-mining-energy-transition-minerals-environmental-laws-uranium-rare-earth-toxic-waste-investor-state-dispute-settlement-isds-aoe.

10 Jacob Gronholt-Pederson, "Uranium Ban Repeal in Greenland Could Revive Massive Rare Earth Project, License Holder Says," *Reuters*, February 11, 2025, https://www.reuters.com/markets/commodities/uranium-ban-repeal-greenland-could-revive-massive-rare-earth-project-licence-2025-02-11/.

than 50' wide and moving so quickly that our only choice is to try and follow it upstream, toward the primary intake for Narsaq, instead of wading across. This is what climate change looks like in the Arctic—massive flooding and landslides as the remnants of a hurricane (a common occurrence each fall) overload a system already operating past its capacity. We are watching the hydrology of Narsaq be reorganized in real time on our walk.

After an hour of slipping and sliding our way to the first ridge north of Narsaq, we reach the settlement's intake system. Students wander through the area, once again collecting materials, taking photographs, and otherwise engaging in their various fieldwork practices while Zane and I watch the icebergs drifting through the outlet bay west of town. They occasionally collide and when they do, the booming crash echoes up and down the fjords around us. The rain begins to pick up, so we head back to the hotel to change and warm up before enjoying dinner at the Ulu Netcafe—one of three restaurants in Narsaq. Over local lamb, root vegetables, and beers, we debrief from the day and discuss the rest of our itinerary for the week.

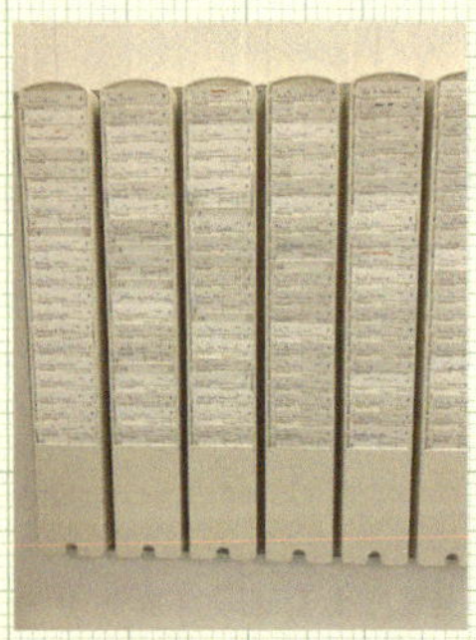

26 September 2022
9:57 am WGST
Elevation: 121'
Temperature: 12C
Latitude/Longitude: 60.909204°, -46.042913° (Neqi Slaughterhouse)
Modes of Transport: Feet

I wake up and stumble into the hotel's restaurant for coffee and pastries. My table looks out over the town, and I watch the fog and mist slink through the alleys and gravel roads as I wait for my caffeine fix to take over. As I do, a queue of sorts begins to form on the road that connects our hotel to the docks below. Local residents are heading to work—some to crew the fishing and ferry vessels outside; some to begin their own forms of subsistence labor (farming, artisanal mining, and textile making); but most march toward the Neqi slaughterhouse facility beside the docks. That's where we're heading.

Work in Narsaq looks much different than it does for most of the people in graduate architecture school. It's highly seasonal—there are times when the fish and meatpacking plants are staffed in the hundreds and times when they are all but closed; there are times when the commercial fishing vessels are desperate for more sailors and times when they're moored for months; there are times when the settlement's hunters are at sea in search of seal and whale and times when they are in the highlands in search of musk ox, mule deer, and polar bear; and there are times when little, if any, work is to be done outside of one's own home. The rhythms here are unique to the seasonality of South Greenland—a political ecology that is rapidly changing as the Arctic thaws.[11]

We're here in late fall, and that means the annual sheep roundup is upon us. This is a miraculous moment in the region—the time when sheep herders, many of whom live in almost total isolation, at elevations and in weather conditions that seem impossible, converge on Narsaq to have their flocks shorn and slaughtered. All of the meat stays in Greenland (a requirement of their colonial rulers, Denmark), but much of the fur is sold in international markets. It, along with seafood, is one of the nation's only true exports.

We arrive and meet the head of the facility, Lars. He's a middle-aged Danish man, a theme that becomes clearer the more time we spend in the region—he even discloses to us that, not only are all of the best-paying managerial jobs in Greenland held for Danes keen to work abroad, but that they're justified in their own legal system through a series of incredibly racist tropes about the ability and work ethic of Greenlanders. This is all to say that wealthy Danes like Lars get to manage these kinds of facilities, and local Greenlanders have to operate them. As we move through our slaughterhouse tour, that visceral reality is quickly on display.

As Lars tours us around, there is a constant buzz and whirr—we see skinless sheep being gutted; we see a station for decapitation and a station for carving the loins, chops, and other choice cuts; we see live sheep being marked for either shearing or slaughter; and we see a small cadre of cleanly dressed Danes moving through a space full of frenetic and bloodied Greenlanders as they saw and slice their way through thousands of sheep per day. It's a gruesome sight to anyone unfamiliar with the workings of an industrial agriculture operation—which is to say, most of my students (for better and for worse, I worked for a small stretch in a meatpacking plant, so this is all too familiar for me). The tour ends where we began—in their mess hall, a strange circular route that ends with us eating some of what we just saw in the slaughterhouse below our feet. Most students decide to forego lunch.

We walk back to the hotel to gather our packs and regroup for our longest hike of the trip. The rain has finally relented.

-----

11 For more on the rhythms and structure of Greenland's workforce, see: Cecile Krogh and Laust Høgedahl, "Mapping the Greenlandic Labour Supply: Evidence from Greenland's First Labour Force Survey," in *Greenland's Economy and Labour Markets* (London: Routledge, 2021).

26 September 2022
1:22 pm WGST
Elevation: 2,422′
Temperature: 14C
Latitude/Longitude: 60.978888°, -45.991417° (Kvanefjeld Mining Site)
Modes of Transport: Feet

Zane and I have organized this hike for a few reasons. One is that our ultimate destination is *the* key site in Greenland's fight for climate justice and in the broader global fight for a just transition. The Kvanefjeld mine was a key strategic site in the nuclear weapons and energy programs of the U.S. and Europe after World War II. Its closure in the early 1990s played an important symbolic role for Urani? Naamik! activists—and a vital health role for the residents of Narsaq. Their efforts pushed a uranium mining ban through parliament, effectively ending all industrial mining operations in the country (nearly all open-pit mines expose or extract radioactive material, whether as a byproduct of securing other minerals or as a goal unto itself).[12]

Now, nearly thirty years after securing its closure, the Kvanefjeld is once again a site of contestation. It is being reimagined by multinational, industrial-scale mining operators as a key strategic site in the global energy transition—and exploratory drilling in the region during the 2010s revealed exceptionally dense deposits of terbium, praseodymium, neodymium, and dysprosium (amongst others)—a source of wealth so great that it could support the full independence of Greenland. Of course, this is why Denmark is so reluctant to cede the territory, and why multinational mining operators are so keen to secure licenses to dig there—the Kvanefjeld is projected to be the most productive site in the region.

Greenland Minerals' initial application was quickly denied—according to Greenland's environmental regulators, it clearly violated the nation's uranium mining ban. That did not stop the company from filing an appeal and launching a political campaign to pressure elected officials into intervening on their behalf. But their plan largely backfired—the public nature of their campaign placed mining at the center of Greenland's 2019 parliamentary elections, and a new wave of anti-mining members were swept into office, effectively killing the permit and ensuring the Kvanefjeld would remain closed. But Greenland is not truly sovereign—it remains a protectorate of Denmark and thus subject to the Danish court system. And that's where Greenland Minerals is now waging their fight for an exploitation license—by suing Greenlanders in Danish court. I came to this work via Zane and quickly learned that one of our most valuable contributions here would be one of simple documentation—to detect and share the various radiological hotspots in and around the mining zone, and to make those data accessible to Greenlanders in their ongoing legal fight in Denmark.

There is a massive data asymmetry between executives at Greenland Minerals and locals in Narsaq. As the subsidiary of a highly profitable, multinational mining company, Greenland Minerals enjoys nearly infinite access to the kinds of things one needs to prevail in court—robust datasets, massive in-house technical processing capacity, high-priced legal teams, and the ability to wage their fight indefinitely if so desired. In fact, one of their primary arguments for approving the exploitation license is that previous uranium mining operations did little, if anything, to increase ambient radiation in and around Narsaq. In their

12 For more on the history of Greenland's radioactive politics, see: Mark Nuttall, "Zero-Tolerance, Uranium, and Greenland's Mining Future," *The Polar Journal* 3, no. 2 (2013): 368–383.

telling, bolstered with some questionable data, there is nothing to suggest that the radioactive material pulled out of the Kvanefjeld ever posed—or ever could pose—a public health risk to the people living in the dust and wastewater tailings such an operation would produce.[13]

We're taking this hike in part to test their argument—that the ambient radiation in the region is within a normal range, and that there are not extant radioactive hotspots from prior mining operations in the area. We've brought Geiger counters and pre-planned routes so we can link the GPS coordinates of a walk (via our Apple Watches) and the measurements we take as we go. The counters are fixed to our backpacks and set to track our data throughout the hike—and to beep loudly when we encounter an area with dangerous levels of radioactivity. Before arriving, we asked our key local interlocutors (Greenlanders and Danes in Narsaq, as well as a few British scholars who have established a modest research station in town) if anyone had ever measured radioactivity in the region before. Other than a few government surveys nearly a century ago, the consensus seems to be that we are the first people to ever bring Geiger counters to the region. Before we depart, we agree to collect our data, share it with our local partners, and to leave the Geiger counters with the local research station. I also promise to return to begin scoping out a broader monitoring system—essentially a mesh network of Arduino-based Geiger counters that could passively detect ambient radiation levels, store them on a secure server, and otherwise be accessible to Greenlanders in perpetuity.

But today we begin by retracing our water infrastructure tour—through the semi-paved roads rimming Narsaq, past the local football pitch and Greenland Minerals outpost, and, eventually, back to the flooding tailwaters that are roaring through a field of slag ore and abandoned machinery. As we approach the decommissioned sifter and a field of cylindrical core samples, the Geiger counters begin to beep. We check our Geigers and see a steadily climbing level of radiation that we later realize is equivalent to readings taken in Fukushima immediately after the disaster. We quickly move on, though Zane grabs a cylindrical core sample to take back for further testing. (It would eventually be lost in transit.)

Here, we split into two groups: one, led by Zane, takes the standard route, wading through the river (no small feat in these conditions) and then following a lowland trail around the bay, up a gentle incline, and to the mining site. The other, led by me, heads straight up the fjord's ridgeline, on a narrow footpath stippled with slippery lichen and moss, until we reach a gorge with a narrow enough crossing to scale down to the water and to climb back out of the gorge and into the alpine plateau that eventually leads to the mining site.

This all sounds simple and straightforward. It was not. Zane's party takes more than an hour just to get everyone across the river—the water is high and fast, and several students packed questionable gear for these conditions. As they do, their Geiger beeps constantly—the ambient radioactivity here is so high that its creating an exposure alert for us (granted, it would take significant and sustained time in the area to bear any real risk; this is fine for us on the hike, but obviously

13 For more on ambient radiation due to mining waste in Narsaq, see: Sven Poul Nielsen, Per Roos, and Kasper Grann Andersson, "Predicted Radiation Exposure from Mining at Kvanefjeld: Introduction to Radiation, Review of Baseline Information, and Predicted Radiation Exposures from Kvanefjeld Mining, Mineral Processing, and Refining," DTU-Nutech-R, no. 11.

not for the Greenlanders living and working here). It's nearly 3 pm before they're truly on their way to the mine.

My route proves significantly more difficult. Not only do we gain and lose well over 2,000' in elevation in just a few short miles, but we have to do so while traversing some of the slickest terrain I've ever encountered. We also have to climb down a relatively sheer rock face, cross at a narrow but very deep point in the river, and then ascend back out of the gorge. Every time we approach a large boulder or an exposed bit of rock on the mossy slope of the gorge, or spend time in the channel at its nadir, our Geiger counter beeps. The radioactivity here is widespread and exceptionally high—well beyond the Fukushima levels we found at the extant mining waste site. Other than the beeping, the walk itself is surreally quiet. There are no bird or insect sounds. Unless we peer over the edge of the gorge, there is no sound from the river. We can only really hear the faint drops of rainwater splatting around us, the squish of our feet in the spongey plants and soil, and the occasional grunt or groan when one of us slips. It's easily the most secluded I have ever felt in my life. This traverse takes us nearly three hours to complete. By the time we're on the alpine plateau—where we meet a large herd of cows and an arctic hare—it's nearly 4 pm and we still have most of the hike to go.

As Zane's party walks the edge of the bay, past a farmhouse and toward the Kvanefjeld trailhead, we scurry through a field of brilliantly colored succulents, moss, lichen, and other alpine plants—they squish and slide beneath our feet, delivering a satisfying "thwock" with each step. We pass large boulders, scramble through loose rock fields, and occasionally climb up and down the granite shelves that pock the plateau—and each time we do, our Geiger counter alerts us to the presence of exceptionally high radioactivity. We finally catch up to Zane around 5 pm at the bank of yet another washed out waterway. Their Geiger counter has been beeping nonstop for over an hour at this point—we initially think it's a malfunction, but the second counter begins to do the same. This water crossing is too much to risk with a team of students—it's deep enough that I can't feel the bottom more than a few feet past the bank, and it's flowing fast enough to take any of us off our feet. We regroup together and discuss the conditions—how Zane has made this hike several times before and never seen it in such bad shape; how one of our interlocuters, a local activist named Jan Rehtmar-Petersen and leader of Urani? Naamik! (who we'll meet later), warned us against trying it while water levels were so high; and how this reshapes our understanding of what sustainability looks like in the Arctic and what it means to truly be on the front lines of the climate crisis (the literal only road in town is washed out, it may never be repaired, and there's no way out of Narsaq beyond emergency helicopter when it's like this).

Stumbling back into town, we are cold, wet, tired, and hungry—the precipitation ranged from a fine mist to a moderate downpour throughout the hike. That, coupled with a few water crossings, has us thoroughly exhausted. As we return to the hotel, I crank the heat in my room, put my clothes to dry on the radiator, take a shower, and meet the class at the hotel's restaurant for Kayaq beers and musk ox steaks. We trade stories about the day's hike. I barely make it to 10 pm before collapsing in my room.

-----

27 September 2022
8:09 am WGST
Elevation: 3,082′
Temperature: 16C
Latitude/Longitude: 60.464590º, -45.142727º
(Qorlortorsuaq Hydroelectric Dam and Power Plant)
Modes of Transport: Boat, Feet

The morning comes quickly and we spill out of our rooms and back toward the docks. We're set to visit the 7.2 MW hydroelectric dam and power plant in Qorlotorsuaq and have a long boat ride ahead of us. For most of Greenland's postcolonial life, its small settlements have been relatively isolated from one another. Without roads to connect them, the only real way to get from one place to another is by boat, aircraft, snowmobile, dogsled, or foot. Because of this, not only were trade and other forms of exchange somewhat rare, but as Greenland's energy system developed, it became impossible to justify the cost and management expense of a national and interconnected grid to power its smattering of small coastal villages. Instead, most of Greenland's legacy

energy system is organized around small, modular diesel plant operations in each settlement—even it's smallest ones clinging to the salt-sprayed rock outcroppings of the coast. Its purpose-built hydroelectric plants, designed and built by Icelandic engineers, began to come online in the late 1990s—and the full system will not be complete until well into the 2030s. This became one of Iceland's most lucrative exports—the transmission of hydroelectric engineering expertise to other Nordic and island nations where their technology could be readily deployed.[14] Each settlement in Greenland operates on a micro-grid—a grid that serves only its respective community, with no real connection to some larger power pool where backup electricity can be bought or sold as needed. Instead, settlements like Narsaq rely on far-flung and remote hydroelectric plants like the one we're going to visit—a majestic high-altitude dam in the settlement of Qorlortorsuaq—and often keep their decommissioned diesel plants around as a backup.

In South Greenland, this energy transition is hard to fathom—one of the most remote valleys on the planet is home to one of the most sublime and immaculate hydroelectric generation, storage, and transmissions networks I've ever seen. As we pile into our ship for the three-hour ride to the plant, most students quickly drift back to sleep and miss an important announcement from our captain: we'll be sailing through a bit of open water, unprotected by the fjords, where we can expect to hit some rough water. We shove off and begin trolling our way down the channel, drifting past icebergs, seabirds, and flotsam.

The first hour or so is uneventful. We push out past any signs of settlement—the colored homes give way to fields and farmhouses. Before long, they're replaced by rock outcrops and battered coastlines—some of the only rocks we'll see in Greenland that are completely stripped of vegetation. As we sail past a beacon and lighthouse, I see the first large wave moving toward the front of our ship. It hits with enough force to rock the captain's coffee mug over, toss a few bags out of the overhead, and wake up most of our students. Within minutes, we are firmly in the chop, jolting up and down, left and right, lurching in sudden, jarring

14 In addition to tourism, the export of energy systems engineering is now one of Iceland's primary economic development priorities. In addition to Arctic hydroelectric power plant development, its geothermal industry is developing new plants in Canada, portions of the Rift Valley in Africa, and China. For more on this, see Maria Gallucci, "Magma and Hot Rocks: Iceland Seeks the Future of Geothermal Energy," *Canary Media*, July 22, 2024, https://www.canarymedia.com/articles/geothermal/magma-and-hot-rocks-iceland-seeks-the-future-of-geothermal-energy#:~:text=By%20gathering%20data%20and%20developing,lava%20and%20seething%20hot%20-springs.

movements every time we hit a large swell—and we hit what feels like a million of them.

This goes on for at least forty-five minutes. By the end, most of us are at the stern of the ship, hoping the fresh air will calm our stomachs. As we begin to turn back into protected water, the clouds begin to lift and we're treated to views of glistening, brilliantly colored coastal landscapes. Soon, those landscapes are peppered with the same pastel-colored Dutch colonial homes we saw in Narsaq as we pass Qaqortoq and Alluitsup Paa, then Saarloq and Lichtenau, Ammassivik and Qagdlumiut. We also drift past a few rocky outcrops with a stray home or two clinging to their sides—settlements of 3–6 people that pock the otherwise uninhabited Arctic coastline between the larger, 2–300 person named settlements. I can't imagine what life in one of those tiny, remote homes on an outcrop must be like—or perhaps I just don't want to know.

Around the 2.5 hour mark, we finally make the turn towards Qorlortorsuaq—a northern strait running up the fjord and toward an outflow glacier and moraine nestled between the largest mountains we've seen on this trip. I'm sure they have a name, but neither our captain nor our smartphones call tell us what they're called. We simply enjoy their presence as we begin to dock and disembark. Our captain tells us to be back here around 3 pm (it's just a touch before noon now) and motors away as we begin the mile or so hike to the dam. Because we're in Greenland as the remnants of Hurricane Fiona blow through, this means that the only road from the dock to the plant is completely washed out—we're forced to do a few river crossings on our way in, though nothing like the Kvanefjeld hike. A few students unpack the drone we've brought to capture some footage of this regional power system, and it whirs by overhead as we crest our last hill and glimpse our host: a jovial Greenlandic man named Otto, accompanied at all times by his large dog, Hund.

He welcomes us and walks us into the bright blue building where the turbines are spinning—we're so close we can touch them and any other instruments we want. It's a far cry from the militarized energy infrastructure of the U.S., where we'd never be able to get this close to such a critical piece of technology. Otto walks us through the plant, up the mountainside, and into the access tunnel that was drilled to build the intake system for the dam. It's full of control panels, emergency valves, and other critical procedural equipment, and dead ends beneath the alpine lake that now powers Narsaq.

The official tour is over quickly—just thirty minutes or so. We all then scatter—some students head back toward the power plant to regroup and conduct some fieldwork; I and a few others continue up the mountain, in search of the control structure, a summit or two to scale, or perhaps both.

I take a small group up to the control structure. There, we're treated to panoramic views of the alpine lake, mountain range, and valley floor, each cloaked in the misty spray drifting up from the riprap below. They stay behind to take photographs, collect materials for their fieldwork, and eat lunch. I take Hund up the mountain in search of a summit—or at least to see the point at which the plant's powerlines crest the peaks encircling our group.

The hike is rough—it's mostly a scramble up a field of lichen and moss-covered boulders. As I slowly make my way to the summit, Hund races up and down the mountain, thrilled to have someone tagging along on what soon becomes clear is a regular hunting trip for the dog. As I get to the saddle that will take me up to the nearest peak, I see hund race up the mountainside and out of sight, emitting a low growl—the first sounds I've heard him make. A few moments later, I see him bounding back down the incline, racing after a darting tuft of white I soon realize is an arctic hare—a massive rabbit that must weigh at least fifteen pounds. Hund manages to catch and kill the hare just a few yards below me. I keep moving upward, and soon he zips past me again, in search of another hare.

As I reach the top and sit down, the dramatic landscape overwhelms me—the quiet (all I can hear is a slight rustle from the wind), the remoteness (I can see the control structure and power plant, but no people), and the stark landscape (the mountains' relief is as sharp as I've ever seen). I feel odd, almost stoned, as I take it all in. But my watch begins to buzz—I forgot that I set an alarm for myself, so I'd know when to turn back in time to make our boat ride home. That time is now, so Hund and I begin the slow walk back to port.

I notice a few additional features en route—a small conference center that can host up to 20 people (note for the next trip here), a small cemetery where the initial crews that built and maintained the system are buried, and an old machine shop where the excavators and other earth-moving technology was maintained. Hund walks us back to the boat and then turns to sprint home.

The return trip is much warmer and smoother. We can sit outside, taking photographs and glimpsing the scattered settlements that rim South Greenland's rocky shoreline. As we pull into port, Zane reminds us that we have about an hour before our next meeting with the local leader of Urani? Naamik!

GAS FIELDS
Estimated Ultimate Recovery – Millions of Barrels of Oil Equivalent
100
1,000
5,000
10,000
100,000
SEA ICE EXTENT
1
Number of Septembers Covered in Ice from 1979 – 2021
INDIGENOUS DIALECTS
CRITICAL MINERALS
Occurrence
Deposit
SHIPPING ROUTES
Anchorage
Norton Sound
Juneau
Whitehorse
Fairbanks
Barrow
Yellowknife
Tłįchǫ
North Dene
South Dene
Copper People
Dene Sųłiné
Beaver / Dunne-za / Dane-zaa
Tsuut'ina
Kaska
Mountain
Gwich'in
Inupiaq
Inuktun
Inuktitut
Iqaluit
Qaanaaq
Thule Greenlandic
West Greenlandic
East Greenlandic
Nuuk
Tasiilaq
Reykjavik
180th Meridian Line
Siberian Yupik
Central Alaskan Yup'ik
Koyukon
Holikachuk
Deg Xinag
Dena'ina
Ahtna
Upper Tanana
Lower Tanana
Tanacross
Hän
Southern Tutchone
Northern Tutchone
Tagish
Tahltan
Nisga'a
Gitksan
Witsuwit'en / Nedut'en
Dakelh
Tsek'ene
Nicola
Haida
Coast Tsimshian
Eyak
Alutiiq
Kuskokwim

27 September 2022
7:31 pm WGST
Elevation: 77′
Temperature: 13C
Latitude/Longitude: 60.9125020°, -46.048809°
(Narsaq Atufarfia [school])
Modes of Transport: Feet
Southern Yukaghir
Even
Yakut*
Evenki
Yakutsk
Evenki
Tura
Tofa
Tuvan
Dolgan
Nganasan
Chulym
Ket
Evenki
Shor
Dudinka
Forest Enets
Tundra Enets
Northern Selkup
Tomsk region Selkup
East Khanty
Forest Nenets
Salekhard
Tundra Nenets
North Khanty
Khanty-Mansiysk
North Mansi
East Mansi
South Mansi
West and South Mansi
Naryan-Mar
Komi Zyrian*

There's a dense carpet of fog settling over Narsaq. As Zane leads students toward the local school for our meeting with leaders of Urani? Naamik!, I wait outside to help direct folks inward. We've invited the entire town via Facebook—all forms of news, be it government, media, or otherwise, are delivered to residents via Facebook here—though few actually show up. I wait until 7:45 pm or so and then head inside just as introductions are being made.

Inside, we have a rather complicated translation system underway. The leaders of Urani? Naamik! are there and most only speak Greenlandic—an Inuit language considered "dead" by most linguistic scholars, and thus for which there are no real translational books or software to assist us—though one member also speaks some Danish. We're also joined by Jan, a Dane who curates the local museum (one that focuses on the Viking history in the region). He speaks Danish, Greenlandic, and English, so he's our primary means of communication tonight.

In all, our meeting lasts about an hour. Some of the details have been omitted here—we received consent to record and use the meeting notes in our work, but I do not feel comfortable disclosing identifying details in a place where a multinational mining operator with a history of extralegal acts is setting up shop. We heard about the intergenerational history of Urani? Naamik!, especially their victories in the 1990s—a decades-long campaign to ban uranium mining that was won through a parliamentary referendum—and the group's spread from Narsaq to Qaqortoq to Nuuk and, eventually, across the whole of Greenland. They're beginning to link up with anti-extraction activists in Canada and Europe, too—a sort of pan-Arctic network of Indigenous people pushing back against the swell of industrial-scale mining in their thawing landscapes.[15]

We also learned some shocking news about their contemporary struggle against Greenland Minerals. For more than a year, Zane and I have been reading some of the technical reports that Urani? Naamik! has produced to show the effects of uranium mining on public health, air quality, and water quality in the region. But we learn over the course of this conversation that none of those reports were developed using primary data. Rather, the group has been forced to scour the internet for studies of similar phenomena in comparable resource communities, and to then extrapolate those findings to Narsaq and South Greenland. We also learn that we've brought the first few Geiger counters to the region that have not been controlled by a mining company, and we decide then and there to leave them behind when our trip concludes. This thread becomes a key insight to the rest of the studio, and to my own research goals in the region: that one way we might make ourselves useful would be to provide the kind of materials (Geiger counters), technical skills (the ability to make and maintain them, perhaps via Arduino machines we could train locals to work with), and digital infrastructure (the ability to gather, store, and process data regarding the ambient radiation of the region so that Greenlanders might prevail in court battles now underway). The stakes feel real for perhaps the first time on this trip.

The meeting ends and we scatter—some students head back to the hotel, others to Café Ingussuk for darts, local beers, and a bit of live music.

-----

15 For more on the recent efforts of Urani Naamik, see: Amanda Boetzkes and Jeff Diamanti, "Geofetishism and the Tender Violence of Rare Earths," *SubStance* 163, no. 3 (2021): 9–30.

28 September 2022
9:49 am WGST
Elevation: 48′
Temperature: 12C
Latitude/Longitude: 60.910899°, -46.050574° (Narsaq Museum)
Modes of Transport: Feet

It's our final day in Narsaq. We check out of our hotel and lug our bags 1,000' or so feet to the Narsaq Museum—a small campus of buildings on the waterfront, just west of the docks we arrived on a few days ago. It has two main attractions—a large building where the Indigenous, Viking, and other Settler histories in the region are told through a rather impressive collection built by Jan over the last few decades (including some original historiographies of Eric the Red, who Jan believes actually landed in Narsaq first and is why he set up the museum here); and a smaller turf house building that Jan has restored and opened up to visitors.

I pick up a few books here—one is a mineralogy book of Greenland (seems important) and another is an illustrated text titled *Beasts of Greenland* that tells the story of the mythical creatures (think, Arctic cryptids) inhabiting the region, and their various origin stories. We all lose track of time exploring the exhibits.

We finish up, grab our bags, and head to the boat. We're running a bit behind, so once the final group is ready to go we're forced to run—and we make it just as our captain is preparing to shove off without us.

28 September 2022
1:44 pm WGST
Elevation: 23′
Temperature: 13C
Latitude/Longitude: 61.162967°, -45.417728° (Narsarsuaq Airport)
Modes of Transport: Boat, Bus, Feet

Our boat whips around and begins trolling back to Narsarsuaq. We're moving slow, as heavy fog and low visibility—along with the usual iceberg mélange drifting through the fjords—make it so that we can't see the shore from the deck (less than 30' away). The ride out is far more harrowing than the one in. We're sloshing around icebergs, ramming into a few small ones, and otherwise moving at a deliberate pace as the conditions worsen.

We eventually pull into port and make our way to the airfield around 3 pm. Students grab a table in the café and I walk over to the gate to check our flight status: delayed until 4:30 pm. I let everyone know and then offer to grab some snacks while we wait at the local Pilersuisoq (the local post office and grocery store). Zane comes along and we buy the strangest collection of snacks I've ever seen in one place. We return to the airport to discover that our flight has been cancelled, so I send everyone down the road to the Hotel Narsarsuaq while I gather room and meal vouchers, ask about rebookings, and let the travel staff at Yale know what happened.

Things begin to spiral out of control from here. For those of us temporarily stranded in Narsarsuaq, the additional time in the country feels like a gift. We organize hikes to the ice sheet, tours of the abandoned military outpost that is now partially inundated by a melting outlet glacier, and otherwise catch up on work and with loved ones from our hotel. The situation there is as good as we've had it in Greenland—our meals are all hearty and free; our rooms are spacious if spartan; our Wi-Fi and cellular signals are exceptional; and our ability to rest and process all we've seen is heightened.

After several days of trying and failing to convince the gate agent to create a flight for us out of thin air that could start our journey home, I managed to finally join my students in the hotel mess hall for a meal. There, I met a Danish construction crew that was flying out the next day on a private charter. I relayed this information to my contacts at Yale. So, the next morning I walked back to the airport, met the crew and their pilots, exchanged contact information, and forwarded it on to the bosses in New Haven where it went through a rigorous risk analysis process led by their general counsel.

From there, our exit plan came into focus. A key facet of it, according to the leadership at Yale, was this: that our charter flight would take us back to Iceland, and from there we'd fly out in waves back to the states over the course of two days. It was imperative to Yale that their students be back in studio on Monday—though I had no plans to hold the studio that day (I would later learn that they knew this, and wanted students back so they could be cornered in class and interrogated about the trip). In fact, the school's administrators went so far as to include a note in our final departure instructions that, if for any reason our flight was near its weight limit (heavy bags, too many people, etc.) that Zane and I would have to remain in Greenland and they weren't sure when, or even if, Yale would help us return to the states (one date they floated was around Halloween, though they noted we may be responsible for buying our own tickets in that case).

When that departure email hits all of our inboxes, our students become apoplectic—over the school's handling of the situation and for threatening to maroon Zane and I in the Arctic during winter indefinitely. At this point, I just want to get home. So I ask students to gather up all of their non-essential items and walk to the local post office with me—we are mailing their hair straighteners, books, and other items home; I am not waiting around until Halloween to get back to the states. As we package up our belongings, the postal staff cackle in disbelief—we have nearly 8 full boxes of materials, at a cost of almost $2,000 to ship. Most of our items never make it back.

Later that day, we walk to the airport together and board our tiny charter flight home. Takeoff is a blur and the rest of the flight is a dream—I wake after touchdown in Reykjavik and begin the process of shepherding students to and from Keflavik International for their return flights. When I finally board my flight home, I realize that I've been booked on a flight to New York (not Philadelphia, where I live) with an arrival time so late that I won't be able to catch a train home. As I start searching for hotels on the flight home, my partner offers to drive up and get me—a godsend, even though we don't get back to Philadelphia until the wee hours of the next morning.

I'm too exhausted to be enraged—though that would change over the next few days. I won't detail much of what happens next, except to say that there is no circumstance on Earth that could compel me to set foot in New Haven ever again.

Coda

I am preparing to return to Greenland in the fall of 2024 (and likely several other times after). I'll lead another studio (this time back home with landscape architects at Penn) and the rest of my work will likely fit into one of three tracks: (1) a set of short-term interventions aimed at building a robust radioactivity monitoring network. We are planning to test a set of Arduino-based Geiger counters there this fall, and to return in 2024–2025 with enough materials to build a full mesh network and enough time to organize a series of workshops that can train local Greenlanders to build and maintain these technologies. Our plan is to wrap our end of that work by 2026 and leave a self-sustaining system of monitoring technology and trained Greenlanders in place; (2) a set of medium-term interventions aimed at diversifying the food system in South Greenland. Like much of the Arctic, Greenlanders rely on root vegetables, lamb, musk ox, the occasional seal or whale, and commercial fishing for their dietary needs. But most homes are set on large plots of land that are ideal for small and community-scale greenhouse operations. We're hoping to begin piloting that bit of work in 2025; and (3) a set of long-term interventions aimed at harnessing Greenland's ample glacial rock flour (GRF) resources. GRF is just what it sounds like—a fine powder of rock and soil spewing out of the bottom of glacial tongues (the result of millions of years of crushing motion). It also happens to be the most nutrient-dense soil medium in the world, one that some experts believe could completely supplant all synthetic fertilizer. That vast amount of potential wealth may be the key to securing Greenland's independence.

BIBLIOGRAPHY

Bertelsen, Rasmus Gjedsø, Jens Christian Justinussen, and Coco Smits. "Energy as a Developmental Strategy: Creating Knowledge-Based Energy Sectors in Iceland, the Faroe Islands, and Greenland." In Exner-Pirot, Heather, and Heather Nicol (eds.), *Handbook of the Politics of the Arctic*, 3–25. London: Elgar, 2015.

Boetzkes, Amanda, and Jeff Diamanti. "At the Moraine." *e-flux: Accumulations*, September 2020. https://www.e-flux.com/architecture/accumulation/341644/at-the-moraine/.

Boetzkes, Amanda, and Jeff Diamanti. "Geofetishism and the Tender Violence of Rare Earths." *SubStance* 163, no. 3 (2021): 9–30.

Cambou, Dorothee, and Greg Poelzer. *Enhancing Energy Justice in the Arctic: An Appraisal of the Participation of Arctic Indigenous Peoples in the Transition to Renewable Energy*. London: Routledge, 2021.

Carruth, Susan. "Developing Renewable Energy in Discontiguous Greenland: An Infrastructural Urbanism of 'Material Practices.'" *Journal of Landscape Architecture* 11, no. 1 (2016): 66–79.

Carvajal Gamboa, Jose Ricardo. "Rare Earth Mining and Sovereign Statehood in Greenland: 'Sustainable Engagement' with Extractivism in Kvanefjeld." Thesis. University of British Columbia, 2022.

Cooper, Zane Griffin Talley. "Bit/Coin, Rare/Earth: Data, Energy, and Extraction Across the Arctic." PhD diss. University of Pennsylvania, 2024.

Dahl-Jensen, D., et al. "Past Temperatures Directly from the Greenland Ice Sheet." *Science* 282, no. 5387 (1998): 268–271.

De Jonghe, Bert. *Inventing Greenland: Designing an Arctic Nation*. New York: Actar, 2022.

Dzik, Anthony. "Interplays of Site and Situation along Tunalliarfik Fjord, South Greenland." *Journal of Settlements & Spatial Planning* 5, no. 2 (2014): 67–89.

Frederiksen, Mads Qvist. "If We Want an Energy Transition, We Must Have More Mining." Arctic Economic Council. May 21, 2025. https://arcticeconomiccouncil.com/news/if-we-want-an-energy-transition-we-must-have-more-mining/.

Gallucci, Maria. "Magma and Hot Rocks: Iceland Seeks the Future of Geothermal Energy." *Canary Media*. July 22, 2024. https://www.canarymedia.com/articles/geothermal/magma-and-hot-rocks-iceland-seeks-the-future-of-geothermal-energy.

Garbis, Zoe, et al. "Governing the Green Economy in the Arctic." *Nature Climate Change* 176 (2023): Article 33.

Goelzer, Heiko, et al. "Recent Progress in Greenlandic Ice Sheet Modelling." *Glaciology and Climate Change* 3 (2017): 291–302.

Greenfield, Patrick, and Phoebe Watson. "Fearing Toxic Waste, Greenland Ended Uranium Mining. Now, They Could Be Forced to Restart—or Pay $11 Billion." *The Guardian*. March 5, 2025. https://www.theguardian.com/environment/2025/mar/05/greenland-mining-energy-transition-minerals-environmental-laws-uranium-rare-earth-toxic-waste-investor-state-dispute-settlement-isds-aoe.

Grenoble, Lenore, Hilary McMahan, and Alliaq Kleist Petrussen. "An Ontology of Landscape and Seascape in Greenland: An Encoding of Land in Kalaallisut." *International Journal of American Linguistics* 85, no. 1 (2019): 1–43.

Gronholt-Pederson, Jacob. "Uranium Ban Repeal in Greenland Could Revive Massive Rare Earth Project, License Holder Says." *Reuters*. February 11, 2025. https://www.reuters.com/markets/commodities/uranium-ban-repeal-greenland-could-revive-massive-rare-earth-project-licence-2025-02-11/.

Heikkinen, Hannu, Lill Rastad Bjorst, and Albina Pashkevich. "Challenging Tourism Landscapes of Southwest Greenland: Identifying Social and Cultural Capital for Sustainable Tourism Development." *Arctic Anthropology* 57, no. 2 (2020): 212–228.

Huybrechts, Phillip, and Jan de Wolde. "The Dynamic Response of the Greenland and Antarctic Ice Sheets to Multiple-Century Climactic Warming." *Journal of Climate* 12, no. 8 (1999): 2169–2188.

Krebs, Damien. "Greenland's Unique Kvanefjeld Multi-Element Rare Earth Project." *Technology Review* 61, no. 2 (2000): 154–164.

Krogh, Cecile, and Laust Høgedahl. "Mapping the Greenlandic Labour Supply: Evidence from Greenland's First Labour Force Survey." In *Greenland's Economy and Labour Markets*. London: Routledge, 2021.

Nielsen, Sven Poul, Per Roos, and Kasper Grann Andersson. "Predicted Radiation Exposure from Mining at Kvanefjeld: Introduction to Radiation, Review of Baseline Information, and Predicted Radiation Exposures from Kvanefjeld Mining, Mineral Processing, and Refining." DTU-Nutech-R. Technical University of Denmark (DTU). 2015.

Nuttall, Mark. "Zero-Tolerance, Uranium, and Greenland's Mining Future." *The Polar Journal* 3, no. 2 (2013): 368–383.

Priebe, Janina. "Glacial Energy Futures? The History of Unbuilt Hydropower in Greenland from the 1950s to the 1970s." *Water History* 16 (2024): 271–290.

Sorensen, Henning, et al. *The Uranium Deposit at Kvanefjeld, the Ilimaussaq Intrusion, South Greenland*. Copenhagen: The Geological Survey of Greenland. 1974.

IMAGE CITATIONS + CREDITS

192 Map sources: Lars Lund Sørensen, Per Kalvig & Diogo Rosa, GEUS. "The Rare Earth Element Potential in Greenland." Geology and Ore, Exploration and Mining in Greenland. Ministry of Mineral Resources (MMR), Government of Greenland, 2018; MacGregor, J. A., et al. 2016. "A synthesis of the basal thermal state of the Greenland Ice Sheet," *J. Geophys. Res. Earth Surf.*, 121, 1328–1350. doi:10.1002/2015JF003803.

194 Photo: Billy Fleming.

198 Photo: Billy Fleming.

200 Photo: Billy Fleming.

Map sources: Nukissiorfiit and Asiaq. Hydro Power Potentials in Greenland. Dataset including multiple data sources indicating hydro power generation potentials throughout Greenland. Metadata compliant with ISO 19139 (2007). Last updated June 4, 2021. Dataset language: Danish. Accessed June 8, 2025. https://www.arcgis.com/home/item.html?id=32dd665060774772b20567bda7d15993.

204 Photo: Billy Fleming.

206 Photo: Billy Fleming.

210 Photo: Billy Fleming.

214 Photo: Billy Fleming.

Map sources: G. J. Orris, Y. Seo, D. A. Briggs, and M. D. Cooker., 2018. *Global Rare Earth Element Occurrence Database*: U.S. Geological Survey data release. https://doi.org/10.5066/F7DR2TN4; "Arctic Sea Routes." Accessed May 17, 2025. https://arctic-nga.opendata.arcgis.com/datasets/nga::arctic-sea-routes/about; F. Fetterer, K. Knowles, W. N. Meier, M. Savoie, and A. K. Windnagel. 2017, updated daily. Sea Ice Index, Version 3. Boulder, Colorado USA. NSIDC: National Snow and Ice Data Center. doi: https://doi.org/10.7265/N5K072F8; "Arctic Indigenous Languages and Revitalization: an Online Educational Resource." 2023. https://doi.org/10.18710/Y21DH0; DataverseNO, V2. "Giant, Supergiant & Megagiant Oil and Gas Fields of the World." Accessed May 17, 2025. https://www.arcgis.com/home/item.html?id=06e2ad891f1648259fb9dc7e6033f9a5; Esri; TomTom; Garmin; FAO; NOAA; USGS; OpenStreetMap contributors, and GIS User Community.

217 Photo: Billy Fleming.

218 Photo: Billy Fleming.

LITHIUM

White Gold and Black Geographies of Resistance in Brazil

by Fitsum Areguy

Lithium Mine Capacity (Mt/Y)
270,000
40,000
Quilombola Territory
Indigenous Territory
Feldspar Mining Concessions
Spodumene Mining Concessions
Authorized Lithium Research
Lithium Extraction Request
Lithium Availability
Lithium Research Request
Highway
State Boundary
Water Body
Salinas
Salinas
USGS
Sigma Lithium's Grota do Cirilo
Mina Da Cachoeira
Bandeira
Araçuaí
Padre Paraiso
Neves
Capelinha
Itamarandiba
Teófilo Otoni
Boca Rica

The overnight bus climbs into the mountains as dawn breaks over Minas Gerais—literally "General Mines"—Brazil's historic mining heartland.[1] Through dust-streaked windows, dawn breaks as a slow bleed of light over terraced hillsides. Each contour bore scars—testament to centuries of extraction that built colonial Brazil's wealth with gold and slave labor, filling Portuguese coffers and bankrolling European empires.[2] In Araçuaí, the bus station buzzes with morning energy. Araçuaí sits in the heart of the Jequitinhonha Valley, a vast region in northern Minas Gerais stretching toward the borders with Bahia and Espírito Santo. Though only 400 km from Belo Horizonte, the state capital, the valley remains largely isolated from major economic centers—a marginalization that has persisted since colonial times. Vendors hawk açaí and tapioca while passengers shuffle toward connecting routes deeper into the Jequitinhonha Valley.

I step into the humid air, my phone immediately losing signal. The region, rich in the minerals needed for our hyperconnected world, remains largely disconnected itself. I wasn't in the town for long; I left searching for a mining village said to lie in the shadow of Sigma Lithium's Grota do Cirilo project. Rounding a bend, I encounter a checkpoint—a chain stretched between posts, with *proibido* (no entry) signs. Beyond it, the dirt road becomes private mining property. A lone security guard steps out of a tin-roofed shack. In a move that, in retrospect, revealed my own naivety about the region's spatial politics, I try to explain that I

1 The name "Minas Gerais" literally translates to "General Mines," reflecting the region's historical importance in Brazil's colonial mining economy.
2 Costa, Marco Aurélio da, and Francisco Javier Rios. "The Gold Mining Industry in Brazil: A Historical Overview." *Ore Geology Reviews* 148 (September 1, 2022): 105005. https://doi.org/10.1016/j.oregeorev.2022.105005.

am a researcher hoping to visit the nearby quilombola community.[3] His face stays stony. "Não pode entrar sem autorização," he states—no entry without company authorization, a phrase whose meaning I grasped with considerable delay. My attempts to negotiate further are unavailing; he merely shakes his head and points back the way I came.

Frustration washes over me... My assumption that access to the community ran through the company's checkpoint was, as an organizer later confirmed, a fundamental misreading of the local geography of power. True access depended not on corporate gatekeepers, but on relationships of trust I had not yet had time to build. This was underscored by the reality that some quilombos are urban. For example, due to drought and work demands, most of the quilombolas from Quilombo do Baú now live in a neighbourhood called Pipoca—itself considered a quilombo—just two blocks from the main square where I had been. On the walk back to town along a dirt road skirting the Jequitinhonha River, I reflected on this barrier. By the water's edge, women washed clothes in brown current, pounding garments against smooth river rocks—waters that now risk drying up, diverted upstream for new lithium mines.

This contemporary struggle over land and water is deeply rooted in the valley's history, a history inextricably linked to the Quilombola communities who have long inhabited these marginalized landscapes. The Jequitinhonha Valley hosts one of the largest concentrations of Quilombola communities in Brazil—settlements originally forged by Africans who escaped enslavement.[4] Their presence here is not accidental; the states that built their wealth on the brutal extraction of gold and the forced labour of enslaved Africans—such as Minas Gerais and Bahia—are today the states with Brazil's largest Black and *parda* (mixed-race)[5] populations.[6] These communities often formed in the marginal landscapes surrounding former mining zones, areas deemed less valuable once accessible gold was exhausted.[7]

3 The term *quilombo* refers to hinterland settlements in Brazil originally founded by people of African origin who escaped enslavement. Historically, these communities were sites of resistance against the colonial slave system. Today, the inhabitants, known as Quilombolas, are recognized by the Brazilian constitution as distinct ethno-racial groups with rights to their ancestral lands, representing an ongoing struggle for cultural preservation and territorial autonomy. The use of "Quilombola" throughout this chapter refers to the contemporary descendants of these communities.

4 Instituto Brasileiro de Geografia e Estatística (IBGE), "2022 Census: IBGE Maps for the First Time Officially Delimited Quilombola Territories," August 24, 2023, https://agenciadenoticias.ibge.gov.br/en/agencia-news/2184-news-agency/news/37785-2022-census-ibge-maps-for-the-first-time-officially-delimited-quilombola-territories. The shapefile for these territories, however, was inaccessible as of June 2025.

5 The term *parda* is one of five official ethno-racial categories used by the Brazilian Institute of Geography and Statistics (IBGE). It is a broad and complex classification encompassing multiracial individuals, typically of mixed European, African, and/or Indigenous ancestry. While its meaning is subject to ongoing social and political debate, its place within the census distinguishes it from the category of *preto* (Black). However, as noted, social movements often group *pardo* and *preto* individuals under the political umbrella of *negro* (Black).

6 IBGE, "2022 Census: Brazil Has 203.1 Million People, 10 Million More Than in 2010," June 28, 2023, https://agenciadenoticias.ibge.gov.br/en/agencia-news/2184-news-agency/news/37225-2022-census-brazil-has-203-1-million-people-10-million-more-than-in-2010; Marco Aurélio da Costa and Francisco Javier Rios, "The Gold Mining Industry in Brazil: A Historical Overview," *Ore Geology Reviews* 148 (2022): 105005, https://doi.org/10.1016/j.oregeorev.2022.105005. This history is a clear instance of racial capitalism, a system where racial hierarchy is instrumental to resource exploitation and capital accumulation. The wealth of Minas Gerais was built on a foundation of gold extracted by enslaved Africans, establishing a logic that continues to devalue Black life and land in the pursuit of mineral wealth.

7 Costa and Rios, "The Gold Mining Industry in Brazil."

Mining History in Minas Gerais

This historical formation can be understood as a form of "fugitive planning" and undercommons.[8] Quilombola settlements became maroon socialities that operated in, but not of, the dominant extractive logic, creating life and culture in the spaces the colonial state disregarded. In these interstices, a deep legacy of cultural creation and self-sufficiency was fostered out of necessity, standing in stark contrast to the region's material poverty.[9] Today, this historical geography directly informs contemporary conflicts over land and resources. Data visualizations reveal a significant overlap between current mining concessions—both industrial and artisanal (*garimpos*)—and the territories of Quilombola and Indigenous peoples.[10] This spatial collision places the pursuit of "critical minerals" like lithium at the centre of ongoing struggles for land rights and environmental justice, disproportionately affecting the descendants of those who were originally dispossessed and exploited by prior extractive booms.[11] This physical encroachment is accompanied by a symbolic one, exemplified by the state-sponsored effort to rebrand the area as "Lithium Valley"—an act local communities identify as an attempt to erase their collective memory.[12]

Now, three centuries later, lithium companies arrive, their prospectuses echoing the same promises of development and progress that gold prospectors once peddled. The new "white gold" rush follows pathways carved by the old, with the potential for similar patterns of displacement and environmental degradation to unfurl across communities that have never fully recovered from the first boom.

I often seek out Black people and Black spaces when I need to ground myself in new and unfamiliar places. The sun hung low as I approached a small barbershop in a plaza near Araçuaí's main square. Two men sat on the curb out front, the tips of their cigarettes glowing as they

8 Stefano Harney and Fred Moten, *The Undercommons: Fugitive Planning & Black Study* (Wivenhoe, UK: Minor Compositions, 2013), 74.
9 Márcio Simeone Henriques, "Sobre o Vale do Jequitinhonha." *Polo Jequitinhonha*, Universidade Federal de Minas Gerais, accessed June 1, 2025, https://www.ufmg.br/polojequitinhonha/o-vale/sobre-o-vale-do-jequitinhonha/.
10 Global Forest Watch, "Brazil Mining Concessions," accessed June 10, 2025, https://data.globalforestwatch.org/search?collection=Dataset&q=mining; U.S. Geological Survey (USGS), "Mineral Resource Data System," accessed June 10, 2025, https://mrdata.usgs.gov/mrds/; Agência Nacional de Águas e Saneamento Básico (ANA), "Garimpos," accessed June 10, 2025, https://metadados.snirh.gov.br/geonetwork/srv/api/records/ec62785d-6559-4f26-9df4-46f3a504bfcd; IBGE, "2022 Census."
11 This claim is a synthesis of the spatial relationship evident from the data in the previously cited sources, reflecting the core issues of land rights and environmental justice that arise from the geographic overlaps.
12 Edson Krenak, "The Violent Cartography of Lithium in Brazil: Indigenous and Traditional Communities Struggle with the Giant of Transition Minerals in Brazil," *Cultural Survival*, June 26, 2023, https://www.culturalsurvival.org/news/violent-cartography-lithium-brazil-indigenous-and-traditional-communities-struggle-giant.

talked. Inside, the rhythmic buzz of clippers and a murmur of conversation mixed with the sound of *Sertanejo* music drifting from a radio. The owner stood behind a young customer, a cell phone tucked into his shoulder as he talked and cut. He acknowledged me with a belated nod. Just like back home, I moved to the couch on the opposite side of the shop to wait my turn. A young woman in jean shorts bounced a bored young boy on her knee, while two teens sporting fresh fades and beat-up Jordans sat close together next to me, sharing a phone.

The owner unclipped the gown on the young man in the chair, who stood up to lean into the mirror. Looking pleased, he rubbed his hands and walked over to pay. The other two teens jumped up to greet him, slapping his neck in approval. It was finally my turn. I settled into the single worn chair, swiveling to scan the room. An old fan rattled in a corner while a television flickered above, the news on mute. The scent of talc and something vaguely antiseptic hung in the air. As the owner worked on my hair with methodical precision, the conversation flowed around me. I asked what they thought of the mining in the valley.

"The mining companies promise jobs," one elderly man observed, his voice raspy, "but the good jobs, they go to people from outside. We get dust and noise." Outside, a heavy truck, laden with unidentifiable equipment, rumbled past. The skepticism in his voice reflected what scholar Tiffany Lethabo King terms "friction".[13] For King, friction is the generative force that emerges when Black life confronts and disrupts the logics of colonial control. It is the practice of reading corporate promises through the lens of historical experience and, as the owner's comment about his grandmother illustrates, affirming the very knowledge systems that extraction threatens to erase. As if to underscore this point, he mentioned his grandmother as he finished up my haircut, boasting about how she knew every plant in the valley.

There is a pervasive belief that the valley is a savanna with dry vegetation, and has little to no biodiversity to be preserved.[14] This misconception belies the ecological richness of the Cerrado biome, which transitions into the Atlantic Forest in parts of this region, and the deep, often unacknowledged, dependence of local communities, particularly quilombolas, on this very biodiversity. This profound biodiversity, stewarded by Quilombola communities, can be conceptualized as the "unregulated wildness" of the undercommons—an alternative, life-affirming order that stands in stark opposition to the monocultural logic of

13 Tiffany Lethabo King, *Black Shoals: Offshore Formations of Black and Native Studies* (Durham, Duke University Press, 2019).

14 Laura Rojas, conversation with author, June 1, 2025, Toronto.

extraction.[15] Across Brazil, studies have documented an incredible pharmacopeia: one review identified 297 plant species from 80 different families utilized by Quilombola communities for a wide array of therapeutic purposes.[16] The most frequently cited plant families include Asteraceae (the daisy family), Lamiaceae (the mint family), and Euphorbiaceae (the spurge family), with species like lemon balm (often used for calming teas), anamu (with various traditional uses) and bitter melon being commonly cited. The preparation methods are as diverse as the plants themselves. Decoction, alcohol maceration, water infusion—each method draws on understanding that leaves are potent sites of active compounds like alkaloids and tannins.[17] This intricate knowledge system, where specific plants are used for ailments ranging from digestive issues and respiratory complaints to nervous system disorders and general inflammation, forms a cornerstone of community health and well-being.[18]

This ecological interdependence is now under direct threat from the burgeoning lithium industry. Canadian company Sigma Lithium holds permits to extract 3.8 million liters of water per day from the Jequitinhonha River and its tributaries for its Grota do Cirilo project—enough to supply 34,000 families. Many rural communities rely on communal water tanks holding 6,000 litres, meant to last for eight months during the dry season. The Araçuaí River already runs discolored from mining activities upstream, a visible wound on the landscape.[19] The massive diversion of water for lithium processing and the potential contamination of remaining sources have cascading consequences. They decimate the very habitats where these crucial medicinal plants thrive, undermine local agriculture, and jeopardize the primary water source for drinking and daily life. The conflict over water, therefore, is not merely about resource allocation; it is an assault on an entire cultural-ecological system. When the rivers shrink and the plants wither, it is not just biodiversity that is lost, but also the health autonomy of communities, their ancestral knowledge, and a way of life intricately woven with the land.
The plant knowledge of the Vale do Jequitinhonha—generations of sustainable relationships with the Cerrado—faces erasure. This is not only a tragedy for the Quilombola communities who are its custodians but also a loss for broader ecological understanding and, as some research suggests, a potential loss of knowledge that could contribute to future pharmacological discoveries.[20] Thus, the "green" credentials of the electric car are underwritten by the destruction of genuinely sustainable, locally rooted knowledge systems—a direct contradiction of the ecological principles the transition purports to uphold.

-----

15 Harney and Moten, *The Undercommons*, 28.
16 Ramos, Letícia Francine Silva, et al, "Ethnobotanical Surveys of Plants Used by Quilombola Communities in Brazil: A Scoping Review." *Life* 14, no. 10 (September 24, 2024): 1215. https://doi.org/10.3390/life14101215.
17 Ramos et al, "Ethnobotanical Surveys," 39.
18 Ramos et al, "Ethnobotanical Surveys," 38.
19 Guatelli, Caio. "Lithium Mining Brings Sickness to Jequitinhonha Valley Communities," *Mongabay Environmental News*, November 18, 2024, sec. Environmental news, https://news.mongabay.com/2024/11/lithium-mining-brings-sickness-to-jequitinhonha-valley-communities/.
20 Ramos et al. "Ethnobotanical Surveys," 39.

A narrative of progress and prosperity is often deployed to justify extractive projects.[21] In May 2023, Minas Gerais' Governor Romeu Zema traveled to New York to launch the "Lithium Valley" project on the NASDAQ stock exchange.[22] This initiative, promoted as the "economic and social salvation for the Jequitinhonha Valley,"[23] aimed to rebrand the region as a hub for lithium extraction, attracting international investment. Yet, for many local communities, this rebranding felt less like salvation and more like an act of erasure. The very name "Jequitinhonha," meaning "long river of fish" in the dialect of the maxacalis, evokes a rich ecosystem and a deep ancestral connection to the land and water.[24] "Lithium Valley" reduces a complex socio-ecological tapestry to a single commodity, an identity legible and attractive to global capital, but one that silences centuries of local history and meaning.[25] This renaming prepares the ground for extraction, erasing local meanings and connections to make way for corporate claims.[26]

At the forefront of this transformation is Sigma Lithium Resources, a Canadian company that holds the largest mining presence in the region.[27] Sigma promotes its operations as "sustainable" and its product as "green lithium," aligning itself with the environmentally conscious branding of the global energy transition. The company's Grota do Cirilo complex is projected to produce 440,000 tons of high-purity lithium concentrate per year.[28] However, the reality on the ground, as reported by local communities and researchers, starkly contrasts with this sanitized image. Sigma Lithium received its operating license in March 2023 amidst serious allegations of violating the rights of local populations by failing to obtain their Free, Prior, and Informed Consent (FPIC)—the right of Indigenous and traditional communities to approve or reject projects affecting their territories—a right enshrined in international conventions to which Brazil is a signatory, such as ILO Convention 169, the International Labour Organization's binding treaty on Indigenous and tribal peoples' rights, which Brazil ratified in 2002.[29 30] Disturbingly, reports indicate that Sigma commenced mapping and earthmoving operations years before securing the necessary environmental

21 Kathryn Yusoff warns that contemporary eco-optimism often disguises colonial power relations.
22 Sigma Lithium Corporation, "Sigma Lithium and Brazilian Government Officials Ring Nasdaq Opening Bell to Celebrate the Launch of Lithium Valley Brazil Initiative," Nasdaq, May 11, 2023, https://www.nasdaq.com/press-release/sigma-lithium-and-brazilian-government-officials-ring-nasdaq-opening-bell-to.
23 Krenak, "The Violent Cartography of Lithium in Brazil."
24 Henriques, "Sobre o Vale do Jequitinhonha."
25 This process is a hallmark of "green extractivism," or "green colonialism," where the urgent, global demand for transition minerals is used to justify and perpetuate colonial patterns of resource appropriation. It outsources the profound environmental and social costs of the energy transition to racialized communities in the Global South, all under a deceptive "green" narrative.
26 Krenak, "The Violent Cartography of Lithium in Brazil."
27 Sigma Lithium Corporation, "Sigma Lithium Starts Construction of Production Plant Foundation and Infrastructure and Orders Long Lead Items, on Schedule for Production in 2022," Sigma Lithium Corp (blog), December 6, 2021, https://sigmalithiumcorp.com/sigma-lithium-starts-construction-of-production-plant-foundation-and-infrastructure-and-orders-long-lead-items-on-schedule-for-production-in-2022/.
28 Sigma Lithium Corporation, "Sigma Lithium Starts Construction Of Production Plant."
29 LIQUIT: Vozes Dos Territórios, "Researchers Report Serious Social and Environmental Harm in Minas Gerais Caused by the Activities of Sigma Lithium," *Cultural Survival,* April 30, 2025. https://www.culturalsurvival.org/news/researchers-report-serious-social-and-environmental-harm-minas-gerais-caused-activities-sigma.
30 These rights are under increasing threat. In March 2025, a draft bill advanced in Brazil's Supreme Court that, if passed, would severely undermine Indigenous land rights and the protective power of international conventions like ILO 169, potentially legitimizing land grabs for extractive projects. https://news.mongabay.com/2025/03/unprecedented-supreme-court-bill-threatens-indigenous-rights-in-brazil/.

licenses, suggesting a disregard for due process and community rights from the outset. More recently, in May 2025, the company was accused of harming the local water supply and instigating community conflicts.[31] In June 2025, the Business & Human Rights Resource Centre reported that Sigma Lithium had not responded to allegations of attempting to intimidate the Mining Observatory, an investigative journalism centre critical of its operations.[32]

The response from the affected communities has been one of growing alarm and organized resistance. In July 2022, a collective of traditional and Indigenous communities from the Jequitinhonha Valley issued a powerful letter of protest, a cry against the onslaught of destructive industries:

> *We denounce... and no longer tolerate... mining companies that: open prospecting holes and leave the craters of greed in our soil, imprison and kill our rivers, steal our plateaus and dry up all our fountains of life; who invade our territories and spray poisons, contaminating our Mother Earth and our people... That is why we cry out in one voice:...Sigma, out!*[33]

This resistance has continued to evolve. In early 2025, *Cultural Survival* reported on ongoing efforts by Quilombola communities to strengthen food sovereignty as a direct response to mining threats and on new alliances being forged between Pankararú and Pataxó peoples to protect their territories from extraction.[34]

This local struggle is situated within a broader context of Canadian mining activities across Latin America. Canadian stock exchanges, particularly the Toronto Stock Exchange (TSX), are global hubs for mining finance, listing 47% of the world's public mining companies.[35] According to the latest data for year-end 2023, Canadian mining assets abroad totaled $220.4 billion, with $113 billion concentrated in Latin America and the Caribbean.[36] Reports from organizations like MiningWatch Canada detail a persistent and widespread pattern of harm associated with Canadian mining operations overseas, including severe environmental degradation,

31 Business & Human Rights Resource Centre, "Brazil: Sigma Lithium Accused of Harming Water Supply and Driving Community Conflicts in and among Vale do Jequitinhonha," May 13, 2025, accessed June 1, 2025, https://www.business-humanrights.org/en/latest-news/brazil-sigma-lithium-allegedly-affected-the-water-supply-and-caused-conflicts-among-vale-do-jequitinhonha-communities/.

32 Business & Human Rights Resource Centre, "Sigma Lithium Did Not Respond to Allegations of an Attempt to Intimidate the Mining Observatory," June 3, 2025, accessed June 10, 2025. https://www.business-humanrights.org/en/latest-news/sigma-lithium-did-not-respond-to-allegations-of-an-attempt-to-intimidate-the-mining-observatory/.

33 Committee of Traditional Communities of Vale do Jequitinhonha, "Carta Coletiva das Comunidades Tradicionais do Vale do Jequitinhonha," July 2022, translated excerpt in Edson Krenak, "The Violent Cartography of Lithium in Brazil: Indigenous and Traditional Communities Struggle with the Giant of Transition Minerals," *Cultural Survival*, 2023.

34 Xiting Tong, "Resisting Lithium Mining and Strengthening Food Sovereignty: Quilombola Community in Girau, Brazil," *Cultural Survival*, January 15, 2025, https://www.culturalsurvival.org/news/resisting-lithium-mining-and-strengthening-food-sovereignty-quilombola-community-girau-brazil.

35 Barry, James J. *The Mineral Industry of Canada*. 2019 Minerals Yearbook–Canada [Advance Release]. (U.S. Department of the Interior, U.S. Geological Survey, April 2024): 5.1. https://pubs.usgs.gov/myb/vol3/2019/myb3-2019-canada.pdf.

36 Natural Resources Canada, "Canadian Mining Assets," Government of Canada, January 2025, https://natural-resources.canada.ca/minerals-mining/mining-data-statistics-analysis/minerals-mining-publications/canadian-mining-assets.

Grota do Cirilo Mine Expansion

human rights abuses, violations of Indigenous rights, and negative economic impacts at local and national levels.[37] The International Rights of Nature Tribunal has highlighted cases like the Belo Sun mining project in Brazil's Xingu territory, backed by Canadian foreign investments, as an example of these destructive dynamics.[38] Therefore, the rhetoric of "green lithium" can be understood not as a commitment to sustainability, but as the latest iteration of a well-documented colonial playbook used by Canadian extractive firms to legitimize dispossession in Latin America.

In lieu of access to the mining villages themselves, the remainder of what was a very brief research trip—only four days on the ground—shifted. My role on this trip was that of a witness, informed by my background as a journalist, focusing on the forms of resistance I could observe from public spaces. Two scenes from those final days stand out: the first an experience in a small church in Araçuaí, and the second on a bus ride in Belo Horizonte, Minas Gerais' metropolitan capital.

Saints and Spirits

One evening in Araçuaí, drawn by a faint candlelight from propped-open wooden doors, I slipped into a small church. The air inside was cool, smelling of old wood, beeswax, and something indefinably sacred. As my eyes adjusted to the dimness, I was surprised to see along the side altars the familiar iconography of Catholic saints was rendered with a profound difference: their skin was dark, their features unmistakably African. There stood São Benedito (St. Benedict of Palermo), the 16th-century Black Franciscan friar, his carved form radiating a quiet dignity. Nearby, a regal figure with a crown was São Elesbão, the 6th-century Ethiopian king who became a saint. An icon of Santa Ifigênia, traditionally believed to be a Nubian princess, graced another altar. Even the cherubic figures painted in a faded mural above the main altar bore brown skin.

During the centuries of enslavement in Brazil, Catholicism, the religion of the colonizer, paradoxically became one of the few sanctioned spheres where Africans and their descendants could gather. They did not simply adopt it; they infused it with their own spirits, their own cosmologies. Black lay brotherhoods (*confrarias*) formed, dedicated to saints whose stories of humility, perseverance,

37 Coumans, Catherine. "Canada's Mining Dominance and Failure to Protect Environmental and Human Rights Abroad: Brief Accompanying Testimony Provided by Catherine Coumans of MiningWatch Canada on February 6, before The House of Commons' Standing Committee on International Trade (CIIT)." MiningWatch Canada, February 14, 2023. https://www.miningwatch.ca/publications/2023/2/23/canada-s-mining-dominance-and-failure-protect-environmental-and-human-rights.

38 *'Rights of Nature Tribunal Finds Canadian Mining Companies Guilty for the Violation of Rights of Nature in 14 Cases in South America and Serbia' Recommendations of the 6th International Rights of Nature Tribunal,* Toronto, Canada: 6th International Rights of Nature Tribunal, February 8, 2025, https://ecojurisprudence.org/wp-content/uploads/2025/03/Rights-of-Nature-Tribunal-Final-Recommendations.pdf; University of Toronto Faculty of Law I International Human Rights Program. "Witnesses Confront Mining Industry at the 6th International Rights of Nature Tribunal: 'The Impacts of Mining and the Post-Extractivism Era,'" February 28, 2025, https://ihrp.law.utoronto.ca/news/title-ihrp-clinic-student-highlights-first-nations-perspectives-mining-development-ontario-2025.

or royal lineage resonated with their own experiences and aspirations.[39] Figures like São Benedito and Nossa Senhora do Rosário (Our Lady of the Rosary) became focal points of Black devotion. As historian Ana Lucia Araujo notes, "Black saints and African deities are closely related in Brazil… Black Catholicism was not dissociated from West African religions such as Vodun and Orisha,"[40] mapping *orixás*—the sacred spirits of Yoruba and other West African traditions—onto Catholic saints. A feast day for a saint could thus become a clandestine occasion to honor an ancestral deity, a way to keep African spiritual memory alive within the very structures of colonial power. This spiritual alchemy is evocative of what Fred Moten conceptualizes as "anarrangement:"[41] the ongoing, disruptive vitality that pressures colonial impositions, in this case transforming the colonizer's religion into a vessel for cultural memory.[42]

After murmuring Rosaries, an elderly woman, her head wrapped in a blue scarf, approached the statue of São Benedito. She gently touched its foot, her lips moving in silent prayer, then turned to me with a gap-toothed smile. On a bulletin board by the entrance, photographs displayed scenes from past Congado festivals: men and women in vibrant costumes of royal blue and white, their bodies alive with rhythm, dancing with drums through the streets. The Congado, a centuries-old Afro-Brazilian ritual, blends Catholic devotion with African musical traditions and royal symbolism, crowning a King and Queen of the Congo in an echo of lost African sovereignties. Born in the crucible of slavery, it was, and remains, a powerful affirmation of dignity, community, and cultural continuity—an act of profound resilience performed across generations.

Stepping out into the warm embrace of the night, the Southern Cross hanging low and brilliant in the ink-black sky, I reflected on the spiritual geography of this valley. The saint devotions, the drum-driven processions, the offerings made to river orixás for rain and bounty—these were all quilombo cosmologies. It's a worldview shaped by those who fled bondage to create free communities in the hinterlands, carrying with them ancestral memories and forging new ways of being in relationship with the Brazilian land.[43] The church I had just left stood on colonized ground, its stones perhaps laid by enslaved hands. Yet, within its candlelit interior, an otherwise world[44] flickered, a spiritual sanctuary nurtured by Black faith, a quiet, persistent resistance against erasure.

39 Sweet, James H. *Recreating Africa: Culture, Kinship, and Religion in the African-Portuguese World, 1441–1770.* (Chapel Hill: University of North Carolina Press, 2003) 206–208 http://archive.org/details/recreatingafrica0000swee.
40 Araujo, Ana Lucia. "Brazil Is an African Country." *Africa Is a Country*, December 5, 2024, https://africasacountry.com/2024/12/brazil-is-an-african-country.
41 Fred Moten, "Resistance of the Object: Aunt Hester's Scream," in *In the Break: The Aesthetics of the Black Radical Tradition* (Minneapolis: University of Minnesota Press, 2003), 1–24, http://www.jstor.org/stable/10.5749/j.cttts6jk.4.
42 Sweet, *Recreating Africa,* 207, 217–218.
43 Shore, "Geographies of Resistance," 59, 62, 69, 71–72.
44 *Otherwise World* is a term used by Tiffany Lethabo King, Jenell Navarro, and Andrea Smith in their book of the same name, describing real-world practices and solidarities that imagine and enact joyful, dynamic futures beyond the violences of settler colonialism and anti-Blackness.

Rhythm and Refusal

The overnight bus to Belo Horizonte pulled away from Araçuaí as the last light faded behind the hills. Inside, fluorescent strips cast harsh angles across tired faces. The air conditioning wheezed against the humid night, mixing diesel fumes with the scent of someone's packed dinner—rice and beans wrapped in foil. Passengers settled into worn vinyl seats, adjusting thin blankets and pillows that had seen countless journeys. Arriving in Belo Horizonte, the city's cadence jarred against the valley's slower pace.

Later, I found myself on a crowded municipal bus winding through the city's periphery toward the centre. The bus carried predominantly Black youth, laughter and conversations animated, heading to claim space in areas that decades of inequality and racialized policing had sought to restrict.[45] As our bus approached a major intersection, the vehicle slowed. Ahead, the flashing red and blue lights of a police checkpoint pulsed in the twilight. A familiar tension rippled through the aisle; conversations died down, laughter ceased. My own heart began to thud, witnessing the residue of long battles for the right to move freely within one's own city.[46] Then, from the back of the bus, a sharp clap rang out, followed by another, establishing a beat. A couple of teenagers, grinning ear to ear, launched into an impromptu funk chant, drumming on plastic seatbacks. The bus, moments before hushed with apprehension, erupted into a defiant, joyous carnival on wheels. The officers, clearly bewildered, exchanged glances. They had perhaps anticipated sullen compliance or nervous silence, but not this—not a busload of young people transforming a moment of potential intimidation into exuberant assertion of presence. Eventually, one officer waved the driver on in exasperation and stepped off the bus. A cheer erupted as we lurched forward, propelled by rhythm and laughter, toward the city centre.

This moment exemplified what Tiffany Lethabo King describes as "Black shoals" the congregation of Black life that, in the face of control, generates unexpected formations and possibilities, exceeding the logic of domination itself.[47] In many ways, the moment of resistance on that bus mirrored the struggles against mining harms in the Jequitinhonha Valley. Black and Indigenous peoples articulating the same fundamental insistence: the right to exist, to flourish, to protect ancestral lands and waters from extractive violence.

Throughout Brazil's history, Black and Indigenous communities have often found their struggles linked. Maroon societies like Palmares, the great 17th-century quilombo, forged alliances with Indigenous neighbors.[48] Today, contemporary

45 Weber, Gabriel. "474 JACARÉ – COPACABANA: O auto da linha do inferno." *assossiassão* Índice Temporada 2: Viu (July 28, 2023), https://assossiassao.pt/474-jacare-copacabana-o-auto-da-linha-do-inferno.
46 Weber, "474 JACARÉ – COPACABANA." Weber writes about how, in Rio de Janeiro, Brazilian authorities subjected buses coming from poorer, predominantly Black northern neighborhoods, to "Operation Summer" blitzes. These operations involved police checkpoints, often at the mouths of tunnels leading to the affluent, where young Black men would be stopped, questioned, and required to show identification, proof of return fare, or even work permits—a blatantly racist protocol aimed at controlling their access to the city's prime leisure spaces. This practice was eventually curbed by court order.
47 Tiffany Lethabo King, *Black Shoals: Offshore Formations of Black and Native Studies* (Durham, Duke University Press, 2019).
48 Hertzman, Marc A., "The Indians of Palmares," in *After Palmares: Diaspora, Inheritance, and the Afterlives of*

coalitions of quilombola groups and Indigenous nations unite against destructive mining, logging, and agribusiness, sharing knowledge of the land and strategies of protest. These histories of co-resistance, though seldom centered in official narratives, persist in grassroots alliances and transnational networks. The fight against mining concessions, which since the late 1990s have directly correlated with the forced displacement of rural communities in Brazil, is one such arena where these solidarities are crucial.[49] Victories, though hard-won, do emerge, as seen in Panama in November 2023, where mass mobilizations against a Canadian copper mining project led to a Supreme Court ruling declaring the mining contract unconstitutional.[50]

Like the youth whose singing and dancing transformed a potentially violent police stop, these moments reminded me that organizing and resisting must be rooted in creative practices that exceed the logic of extraction itself, generating new possibilities from colliding forces.

Weeks after I left Minas Gerais, a community organizer shared some updates from the field.[51] Public hearings in local city chambers had become battlegrounds, with mining proponents and land defenders facing off in increasingly heated exchanges. The Minas Gerais State Prosecutor's Office had contested Sigma's operations based on Definitive Land Possession documents that showed the mining violated required distance buffers from quilombo territories. Yet simultaneously, legislators pushed through approvals to reduce protected areas, selling lands to Atlas Lithium despite federal prosecutors attempting to halt the process for procedural violations. These attacks intensified beyond legal maneuvering, with fake videos circulated on social media intended to delegitimize community leaders.

This fight, and that of so many others, is for life, memory, and land against this reductionist logic of extraction, struggles that intensify daily. Still, even as I write, communities continue to organize, through political protest but also by planting gardens and singing and teaching children the names of plants and spirits of rivers—acts of persistence in the face of escalating violence.

This fight, and that of so many others, connects quilombola and Afro-Indigenous communities across Brazil and beyond, building solidarity networks that span from the Amazon to the Atlantic Forest, from rural valleys to urban peripheries. What's at stake here is the very possibility of worlds where Black and Indigenous life can flourish, unburdened by extractive violence. Their struggle is both ancient and urgent. In their refusal to be erased, in their insistence on living otherwise, they embody the persistent possibility that another world remains within reach.

Zumbi (Durham, NC: Duke University Press, 2024): 166–168.) https://doi.org/10.1215/9781478059547-008.
49 Arboleda, Martin, Planetary Mine: Territories of Extraction under Late Capitalism (London, Verson Books, 2020): 142.
50 *PBS News*, "Panama's Supreme Court Declares 20-Year Contract for Canadian Copper Mine Unconstitutional," November 28, 2023, https://www.pbs.org/newshour/world/panamas-supreme-court-declares-20-year-contract-for-canadian-copper-mine-unconstitutional.
51 Laura Rojas, conversation with author, June 1, 2025, Toronto.

ACKNOWLEDGMENT

This chapter owes a debt of gratitude to Rachel Thevenard for her geospatial research and initial cartographic production.

BIBLIOGRAPHY

Araujo, Ana Lucia. "Brazil Is an African Country." *Africa Is a Country*. December 5, 2024. https://africasacountry.com/2024/12/brazil-is-an-african-country.

Arboleda, Martin. *Planetary Mine: Territories of Extraction under Late Capitalism*. London: Verso, 2020. https://www.versobooks.com/books/3965-planetary-mine.

Barry, James J. "The Mineral Industry of Canada." 2019 Minerals Yearbook–Canada [Advance Release]. U.S. Department of the Interior, U.S. Geological Survey. April 2024. https://pubs.usgs.gov/myb/vol3/2019/myb3-2019-canada.pdf.

Bledsoe, Adam. "The Negation and Reassertion of Black Geographies in Brazil." *ACME: An International Journal for Critical Geographies* 14, no. 1 (July 7, 2015): 324–343. https://doi.org/10.14288/acme.v14i1.1153.

Building and Bildung Und Blackness: Some Architectural Questions for Fela. NOMAS Lecture. Massachusetts Institute of Technology. 2022. https://www.youtube.com/watch?v=EW1aR3rHeZE.

Business & Human Rights Resource Centre. "Brazil: Sigma Lithium Accused of Harming Water Supply and Driving Community Conflicts in and Among, Vale do Jequitinhonha." May 13, 2025. Accessed June 1, 2025. https://www.business-humanrights.org/en/latest-news/brazil-sigma-lithium-allegedly-affected-the-water-supply-and-caused-conVale do Jequitinhonhaflicts-among-vale-do-jequitinhonha-communities/.

Business & Human Rights Resource Centre. "Sigma Lithium Did Not Respond to Allegations of an Attempt to Intimidate the Mining Observatory." June 3, 2025. Accessed June 10, 2025. https://www.business-humanrights.org/en/latest-news/sigma-lithium-did-not-respond-to-allegations-of-an-attempt-to-intimidate-the-mining-observatory/.

Costa, Marco Aurélio da, and Francisco Javier Rios. "The Gold Mining Industry in Brazil: A Historical Overview." *Ore Geology Reviews* 148 (September 1, 2022): 105005. https://doi.org/10.1016/j.oregeorev.2022.105005.

Coumans, Catherine. "Canada's Mining Dominance and Failure to Protect Environmental and Human Rights Abroad: Brief Accompanying Testimony Provided by Catherine Coumans of MiningWatch Canada on February 6, before The House of Commons' Standing Committee on International Trade (CIIT)." Canada. February 14, 2023. https://www.miningwatch.ca/publications/2023/2/23/canada-s-mining-dominance-and-failure-protect-environmental-and-human-rights.

Darimani, Abdulai. "Impacts of Activities of Canadian Mining Companies in Africa." Accra: Third World Network–African Secretariat. October 31, 2005. https://www.miningwatch.ca/sites/default/files/9africacasestudy.pdf.

Gerais, Assembleia Legislativa de Minas Gerais. "Povos do Jequitinhonha denunciam impactos por exploração do lítio–Assembleia Legislativa de Minas Gerais." Portal da Assembleia Legislativa de Minas Gerais. Accessed June 5, 2025. https://www.almg.gov.br/comunicacao/noticias/arquivos/Povos-do-Jequitinhonha-denunciam-impactos-por-exploracao-do-litio/.

Global Forest Watch. "Brazil Mining Concessions." Accessed June 10, 2025, https://data.globalforestwatch.org/search?collection=Dataset&q=mining.

Guatelli, Caio. "Lithium Mining Brings Sickness to Jequitinhonha Valley Communities." *Mongabay Environmental News*. November 18, 2024, sec. Environmental news. https://news.mongabay.com/2024/11/lithium-mining-brings-sickness-to-jequitinhonha-valley-communities/.

Harney, Stefano, and Fred Moten. *The Undercommons: Fugitive Planning & Black Study*. Wivenhoe, UK: Minor Compositions, 2013.

Henriques, Márcio Simeone. "Sobre o Vale do Jequitinhonha." Polo Jequitinhonha. Universidade Federal de Minas Gerais. Accessed June 1, 2025. https://www.ufmg.br/polojequitinhonha/o-vale/sobre-o-vale-do-jequitinhonha/.

Hertzman, Marc A. "The Indians of Palmares." In *After Palmares: Diaspora, Inheritance, and the Afterlives of Zumbi*, 168–188. Durham, NC: Duke University Press, 2024. https://doi.org/10.1215/9781478059547-008.

Instituto Brasileiro de Geografia e Estatística (IBGE). "2022 Census: Brazil has 203.1 Million People, 10 Million More than in 2010." June 28, 2023. https://agenciadenoticias.ibge.gov.br/en/agencia-news/2184-news-agency/news/37225-2022-census-brazil-has-203-1-million-people-10-million-more-than-in-2010.

Instituto Brasileiro de Geografia e Estatística (IBGE). "2022 Census: IBGE Maps for the First Time Officially Delimited Quilombola Territories." August 24, 2023. https://agenciadenoticias.ibge.gov.br/en/agencia-news/2184-news-agency/news/37785-2022-census-ibge-maps-for-the-first-time-officially-delimited-quilombola-territories.

Instituto Estadual do Patrimônio Histórico e Artístico de Minas Gerais (IEPHA). "Você Conhece a Rota Dos Quilombos No Vale Do Jequitinhonha?" IEPHA Minas Gerais. September 3, 2021. https://www.iepha.mg.gov.br/index.php/noticias-menu/606-voce-conhece-a-rota-dos-quilombos-no-vale-do-jequitinhonha.

King, Tiffany Lethabo. *Black Shoals: Offshore Formations of Black and Native Studies*. Durham, NC: Duke University Press, 2019.

King, Tiffany Lethabo, Jenell Navarro, and Andrea Smith. *Otherwise Worlds: Against Settler Colonialism and Anti-Blackness*. Durham, NC: Duke University Press, 2020. https://www.dukeupress.edu/otherwise-worlds.

Krenak, Edson. "The Violent Cartography of Lithium in Brazil: Indigenous and Traditional Communities Struggle with the Giant of Transition Minerals in Brazil." Cultural Survival. June 26, 2023. https://www.culturalsurvival.org/news/violent-cartography-lithium-brazil-indigenous-and-traditional-communities-struggle-giant.

Lima, Eduardo Campos. "In South America's 'Lithium Triangle,' Indigenous People Defend Sacred Sites." *National Catholic Reporter*. Accessed June 5, 2025. https://www.ncronline.org/earthbeat/justice/south-americas-lithium-triangle-indigenous-people-defend-sacred-sites.

LIQUIT: Vozes Dos Territórios. "Researchers Report Serious Social and Environmental Harm in Minas Gerais Caused by the Activities of Sigma Lithium." Cultural Survival. April 30, 2025. https://www.culturalsurvival.org/news/researchers-report-serious-social-and-environmental-harm-minas-gerais-caused-activities-sigma.

Moten, Fred. "Resistance of the Object: Aunt Hester's Scream." In *In the Break: The Aesthetics of the Black Radical Tradition*, 1–24. Minneapolis: University of Minnesota Press, 2003. http://www.jstor.org/stable/10.5749/j.cttts6jk.4.

Natural Resources Canada. "Canadian Mining Assets." Government of Canada. January 2025. https://natural-resources.canada.ca/minerals-mining/mining-data-statistics-analysis/minerals-mining-publications/canadian-mining-assets.

Passarini, Ígor. "The Greenwashing Behind Brazil's Lithium Boom." *Observatório da Mineração*. October 17, 2024. https://observatoriodamineracao.com.br/the-greenwashing-behind-brazils-lithium-boom/.

PBS News. "Panama's Supreme Court Declares 20-Year Contract for Canadian Copper Mine Unconstitutional." November 28, 2023. https://www.pbs.org/newshour/world/panamas-supreme-court-declares-20-year-contract-for-canadian-copper-mine-unconstitutional.

Ramos, Letícia Francine Silva, Ananda Gomes de Sousa, Rebeca de Siqueira Amorim, Alan de Araújo Roque, Israel Luís Diniz Carvalho, Ana Laura Vilela de Carvalho, Milena Evangelista Dos Santos, et al. "Ethnobotanical Surveys of Plants Used by Quilombola Communities in Brazil: A Scoping Review." *Life* 14, no. 10 (September 24, 2024): 1215. https://doi.org/10.3390/life14101215.

Rights of Nature Tribunal. "Brazil: Amazon Belo Sun Gold Mine."January 31, 2025, https://www.rightsofnaturetribunal.org/cases/belo-sun-gold-mine-in-the-brazilian-amazon/.

Rights of Nature Tribunal. "Rights Of Nature Tribunal Finds Canadian Mining Companies Guilty For The Violation Of Rights Of Nature Across South America And Serbia." March 1, 2025. https://www.rightsofnaturetribunal.org/canadian-mining/.

Rights of Nature Tribunal. "'Rights of Nature Tribunal Finds Canadian Mining Companies Guilty for the Violation of Rights of Nature in 14 Cases in South America and Serbia' Recommendations of the 6th International Rights of Nature Tribunal." Toronto, Canada: 6th International Rights of Nature Tribunal. February 8, 2025. https://ecojurisprudence.org/wp-content/uploads/2025/03/Rights-of-Nature-Tribunal-Final-Recommendations.pdf.

Shore, Edward. "Geographies of Resistance: Quilombos, Afro-Descendants, and the Struggle for Land and Environmental Justice in Brazil's Atlantic Forest." *Afro-Hispanic Review* 36, no. 1 (March 22, 2017): 58–79.

Sigma Lithium Corporation. "Sigma Lithium and Brazilian Government Officials Ring Nasdaq Opening Bell to Celebrate the Launch of Lithium Valley Brazil Initiative." Nasdaq. May 11, 2023. https://www.nasdaq.com/press-release/sigma-lithium-and-brazilian-government-officials-ring-nasdaq-opening-bell-to.

Sigma Lithium Corporation. "Sigma Lithium Starts Construction of Production Plant Foundation and Infrastructure and Orders Long Lead Items, on Schedule for Production in 2022." Sigma Lithium Corp (blog). December 6, 2021. https://sigmalithiumcorp.com/sigma-lithium-starts-construction-of-production-plant-foundation-and-infrastructure-and-orders-long-lead-items-on-schedule-for-production-in-2022/.

Silva, Henrique Moreira de Melo, Luiz Tarcizio Gonzaga de Oliveira, and Natan Ferreira de Carvalho. *Relatório Técnico: Processos de licenciamento ambiental n.º 04515/2024 e n.º 1267/2023: Licenciamento ambiental do empreendimento Grota do Cirilo de responsabilidade da empresa Sigma Mineração S.A. e seu impacto em comunidades tradicionais.* Belo Horizonte/Diamantina, MG: Ministério Público do Estado de Minas Gerais, April 16, 2025.

Sweet, James H. *Recreating Africa: Culture, Kinship, and Religion in the African-Portuguese World, 1441–1770.* Chapel Hill: University of North Carolina Press, 2003. http://archive.org/details/recreatingafrica0000swee.

Tong, Xiting. "Resisting Lithium Mining and Strengthening Food Sovereignty: Quilombola Community in Girau, Brazil." *Cultural Survival.* January 15, 2025. https://www.culturalsurvival.org/news/resisting-lithium-mining-and-strengthening-food-sovereignty-quilombola-community-girau-brazil.

University of Toronto Faculty of Law I International Human Rights Program. "Witnesses Confront Mining Industry at the 6th International Rights of Nature Tribunal: 'The Impacts of Mining and the Post-Extractivism Era,'" February 28, 2025. https://ihrp.law.utoronto.ca/news/title-ihrp-clinic-student-highlights-first-nations-perspectives-mining-development-ontario-2025.

U.S. Geological Survey (USGS). "Mineral Resource Data System." Accessed June 10, 2025, https://mrdata.usgs.gov/mrds/.

Veran, Jean-Francois. "Quilombos and Land Rights in Contemporary Brazil." *Cultural Survival Quarterly* 25, no. 4 (April 28, 2010). https://www.culturalsurvival.org/publications/cultural-survival-quarterly/quilombos-and-land-rights-contemporary-brazil.

Weber, Gabriel. "474 JACARÉ – COPACABANA: O auto da linha do inferno." *assossiassão* Índice Temporada 2 : Viu (July 28, 2023). https://assossiassao.pt/474-jacare-copacabana-o-auto-da-linha-do-inferno.

IMAGE CITATIONS + CREDITS

Listed by Page Number

222 Map sources: Google Earth, CNES, Airbus.

224 Photo: Laura Martello. Map sources: Instituto Prístino. "Atlas Municipios de Minas Gerais." Accessed June 7, 2025. https://institutopristino.org.br/atlas/municipios-de-minas-gerais/. Esri; Garmin International, Inc.; U.S. Central Intelligence Agency (The World Factbook); National Geographic Society; Catálogo de Metadados da ANA. "Garimpos." Accessed June 8, 2025. https://metadados.snirh.gov.br/geonetwork/srv/api/records/ec62785d-6559-4f26-9df4-46f3a504bfcd. "Malha Municipal | IBGE." Accessed June 8, 2025. https://www.ibge.gov.br/geociencias/organizacao-do-territorio/malhas-territoriais/15774-malhas.html. E. J. McFaul, G. T. Mason, W. B. Ferguson, and B. R. Lipin. "U.S. Geological Survey Mineral Databases; MRDS and MAS/MILS." U.S. Geological Survey, 2000. https://doi.org/10.3133/ds52. Geological Survey of Brazil (SGB); Brazil Ministry of Development, Industry, Trade, and Services (MDIC), U.S. Geological Survey (USGS).

227 Photo: Laura Martello. Map sources: Earthstar Geographics; "Brazil Mining Concessions." Accessed June 8, 2025. https://www.arcgis.com/sharing/rest/content/items/f6c60c7157f144858521b92bc73160c1/info/metadata/metadata.xml?format=default&output=html. "Malha Municipal | IBGE." Accessed June 8, 2025. https://www.ibge.gov.br/geociencias/organizacao-do-territorio/malhas-territoriais/15774-malhas.html. Marco Aurélio da Costa, and Francisco Javier Rios. "The Gold Mining Industry in Brazil: A Historical Overview." *Ore Geology Reviews* 148 (September 1, 2022): 105005. https://doi.org/10.1016/j.oregeorev.2022.105005. Geological Survey of Brazil (SGB); Brazil Ministry of Development, Industry, Trade, and Services (MDIC), U.S. Geological Survey (USGS).

228 Image: Senna occidentalis, illustration from Köhler's Medizinal-Pflanzen, 1887; Coffee senna, illustration from the 1927 work by Sheikh Zain al-Din. Public domain via Wikimedia Commons.

233 Image: Ocimum basilicum, illustration by Francisco Manuel Blanco. From Flora de Filipinas, Biodiversity Heritage Library. Public domain via Wikimedia Commons; Petiveria hexaglochim, from the Coleção Brasiliana Iconográfica. Public domain via Wikimedia Commons; Momordica charantia, artwork by Anoniem, from the Naturalis Biodiversity Center collection. Public domain via Wikimedia Commons. Map sources: Google Earth; CNES; Airbus.

36°25'N
36°24'N
36°23'N
36°22'N
36°21'N
95°18'E
36°19'N
36°18'N
95°18'E
95°17'E
95°16'E
22'N
36°21'
COBALT

Eating Congo Caviar at the End of the World

by Ash Duhrkoop

Without art, Congo is just a mine.
– Sinzo Aanza

Introduction

"Kolwezi is a city that's not supposed to be." On a video call with contemporary Congolese artist Sammy Baloji, we discuss his eponymously titled photography series about artisanal mining in the city. Founded during the colonial period as a mining town for its abundant copper, Kolwezi is today a major source of cobalt. Open pit mines scar the landscape, from industrial scale to small, speculative pits dug by artisanal miners—called *creuseurs* in French, which literally translates to 'diggers'—and local city dwellers in search of cobalt in their yards and under their homes. When monsoon rains sweep through the area, even these small holes have been known to swallow men up.[1] Baloji continues, "Somehow it's a non-existent city either way, and ruled by extraction." Kolwezi is less a city than a sacrifice zone, a site of mineral extraction from its inception, the activities of which have laid waste to the natural environment and exploited local labor for nearly one hundred years. It is a place created to be eaten up.

1 Filip De Boeck and Sammy Baloji, *Suturing the City: Living Together in Congo's Urban Worlds* (London: Autograph ABP, 2016), 208-209.

Kolwezi was established as a colonial settlement in 1938—then called Mariapolis—by the Union Minière de Haut Katanga (UMHK) mining conglomerate to serve as the company's headquarters in the Belgian Congo. UMHK razed the acacia and miombo trees that once dominated the landscape to make way for Musonoi Mine, a set of open cut pits active since 1940, though industrialized mining in the area dates back to 1901. The company built low lying cottages for Europeans and barracks-style housing for miners, usually Africans, in an apartheid-style segregated plan. The settlement didn't obtain city status until 1971, well after the Democratic Republic of the Congo (DRC) won its independence from Belgium.

Today, Kolwezi is the capital city of Lualaba Province in southeastern DRC, and it is being consumed anew by artisanal and small-scale mining (ASM) of the region's rich reserves of cobalt. It has been called "the beating heart of the world's device-driven economy and EV revolution."[2] Satellite images reveal that Kolwezi is full of holes, each bearing witness to this hundred-year history of industrial mineral extraction. According to some estimates, mines occupy 80% of developed land in Kolwezi.[3] Eight of the ten largest cobalt mines in the world are found in the city, to say nothing of the tunnels dug by artisanal miners. Throughout the mining region of the country—formerly known as Katanga—colonial-era practices of extraction are being resuscitated to tap the raw materials needed for the energy transition: cobalt, copper, lithium, nickel, and zinc are all critical minerals for clean energy and all of them are sourced from the DRC, among other locales. Historically, copper was one of the biggest exports from the DRC, but today, 70% of the world's supply of cobalt comes from the Congo.[4] This doesn't make Kolwezi a boomtown in the conventional sense. "Kolwezi is by far the most heavily polluted city in the southeastern provinces. Breathing hurts. Looking burns. ... Stand almost anywhere in Kolwezi, and there will be a mine in every direction."[5]

Kolwezi first came into view for me in 2015 from Baloji's photographs. I was working in the Curatorial Department of the Virginia Museum of Fine Arts when images of an ablated and desolate landscape crossed my desk. The first few I saw were mostly large expanses of dirt and rocky debris, the most striking visual element being tarps spread taught over the ground, sometimes coppery orange and sometimes cobalt blue. Not long before this, it had become common to hear smartphone users—a group encompassing just under half of American adults in 2012—condemned for carrying "a little bit of Congo in their pockets," as awareness of global complicity in the conflict minerals trade spread through academic scholarship and mainstream media.[6] Knowledge of our technological

2 Siddharth Kara, *Cobalt Red: How the Blood of the Congo Powers Our Lives* (New York: St. Martin's Press, 2023), 157.
3 Siddharth Kara, *Cobalt Red,* 158.
4 "Cobalt," U.S. Geological Survey, https://pubs.usgs.gov/pp/1802/f/pp1802f.pdf); "Cobalt Life Cycle," Cobalt Institute, https://www.cobaltinstitute.org/about-cobalt/cobalt-life-cycle/cobalt-mining/.
5 Siddharth Kara, *Cobalt Red*, 158.
6 For example, see Anne DeVoe, "Carrying a Piece of Congo in Our Pockets: Global Complicity to Congo's Sexual Violence and the Conflict Minerals Trade," *Seattle Journal for Social Justice* 10, no. 1 (2011); Matthew Sparkes, "Everyone has a little bit of Congo in their pocket," *The Telegraph*, 31 May 2014.

2013
2015

advancements funding war lords and prolonging civil war in a far-flung place didn't stem the tide of progress; in 2024, 91% of American adults owned a smartphone and the average age for a child to own their own smartphone was eleven. Since then, coverage of cobalt mining in the Congo has become ubiquitous. An exposé published in 2016 in *The Washington Post* painted a dire picture of cobalt mining in Kolwezi.

The world's soaring demand for cobalt is met by workers, including children, who labor in harsh and dangerous conditions. An estimated 100,000 cobalt miners in the Congo use hand tools to dig hundreds of feet underground with little oversight and few safety measures, according to workers, government officials, and evidence found by *The Washington Post* during visits to remote mines. Deaths and injuries are common. And the mining activity exposes local communities to levels of toxic metals that appear to be linked to ailments that include breathing problems and birth defects.[7]

Coverage of the issue ramped up in 2021, with extensive reporting published in *The New York Times* and *The New Yorker*. A major book was also published that year, Erik Bruyland's *Cobalt Blues: La sape d'un géant* (though only available to a Francophone audience). This publication was followed by Siddharth Kara's *Cobalt Red: How the Blood of the Congo Powers Our Lives* in 2023.

The material in focus here—cobalt—makes possible my main mode of interaction with my subject—cobalt mining in Kolwezi and the Congolese artists who historicize and critique extractivism in the region. I researched and wrote this chapter on an Apple MacBook Pro, which is powered by a lithium-ion battery containing cobalt from the Congo. I use my Apple iPhone, also containing cobalt from the Congo, to connect with artists, colleagues, and other contacts and to browse various Instagram accounts belonging to individuals and institutions

7 Todd C. Frankel, "The Cobalt Pipeline: Tracing the Path from Deadly Hand-Dug Mines in Congo to Consumers' Phones and Laptops," *The Washington Post*, 30 September 2016.

based in Kolwezi and the mining region more broadly. The networks of extraction and exchange of cobalt, copper, and other minerals from the DRC are reflected in the electronic circuits, digital networks, and technological apparatus that I rely upon for my research on that very extractivism, for I have not yet been to Kolwezi, or any city in the DRC. A planned trip in the summer of 2020 was cancelled due to the COVID-19 pandemic, and security protocols at institutions in the U.S. have prevented me from traveling to the DRC on two other occasions, based in large part on the U.S. State Department's travel advisory system, which perpetually lists the DRC at Level 3, or "Reconsider Travel."[8]

Without the opportunity to do field work on the ground in the mining region of the DRC, my research relies on representations. I see Kolwezi through lenses: Baloji's camera, Google's satellites, a journalist's headline. In light of how widely known the problems surrounding cobalt mining are, the question becomes how do different forms of media effect our consciousness of the situation? Artworks

by Baloji and fellow artists from the region challenge the trope of the exploited African, victim of global capital, presented in the media. The images in Baloji's *Kolwezi* series challenge the utilitarian, profit-driven approaches dominating geopolitical discourse and humanize abstract and nebulous problems.

Born in 1978 in Lubumbashi, the mining capital of the DRC, Sammy Baloji's prodigious body of work represents an investigation into the cultural and industrial heritage of the Katanga region, the effects of Belgian colonization, and the neocolonial conditions of contemporary mining practices. He renders visible the neocolonial conditions of contemporary resource extraction for the green energy transition, alongside archival sources that show, in hindsight, that the Congo

8 As of January 29, 2025, the U.S. State Department listed the Democratic Republic of the Congo at Level 4: Do Not Travel on its Travel Advisory, representing an escalation from its previous designation at Level 3: Reconsider Travel and indefinitely deferring my research trip. https://travel.state.gov/content/travel/en/traveladvisories/traveladvisories/democratic-republic-of-the-congo-travel-advisory.html.

was perceived as always-already an environmental sacrifice zone at the start of the colonial project, which paved the way for the long history of extractivism and rampant exploitation of the Congo's land, natural resources, and people. While many of his works center around the historical extraction of copper, Baloji's *Kolwezi* series focuses on the more recent artisanal extraction of cobalt. It consists of five subsets of images focused on different aspects of life in the city: mining sites, daily labor, portraits of miners and their families, home interiors, and home exteriors.

Artisanal mining began in Kolwezi and throughout southeastern DRC in the 1990s, following privatization of the national mining company, Gécamines, which had gone bankrupt due to corruption and mismanagement. Unemployed miners and young men looking to earn some money for their studies, Baloji explains to me, began digging by hand in the defunct industrial mines. Some of these industrial mines have been reopened—run mostly by Chinese companies, a presence that Baloji sought to document when he began photographing Kolwezi in late 2008.

Around 2009, artisanal mining took on a different tenor in Kolwezi, when locals discovered mineral reserves under their homes. According to news reports and local lore, a man was digging a hole in his backyard in the Kasulo district of the city when his shovel struck a vein of heterogenite, unmistakable with its shiny black protuberances streaked with the bright turquoise of oxidized copper.[9] His find led to a digging frenzy visible in satellite images. Heterogenite, the rich mineral blend of copper and cobalt found in the region, and the world's most important source of refined cobalt, was famously called "Congo caviar" by a U.S. ambassador in the 1960s and the moniker stuck amongst some circles of investors and speculators.[10] It is a fitting appellation for the mineral, which bears a visual resemblance to spoonfuls of the slippery black delicacy. The phrase also resonates with the elitism and inequity inherent to the circuits of extraction, exchange, and consumption of the ore, in which foreigners are the benefactors of the DRC's mineral riches, while local populations face environmental, public health, and social crises. Once extracted, heterogenite ore is refined into cobalt and copper, two important resources for many electronics, such as laptop computers and smart phones, as well as so-called green technologies. Cobalt, in particular, is a critical element in the production of lithium-ion batteries, the key component in electric vehicles.

---

9 Nicolas Niarchos, "The Dark Side of Congo's Cobalt Rush," *The New Yorker*, 24 May 2021; William Clowes and Thomas Wilson, "Never Mind the Mines. In Congo, There's Cobalt Under the House," *Bloomberg*, 28 March 2018.

10 Tom Zoellner, *Uranium: War, Energy, and the Rock that Shaped the World* (New York: Viking, 2005), 11; Michela Wrong, *In the Footsteps of Mr. Kurz* (New York: Harper Collins, 2001), 108; "Telegram from the Embassy in Belgium to the Department of State," July 10, 1960, FRUS 1958–1960 vol. XIV: 286–287, https://history.state.gov/historicaldocuments/frus1958-60v14/d111.

Critical Minerals, Sacrifice Zone

In Kolwezi, and throughout the mining region, Chinese investors have made deals with the government of the DRC to redevelop the industrial mines there. In exchange for "Congo caviar," the Chinese investors promised to rehabilitate infrastructure. Extraction takes place at sites previously drilled by industrial mining machinery, which have left slopes as steep as 100 meters in height. In a statement about the series, Baloji describes the process of artisanal mining that takes place at these sites:

> *Armed with picks, hammers, lamps, and bags of raffia, miners climb these slopes in search of heterogenite. They must excavate tunnels of 60 to 100 meters into the earth before reaching the vein. Then they go up and down the slopes several times carrying over 50 pounds to accumulate sufficient tonnage for industrial sale. Often miners are victims of landslides, but these losses do not stop the march to gold.*[11]

Around the mines, makeshift tarpaulin tent cities have been established to house the miners, many of whom migrated from other parts of the Katanga region and beyond. Baloji's photographs include both landscapes of the tent cities and portraits of mine workers in their shelters. His statement continues: "Given their current territorial and economic instability, artisanal miners live in tented cities of makeshift tarpaulins near the mining areas. These living spaces and operations are temporary. Maps of these exploitative industrial and artisanal sites remain unknown to all but the mining registers held by [the] government."[12]

In another subset of images, Baloji pairs his photographs of mining sites with mass-produced posters imported from China to make diptychs titled *Raccords* (meaning "connections" in French). The posters are purchased cheaply by miners and other city dwellers to decorate their dwellings, at times just a tarpaulin tent in the vicinity of a mining site. These posters illustrate fantastical cityscapes and impossibly beautiful natural vistas, each one a manipulated version of reality meant to be highly desirable, but never actually attainable. In *Raccord #1*, Baloji has juxtaposed a photograph of the city of Kawama, its orange tarpaulin huts in neat rows, with a poster of a luxurious pool resort. The lush palm trees and green grass surrounding the crystal blue water in the poster is starkly contrasted with the sparse vegetation and sense of desolation present in the photograph of Kawama. Baloji explains, "I have chosen to integrate these posters into my work as an extension of a utopian future born of artisanal mining, the loss of human lives, the export of minerals, and the continuing displacement of populations."

Let us look closely at one of the *Raccord* diptychs. For this subset within the *Kolwezi* series, Baloji paired one of his landscape photographs taken in Kolwezi in 2009 with a Chinese poster sold locally. While all of the *Raccord* diptychs are skillfully curated to seamlessly blend into each other, *Raccord #2 Usine de Shituru* (Twin photos #2 Shituru Plant) is an arresting visual allegory for the

11 Sammy Baloji, "Kolwezi," artist statement, Axis Gallery, 2012, https://axis.gallery/artists/sammybaloji/kolwezi/.
12 Sammy Baloji, "Kolwezi."

"dark side" of the green energy revolution. On the left-hand side, we are faced with the wall of a crater-sized hole, the extent of which exceeds the frame of the photograph, lending its scale the sense of infinitude. Rubble in the foreground and smoke in the midground give the image a war-torn and somber tone. The hilly terrain in the background perfectly aligns with the majestic mountains framing the alpine valley featured in the poster on the right-hand side. The valley is a lush verdant green, nestled charmingly under a crystal blue sky. Juxtaposed with the drab browns and greys of Kolwezi, the super-saturated green hue of the valley becomes almost unbearably bright, radioactive even.

Though no people are visible in the image on the right, there are traces of its human inhabitants: tidy stacks of hay, tended gardens, a neat row of utility poles. Looking closely at the image on the left, of Shituru Plant, the tiny bodies of four Congolese creuseurs become visible in the mouth of the cavernous pit. They are performing the daily, and dangerous, labor of digging for heterogenite. It becomes obvious from Baloji's masterful suturing of these two images that one makes the other possible. The phantasmagoric alpine valley on the right, like so many places in the Global North, relies on the "sacrifice zone" of Kolwezi, and other similar sites usually located in the Global South, where land is depleted, extracted, and human labor exhausted, expendable. As Elizabeth Povinelli puts it, "the toxicity of the capitalist form is concentrated in the very places that its wealth is extracted."[13]

When we think of sustainability, the landscape depicted on the right is undoubtedly amongst those environments we seek to preserve. Perhaps unsurprisingly, we find solar panels adorning the roof of at least one of the homes nestled into the valley. Evidence of the value of this environment, and the responsibility exercised by its residents. *They are doing their part*. Taken as metonyms for their socio-economic counterparts elsewhere throughout the world, the unseen inhabitants of the valley get to *be green*, both figuratively in the sense of "going green," that pithy slogan of capitalist environmentalism, and literally, with fertile and verdant land, because of the land and labor thousands of kilometers

13 Elizabeth Povinelli, "Toxic Ruins, Sovereignty, and Relations," in Lucrezia Cipitelli, Dalinda María Benfield, and Christopher Bratton (eds.), Ex Situ. Christopher Bratton (Milan: Kunstverein Publishing, 2023),15.

away. What do solar panels on a chateau in the Alps have to do with a dusty, post-industrial city in the Democratic Republic of the Congo? The *raccords* run deep.

Some versions of solar panels (DSCs) use cobalt, however these are rare today. The vast majority of solar panels in use are photovoltaic (PV), which do not themselves contain cobalt, however, the batteries used to store the energy transferred by the PV panels almost always do. Demand for the element is surging due to the green energy revolution and the associated demand for electric vehicles amongst consumers in the U.S. and Europe. However, there is something often overlooked in the rhetoric surrounding green energy: while the source of the energy might be renewable (i.e. the sun, in the case of solar energy), the technological infrastructure required to render this energy consumable (i.e. solar panels, lithium-ion batteries) is not renewable—and these technologies are currently not sustainably produced either.

A Geographic Fluke, A Geological Scandal

In the early 20th century, the Katanga region was greedily called *un scandale géologique* by Belgian executives of the Union Minière Haut Katanaga (UMHK) mining company, who controlled the mineral resources found in Katanga during the colonial period, due to the rich abundance and high-grade quality of the ores. This region is part of the Central African Copperbelt, a natural geographic region, distinguished by its common geology, which extends southward from the DRC's Katanga and Lualaba provinces to Zambia's Copperbelt Province. Here, about half of the world's reserves of cobalt lie beneath the earth, an estimated 3.5 million tons of it.

Artisanal mining in Kolwezi is made possible by the geological fact of substantial cobalt deposits at shallow depths, accessible by shovel. This is how the gentleman in the Kasulo neighborhood struck cobalt while digging in his yard. In his book *Cobalt Red: How the Blood of the Congo Powers Our Lives*, Siddharth Kara describes this "geographical fluke":

> *...the reason the copper-cobalt deposits in the copper belt are so shallow is because they are uniquely found in "sediment hosted stratiform deposits." This type of deposit indicates that the cobalt-containing ores occur in discrete layers of sedimentary rocks that were initially laid down in water. Such deposits are the only ones with the potential to be pushed upward to the surface by tectonic activity, thereby making them accessible to artisanal miners. The Central African Copper Belt happens to be located on the western shoulder of one of the most spectacular examples of this tectonic activity – the East African Rift.*[14]

In the Copper Belt region, copper-cobalt ores are found both at great depths and near the surface in different mineral forms. Deeper ores combine with sulfur to

14 See Siddharth Kara, *Cobalt Red,* 24–25.

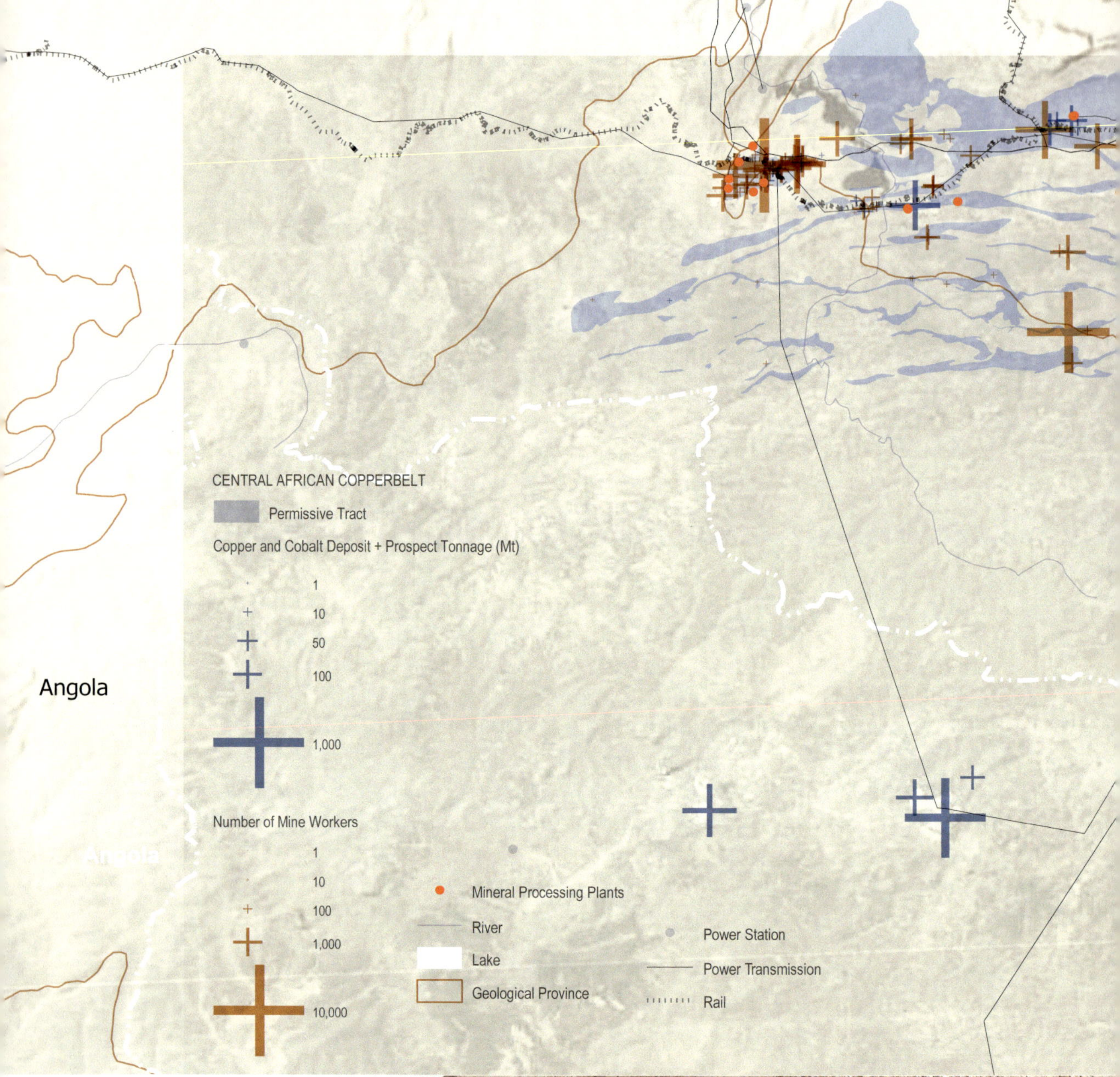
CENTRAL AFRICAN COPPERBELT
Permissive Tract
Copper and Cobalt Deposit + Prospect Tonnage (Mt)
1
10
50
100
1,000
Number of Mine Workers
1
10
100
1,000
10,000
Mineral Processing Plants
River
Lake
Geological Province
Power Station
Power Transmission
Rail
Angola
Zambia

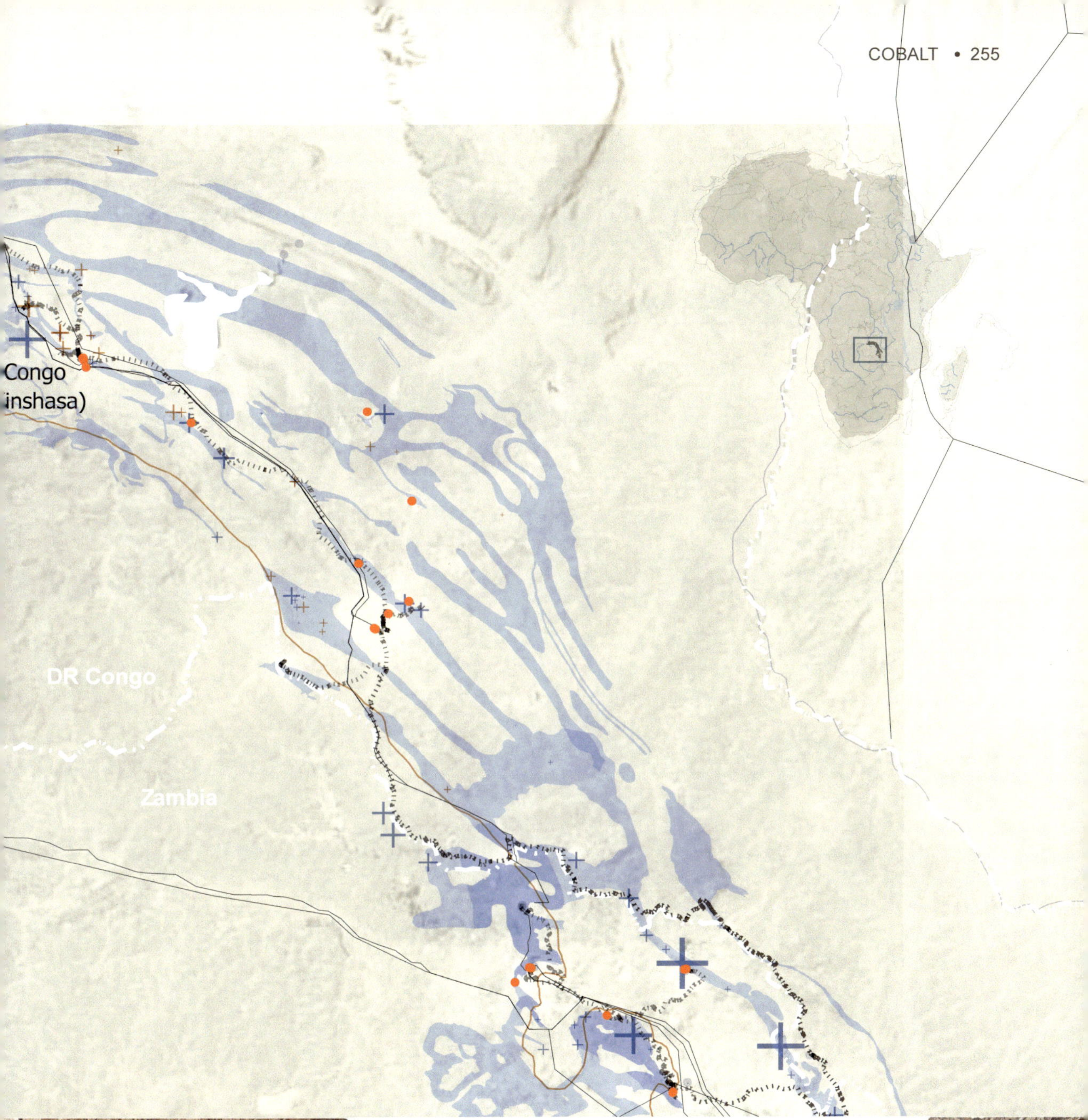
Congo
inshasa)
DR Congo
Zambia

form carrollite, the primary source of industrially mined cobalt. The shallower ores are subject to oxidization, when water combines with sulfur to create sulfuric acid, a geochemical process that forms cobalt-hydroxide in heterogenite, or "Congo caviar." According to Murray Hitzman, expert on Central African geology, "The cobalt-hydroxide ore bodies in Katanga are unique. They form blocks that can be tens of meters to several kilometers in length floating like raisins in a cake."[15]

During the period of the Belgian Congo, lasting from 1908 until 1960, the Congo was already the leading producer of cobalt. According to a Belgian report produced in 1961, "The UMHK [was] the chief world producer of cobalt and produced 8,341 tons of this metal in 1959."[16] At this time, 50 years prior to the invention of the lithium-ion battery, cobalt served several industrial and medical purposes. For example, it is a crucial component of super-alloys used in industrial magnets and airplanes. Befitting of the atomic age, a cancer treatment involving a radioisotope of cobalt—cobalt-60—was developed after World War II. Called cobalt therapy, but nicknamed "the cobalt bomb," the procedure involved an external beam radiotherapy machine, which produced a beam of gamma rays from cobalt-60 directly into the patient's body to kill cancerous tumor tissue. Cobalt therapy was, in fact, made possible by the atomic bomb; the invention of the nuclear reactor for the Manhattan Project led to the creation of artificial radioisotopes for radiotherapy, including cobalt-60. Prior to this, x-ray tubes were used, however, the lower voltage only reached superficial tumors. Due to the "megavoltage" radiation produced by the cobalt isotope (1170–1330 volts, instead of the 50–150 volts of ordinary x-ray tubes), cobalt therapy's penetrating rays could treat tumors deep in the body. In lithium-ion batteries, cobalt allows the battery to stock energy without overheating, effectively rendering the battery rechargeable.

For centuries before its modern technological applications, cobalt has been used as a pigment in glass and ceramics and later as paint. Despite this, it wasn't until 1735 that the element was "discovered" by Swedish chemist and mineralogist Georg Brandt. The language of discovery is endemic to how we speak about the Congo and Africa more broadly. For example, the Congo is said to have been discovered by explorer and journalist Henry Morton Stanley in 1877, when he traversed the territory westward from Tanzania to the mouth of the Congo River on the Atlantic coast, as if the land was a void before Stanley stepped upon it – *terra nullius*.[17] Following this "discovery," natural resources, such as rubber, palm oil, and a cornucopia of minerals and metals were discovered by various explorers and colonists. Belgian geologist Jules Cornet prospected in the Katanga region and found the immensely rich deposits of a variety of valuable mineral resources. His "discovery" led to the establishment of the Belgian corporation UMHK and the start of industrial scale

15 Hitzman, quoted in Siddharth Kara, *Coablt Red*, 25.
16 Jules Gerard-Libois and Benoit Verhaegen, *Congo 1960* (Brussels: Centre de recherche et d'information socio-politiques, 1961). Quoted in Conor Cruise O'Brien, *To Katanga and Back: A UN Case History* (New York: Simon and Schuster, 1962), 87.
17 *Terra nullius* is a Latin expression meaning "nobody's land." It was a principle used in international law to justify claims that territory may be acquired by a state's occupation of it.

mining in the DRC. In 1915, British prospector Robert Rich Sharp "discovered" the radium and uranium orebodies at Shinkolobwe, another site in the Katanga region—an accomplishment that is unironically documented in a 2014 article in *Terrae Incognitae*, the journal of The Society for the History of Discoveries. The rhetoric of discovery inherent to the colonial project undermines the history of local communities' knowledge and use of these resources for centuries before colonialism. For example, African metallurgists have been smelting copper since at least the 4th century.[18] In addition to the artistic, ritual, and social uses of refined copper, heterogenite has long been known and used medicinally by local populations. Despite the ancient history of cobalt's use, Elvira Weeks' classic text *Discovery of the Elements* classifies cobalt as an "eighteenth century metal," as this was when it was isolated and named as an element.

Brandt chose the name cobalt from the German word *kobold*, referring to a subterranean gnome. In ancient German folklore, kobolds are malicious earth sprites who delighted in destroying the work of the miners. Cobalt, before it was known as such, was thought to be silver that was bewitched by kobolds and transformed into a toxic substance. In actuality, the cobalt they mistook for silver released arsenic and sulphur vapors when it was smelted. French chemist Étienne Francois Geoffroy wrote that "it has so strong a Corrosive Quality as sometimes to burn and ulcerate the Hands and Feet of the Miners, and it is a deadly poison for all known Animals."[19]

We can read this history of humanity's relationship with cobalt pictographically in a painting by Congolese painter Edmond Musasa Leu N'seya, who goes by Musasa, in collaboration with Belgian artist Maarten Vanden Eynde. The painting is part of a series of painted charts titled *Material Matters*, each one featuring a different element from the periodic table. Referencing the universality of the periodic table, the chart tells a different story about each element than is found in Mendeleev's table, offering clues to the socio-cultural influences of the element throughout history. There is a matter-of-factness to the placement of the various icons together, linking disparate human creations through a shared material. However, the scientific objectivity of the periodic table of elements is undermined by the tragic irony depicted on the painted schoolroom chart—a young boy, wearing clothes reminiscent of a school uniform, toils in front of a pile of cobalt ore.

18 See Eugenia W. Herbert, *Red Gold of Africa: Copper in Precolonial History and Culture* (Madison: University of Wisconsin Press, 1984).
19 E.-F. Geoffroy, "Treatise of the Fossil, Vegetable, and Animal Substances that are Made Use of in Physick." First published 1736. Reproduced in Mary Elvira Weeks and Henry M. Leicester, *Discovery of the Elements* (Easton, Pa: Journal of Chemical Education, 1968), 94.

"No Congo, No Technology"

Maurice Mbikayi is another Congolese artist whose work questions the technological imperialism and ecological catastrophe of the mining industry in the DRC. Speaking of coltan, he explains, "the world's technology depends on Congo. One of our planet's most precious minerals which is used to make all the technological devices we use – airplanes, rockets, TVs, cars, mobile phones, computers, game devices, guns, etc, etc – and 80 percent of the world's coltan is found only in the Congo. So no Congo – no technology!"[20] Like cobalt, coltan is a critical mineral for the tech sector found in abundance in the DRC. And like cobalt, coltan is extracted through artisanal and small-scale mining, often under dangerous and unregulated conditions. Coltan mining in particular has been linked to armed militias that exploit local populations and use mineral profits to fund conflicts.[21]

Despite all of this, neither coltan nor cobalt is a conflict mineral according to the U.S., a designation which only includes "3TG" minerals (tungsten, tin, tantalum, and gold). As a result, Apple Inc.'s Conflict Minerals Report to the U.S. Securities and Exchange Commission does not disclose any sources for cobalt. Despite the fact that between 60 and 80% of the global supply of cobalt comes from the DRC,

20 Nala Xaba, "No Congo, No Technology: Interview with Maurice Mbikayi," *Chimurenga*, 29 May 2017, https://www.chimurenga.co.za/archives/6058.

21 For more on coltan mining in the DRC, see Alexandra Wexler, "How This Conflict Mineral Gets Smuggled Into Everyday Tech," *The Wall Street Journal*, 6 October 2024, https://www.wsj.com/business/how-this-conflict-mineral-gets-smuggled-into-everyday-tech-5a4b197f; Lydia Todd, "Coltan and Conflict in the Democratic Republic of Congo," SUNY Open Access Repository (SOAR), June 10, 2024, https://soar.suny.edu/handle/20.500.12648/15482; "The Dark Side of Technology: Coltan Mining in the DRC and its Human Rights and Environmental Impacts," 2023, https://globalforestcoalition.org/the-dark-side-of-technology-coltan-mining-in-the-drc-and-its-human-rights-and-environmental-impacts/; "Mining and Illicit Trading of Coltan in the Democratic Republic of Congo," 2021, https://enactafrica.org/research/research-papers/mining-and-illicit-trading-of-coltan-in-the-democratic-republic-of-congo.

major tech companies claim that the cobalt used in their products does not come from the Congo.

> Sony: "We take this issue seriously and have been conducting a fact-finding process. So far, we could not find obvious results that our products contain the cobalt originated from Katanga in the DRC."

> Volkswagen: "To the best of our knowledge, the cobalt in our batteries does not originate from the DRC."

While it is dubious that some of the biggest companies that consume cobalt could avoid Congolese cobalt, given the large percentage of the world supply that the country accounts for, the most egregious responses came from Samsung and Microsoft.

> Samsung: "It is impossible for us to determine whether the cobalt supplied to Samsung SDI comes from the DRC Katanga's mines."

> Microsoft: "We have not traced the cobalt used in Microsoft products through our supply chain to the smelter level due to the complexity and resources required."

Microsoft earned over $72 billion dollars of net profit for the 2022 fiscal year, yet the company purports to not have the resources to investigate whether components in their products are obtained via human rights abuses. Apple Inc. acknowledges the need for transparency in the supply chain for cobalt, in addition to the 3TG minerals, and lists Huayou Cobalt as their supplier, who supplies most of the world's largest battery makers. Huayou Cobalt, based in China, processes cobalt ore from its subsidiary Congo Dongfang Mining (CDM), which has been accused of sourcing its ore from sites using child labor, amongst other human rights abuses.

In light of failed regulatory efforts, the corporate solution has been divestment. Companies like Tesla are investing billions of dollars in developing a cobalt-free battery to divest with the DRC. Indeed, an article published in *Nature* in 2021 reported that "many researchers consider the cobalt problem essentially solved."[22] At least two different labs have successfully produced cobalt-free cathodes that do not show compromised performance—at least, not in the lab. Eliminating cobalt often means replacing that metal with nickel. "We have addressed the cobalt scarcity, but because we're scaling so rapidly, we are heading straight for a nickel problem," said one materials scientist quoted in the article. Removing both cobalt and nickel means switching to a completely different crystal structure for the cathode materials, which means that the battery's energy density could be compromised.[23]

In early 2022, Tesla began producing vehicles with lithium iron phosphate (LFP) batteries, which contain no cobalt or nickel. An alternative to lithium-ion batteries, LFPs were actually developed at the University of Texas in the late 1990s, but were not adopted by the U.S. automobile industry due to their limited range capacity compared to EVs powered by lithium-ion batteries. With improvements in the intervening years, the LFPs in Tesla Model 3 vehicles have a range of 267 miles, which is comparable to the 280-mile range of Volkswagen's ID.4, equipped with a lithium-ion battery containing cobalt and nickel.[24] This range, however, still falls short of the Model 3 powered by a lithium-ion battery, which has a base range of 358 miles.

Removing both cobalt and nickel from EV batteries also diminishes the potential for the circular economy promised by EV advocates, in which spent batteries are recycled alleviating the need for virgin mined materials. However, lithium is cheaper to mine than to recycle; the value of the cobalt and nickel in battery packs is what makes recycling them appealing at all. In a CNN Business report, Ajay Kochhar, CEO of battery recycler Li-Cycle, is quoted, "Recycling is important, but it also has to make sense in terms of the rationale."[25] And that rationale is always economic: if it is cheaper to mine virgin material, there is little incentive to reclaim it from used batteries. Switching to LFPs means we may end up creating many more immense e-waste dump sites, like the infamous Agbogbloshie in Ghana, with millions of spent EV batteries.[26]

Short of eliminating the use of cobalt altogether, exploration geologists are working to find alternative sources of cobalt elsewhere around the world. In 2018, scientists at Canada's University of Alberta examined the cobalt deposits in the DRC with a new dating system and found that cobalt and copper mineralization in the region occurred during a period of mountain building and deformation

---

22 Davide Castelvecchi, "Electric Cars: The Battery Challenge," *Nature* 596 (August 19, 2021), 338.

23 Castelvecchi's article provides a very accessible description of how EV battery packs work, in greater detail than I can reproduce here.

24 Matt McFarland, "The Next Holy Grail for EVs: Batteries Free of Nickel and Cobalt," *CNN Business*, June 1, 2022, https://www.cnn.com/2022/06/01/cars/tesla-lfp-battery/index.html.

25 Matt McFarland, "The Next Holy Grail for EVs."

26 Some analysts argue that once there is a critical mass of spent EV batteries, the economic incentive of scale will make recycling more compelling. Again, the rationale is entirely economic.

between 610 and 470 million years ago. These results suggest that the deposits formed 100 to 150 million years more recently than originally thought.[27] This new information adds crucial insight to the hunt for cobalt to meet the growing demands of the energy transition. As Dr. Robert Creaser, lead geologist on the study, explains:

> *The conundrum is that the western world needs cobalt, and the conditions in some places we currently get it from can be exploitative. The biggest value of this research is opening the possibility of finding more prospective areas worldwide for sources of cobalt. This background information helps exploration geologists develop ideas of where and where not to look.*[28]

However, human rights advocates don't want the world to stop buying Congo's copper, and neither do local Congolese people. If the demand for Congolese cobalt was to cease, hundreds of thousands of local people would lose their one lifeline, however tenuous and volatile it may be.

Conclusion

Tesla's CEO Elon Musk has set out a goal of selling 20 million EVs per year by 2030, an increase of more than 15 times its sales in 2022.[29] Other automotive companies are also scaling up production of their EV offerings. Virta, an automotive research consultancy, forecasts that there will be 140 million EVs on roads worldwide by 2030, up from 16 million in 2021. At around 30 pounds of cobalt per EV battery, four billion two hundred million pounds (2,100,000 tons) of cobalt will have been extracted from the earth to meet this projected demand.

In 2019, Baloji along with Jean Katambayi Mukendi and Daddy Tshikaya, fellow members of the artist collective On-Trade-Off, constructed a 1:1 scale Tesla Model S replica from recycled copper wire for a project titled *Charging Tesla Crash: A Speculation*, which offers a potent critique of the environmental and social consequences embedded in global industrial systems, resource extraction, and the ideals of techno-capitalist progress. The wire construction is reminiscent of the small wire cars made by the children in the DRC. These homemade toys, typically made with second-hand, repurposed materials, are popular throughout

27 University of Alberta, "Earth's Cobalt Deposits Formed Much Later than Previously Believed," *ScienceDaily*, www.sciencedaily.com/releases/2018/12/181212134424.htm. Accessed July 2, 2023,

28 University of Alberta, "Earth's Cobalt Deposits Formed Much Later than Previously Believed."

29 Tesla reportedly sold 1.31 million vehicles in 2022. https://ir.tesla.com/press-release/tesla-vehicle-production-deliveries-and-date-financial-results-webcast-fourth-quarter.

the continent. Through the symbolic construction of a wireframe Tesla model, the artist collective highlights the paradox inherent in green technologies, such as electric vehicles, which are marketed as sustainable yet rely on the unsustainable mining of critical minerals. These resources come at a significant environmental and human cost, leaving local economies static and poor despite their vital contributions to global technological advancement.

Of the project, Katambayi stated:

> *We are now at a time of unbalanced trade between the South and the North. The former offer materials and labor against the later who offer technologies and intelligence. At the end of the construction of the TESLA car in wire at scale 1, we will perform a physical simulation of an accident to express the enrichment of world mineral industry against a local economy still static and poor (in the country where the minerals are extracted). This is when the project becomes speculative…*

By framing the project as speculative, Katambayi invites audiences to reimagine futures beyond the exploitative practices of the present. The speculative and digital elements align with ecocritical aims of envisioning more equitable and sustainable relationships between humans and the environment. Furthermore, the project challenges the greenwashed narratives of companies like Tesla, which present themselves as pioneers of sustainability while perpetuating harmful extraction practices. This critique extends to the North's monopoly on "intelligence" and technology, questioning whether such innovations genuinely promote ecological balance or simply reinforce exploitative systems under the guise of progress.

The wire model is also a potent metaphor for the fragility and incompleteness of modern technological achievements. Its skeletal construction contrasts sharply with the sophistication of an actual Tesla, emphasizing the disconnect between the origins of its raw materials and the polished products consumed in the Global North. The planned crash simulation intensifies this critique, symbolizing environmental and societal collapse as inevitable outcomes of unchecked resource extraction and economic inequality. The crash serves, too, as an allegory for climate change, highlighting the disproportionate burdens faced by vulnerable populations in the wake of the environmental repercussions of industrialization.

Ultimately, *Charging Tesla Crash* demands a reckoning with the global inequalities and ecological violence underlying technological advancement. It emphasizes the interconnectedness of human and environmental systems, calling for an ethical shift toward more equitable and sustainable practices. By linking the material realities of resource extraction to the symbolic fragility of a Tesla crash, Katambayi underscores the need to move beyond exploitative economies and imagine a future grounded in ecological justice and shared global responsibility.

Driving home on Interstate-81 from a family trip to Virginia's Blue Ridge mountains in our gas-powered car, we pass an auto hauler carrying a full load of brand new Teslas. I notice that the placeholder tags confidently proclaim "Zero Emissions," and I can't help but think of all the ways in which the equations of the energy transition don't add up to a net neutral. We don't need complex math to tell us this; artists from the Congo have been making the imbalanced economics visible for a long time. "Kolwezi is a city that's not supposed to be." Baloji's words echo in my mind. Its fate was settled by colonial extractivism at its founding. Kolwezi is, by design, a sacrifice zone. Looking at a geological map of the DRC, a geologist at the Africa Museum (formerly known as the Royal Museum for Central Africa) in Tervuren, Belgium, told me how the geological formation at Kolwezi, Lubumbashi, and other mining cities most likely formed during the Protozoic period. I asked him how the minerals got there. *Where did the cobalt come from, to end up in that particular geological formation in what is known today as the Democratic Republic of the Congo?* He paused thoughtfully before responding, "geologists don't ask about origins." He went on to explain that water brings minerals to gaps in rocks. Yes, but *where do the minerals come from*? I wanted him to tell me about cosmic dust and supernova explosions, but this isn't his domain.

Before I left he added that there are other minerals present in the mining region of the DRC, such as indium, that just do not have known applications yet. But once they do—and if history is any indication, they will—the minerals will be there, no matter how they got there.

When you Google "Where does cobalt come from?," every result is the DRC.

BIBLIOGRAPHY

Baloji, Sammy. "Kolwezi." Artist statement. Axis Gallery, 2012. https://axis.gallery/artists/sammybaloji/kolwezi/.

Baloji, Sammy. *Sammy Baloji: Mémoire/Kolwezi*. Oostkamp: Stichting Kunstboek, 2014.

Bruyland, Erik. *Cobalt Blues: La sape d'un géant. Congo 1960–2020*. Brussels: Racine, 2021.

Castelvecchi, Davide. "Electric Cars: The Battery Challenge." *Nature* 596 (August 19, 2021): 338.

Clowes, William, and Thomas Wilson. "Never Mind the Mines. In Congo, There's Cobalt Under the House." *Bloomberg*. March 28, 2018.

Cobalt Institute. "Cobalt Life Cycle." https://www.cobaltinstitute.org/about-cobalt/cobalt-life-cycle/cobalt-mining/.

De Boeck, Filip, and Sammy Baloji. *Suturing the City: Living Together in Congo's Urban Worlds*. London: Autograph ABP, 2016.

DeVoe, Anne. "Carrying a Piece of Congo in Our Pockets: Global Complicity to Congo's Sexual Violence and the Conflict Minerals Trade." *Seattle Journal for Social Justice* 10, no. 1 (2011).

ENACT Africa. "Mining and Illicit Trading of Coltan in the Democratic Republic of Congo." 2021. https://enactafrica.org/research/research-papers/mining-and-illicit-trading-of-coltan-in-the-democratic-republic-of-congo.

Frankel, Todd C. "The Cobalt Pipeline: Tracing the Path from Deadly Hand-Dug Mines in Congo to Consumers' Phones and Laptops." *The Washington Post*, September 30, 2016.

Geoffroy, E.-F. "Treatise of the Fossil, Vegetable, and Animal Substances That Are Made Use of in Physick." First published 1736. Reproduced in Mary Elvira Weeks and Henry M. Leicester, *Discovery of the Elements*. Easton, PA: Journal of Chemical Education, 1968.

Global Forest Coalition. "The Dark Side of Technology: Coltan Mining in the DRC and Its Human Rights and Environmental Impacts." 2023. https://globalforestcoalition.org/the-dark-side-of-technology-coltan-mining-in-the-drc-and-its-human-rights-and-environmental-impacts/.

Goddeeris, Idesbald, Amandine Lauro and Guy Vanthemsche, eds. *Colonial Congo: A History inn Questions*. Turnhout: Brepols, 2024.

Herbert, Eugenia W. *Red Gold of Africa: Copper in Precolonial History and Culture*. Madison: University of Wisconsin Press, 1984.

Kara, Siddharth. *Cobalt Red: How the Blood of the Congo Powers Our Lives*. New York: St. Martin's Press, 2023.

McFarland, Matt. "The Next Holy Grail for EVs: Batteries Free of Nickel and Cobalt." *CNN Business*. June 1, 2022. https://www.cnn.com/2022/06/01/cars/tesla-lfp-battery/index.html.

Niarchos, Nicolas. "The Dark Side of Congo's Cobalt Rush." *The New Yorker*. May 24, 2021.

O'Brien, Conor Cruise. *To Katanga and Back: A UN Case History*. New York: Simon and Schuster, 1962.

Povinelli, Elizabeth. "Toxic Ruins, Sovereignty, and Relations." In Cipitelli, Lucrezia, Dalinda María Benfield, and Christopher Bratton (eds.), *Ex Situ*. Milan: Kunstverein Publishing, 2023: 15.

Sparkes, Matthew. "Everyone Has a Little Bit of Congo in Their Pocket." *The Telegraph*. May 31, 2014.

"Telegram from the Embassy in Belgium to the Department of State." July 10, 1960. Foreign Relations of the United States, 1958–1960, vol. XIV: 286–87. https://history.state.gov/historicaldocuments/frus1958-60v14/d111.

Todd, Lydia. "Coltan and Conflict in the Democratic Republic of Congo." SUNY Open Access Repository (SOAR). June 10, 2024. https://soar.suny.edu/handle/20.500.12648/15482

University of Alberta. "Earth's Cobalt Deposits Formed Much Later than Previously Believed." *ScienceDaily*. www.sciencedaily.com/releases/2018/12/181212134424.htm. Accessed July 2, 2023.

U.S. Geological Survey. "Cobalt." https://pubs.usgs.gov/pp/1802/f/pp1802f.pdf.

Wexler, Alexandra. "How This Conflict Mineral Gets Smuggled into Everyday Tech." *The Wall Street Journal*. October 6, 2024. https://www.wsj.com/business/how-this-conflict-mineral-gets-smuggled-into-everyday-tech-5a4b197f.

Wrong, Michela. *In the Footsteps of Mr. Kurtz*. New York: Harper Collins, 2001.

Xaba, Nala. "No Congo, No Technology: Interview with Maurice Mbikayi." *Chimurenga*. May 29, 2017. https://www.chimurenga.co.za/archives/6058.

Yusoff, Kathryn. *A Billion Black Anthropocenes or None*. Minneapolis: University of Minnesota Press, 2018.

Yusoff, Kathryn. "Geologic Realism: On the Beach of Geologic Time." *Social Text* 37, no. 1 (2019): 1–26.

Zoellner, Tom. *Uranium: War, Energy, and the Rock that Shaped the World*. New York: Viking, 2005: 11.

IMAGE CITATIONS + CREDITS

Listed by Page Number

242 Diptych: Sammy Baloji, Raccord #5 – Mine à ciel ouvert noyée de Banfora, 2011. Image courtesy of the artist.

Map sources: Maxar.

244 Photo: Sammy Baloji, *Mine à ciel ouvert noyée de Banfora #5 – Lieu d'extraction minière artisanale*, 2010. Image courtesy of the artist.

247 Photo: Sammy Baloji, *Vue cité de Kawama*, 2011. Image courtesy of the artist.

Map sources: Esri; Maxar; Earthstar Geographics, and the GIS User Community.

248 Photo: Sammy Baloji, *Mine à ciel ouvert noyée de Banfora #4 – Lieu d'extraction minière artisanale*, 2010. Image courtesy of the artist.

252 Diptych: Sammy Baloji, *Raccord #2 – Usine de Shituru*, 2011. Image courtesy of the artist.

254 Photo: Sammy Baloji, *Mine à ciel ouvert noyée de Banfora #3 – Lieu d'extraction minière artisanale*, 2011. Image courtesy of the artist.

Map sources: C. D. Taylor, J. D. Causey, P. D. Denning, J. M. Hammarstrom, T. S. Hayes, J. D. Horton, M. J. Kirschbaum, H. L. Parks, A. B. Wilson, N. E. Wintzer, and M. L. Zientek, 2013. "Descriptive Models, Grade-Tonnage Relations, and Databases for the Assessment of Sediment-Hosted Copper Deposits—With Emphasis on Deposits in the Central African Copperbelt, Democratic Republic of the Congo and Zambia." U.S. Geological Survey Scientific Investigations Report 2010–5090–J, 154 p. and data files. http://pubs.usgs.gov/sir/2010/5090/j/. Esri; CGIAR; USGS; TomTom; Garmin; FAO; NOAA; OpenStreetMap contributors, and the GIS User Community.

257 Artwork: Maarten Vanden Eynde and Musasa, Material Matters [Co27], 2018–2025. Photo: Philippe de Gobert.

258 Photo: Sammy Baloji, Détail site d'extraction artisanale #5, 2011. Image courtesy of the artist.

261 Artwork: Jean Katambayi, Sammy Baloji, Daddy Tshikaya, Marjolijn Dijkman, Charging Tesla Crash, a Speculation. Photo: Uriel Orlow.

30°0'20"E
0°12'35"N
30°0'25"E
30°0'30"E
30°0'35"E
0°12'35"N
30°0'15"E
0°12'30"N
30°0'10"E
0°12'25"N
30°0'5"E
0°12'20"N
30°E
Competition Board
Downed Bridge
Super Sack Ore Pile
Mine Entrance

CONCLUSION

Where Histories Are Held and Futures Rehearsed

by Matthew Seibert + Hugo Kamya

Oh what amazing days!
Kilembe Mines, Kilembe Mines
It was powerful
Kilembe Mines, Kilembe Mines
Was flowing with money!
Life in Kilembe was really good
Every beautiful thing was in Kilembe

John and Judith bob their heads and shyly grin as they try to recount the lyrics, standing in front of a shell of a rural hospital complex. Tall grass obscures the view, but you can still make out the dunes of sandy earth rolling through the open structure, a striking juxtaposition of ordered and organic form. Also a tranquil, if eerie, memory of a landscape briefly under raging waters, as the wind now whispers through it.

These materialized memories of a former time, a former landscape, are frequent and many. They each evoke the tension of time in different ways. Some display geologic processes. Just up the dirt road from where we now stand, rests a pile. A pile of piles, really. Massive, disintegrating Super Sacks of woven plastic fabric attempt to hold ore mined from nearby. They drape across one another like bodies almost. Plastic shreds breathe in the wind, but what really catches your eye are the vibrant cobalt blue growths emerging like ornate coral from the drabness of dust and industry. Mineral efflorescence, a slow and mystical process of water-soluble salts–and their mineral neighbors–transporting themselves to the surface, evaporating as accreted protrusions. You can mark time by their growth.

Expensive women
Expensive alcohol
Weekend party dances
All were in Kilembe

The lyrics continue in my head as I take in these geologic beings making such mined ore their home, material once slated to be refined into the metals used in electronics decades ago.

You can also mark time by decay and what once was.

John leads us around and over white boulders, then illustrates how to snake through the steel mess of a downed bridge, to a type of scoreboard where mining totals were once posted, in part to spur competition. Row upon row of living quarters hulk behind us, those that weren't washed away at least. Yet even many of those on high enough ground to escape the torrent, sit empty, gradually falling deeper and deeper under time's entropy.

White and Indian
The Baziba and the Kikuyu
The Baganda and the Acholi
All were in Kilembe

Now very few people are. And those that are, largely long for the glory days, sketch out plans to move elsewhere, scratch at a living, or a combination of the three. Nestled into the foothills of the Rwenzori Mountains in Western Uganda, Kilembe Mines didn't always used to be like this, as the recited song suggests. The country's largest copper mine–with an estimated deposit of 4 million metric tons and an undetermined amount of cobalt–was once a bustling community and a point of national pride. A song was written about its prosperity, after all. Shuttered in the 1980s due to aging equipment and plummeting copper prices, and then devastated by floods in 2020, Kilembe

Mines is a place caught between its memory and its hope, between loss and promise, not unlike its piles of effloresecing ore, its value still withheld. But today it looks more like a sign of ruin than possibility.

It's also caught within the geopolitical winds of the green energy transition as global demand surges for "critical minerals." How should competing interests and value systems first be articulated and then navigated in such a time of great change and amid such a history of exploitation? Within its specific context, Kilembe Mines offers potential universal lessons that cut across many disciplines. Design is one of them. As a powerful tool and practice, design can be employed in versatile and generative ways across "wicked problems,"[1] but, importantly, it is not all-powerful.

First, we must listen.

From Dancing and Drinking to Waiting and Wanting: the Story of Kilembe Mines

The Glory and the Grief

Kilembe was once a name that stirred dreams. Its copper veins promised wealth, its streets echoed with music, and its skies shimmered with possibility. In the 1950s and 60s, Kilembe thrived—a bustling mining town that drew in workers from across Uganda and the world. Canadians, Indians, Chinese, and Ugandans danced to the rhythm of prosperity. Homes were built. Schools opened. Roads stretched wide to accommodate the ambitions of a country newly independent.

The copper mine, inaugurated in 1956, became Uganda's industrial heartbeat, reaching peak production in its inaugural year. But this peak was fleeting. By the 1970s, the music began to fade. Political instability, economic mismanagement, and the brutal regime of Idi Amin turned promise into decay. What remains now are skeletal buildings, polluted rivers, toxic dust, and a people waiting—not dancing—for a revival that never comes.

The Voices in the Dust

Among the ruins, four voices rise—Judith, John, Abraham, and Lillian. Their stories, though deeply personal, echo through the abandoned shafts and contaminated rivers of Kilembe.

Judith, a single mother, embodies the quiet resilience of women who supported Kilembe from the margins. Her backyard poultry farm and vegetable garden are lifelines in a landscape of economic drought. She hasn't been paid in over 18 months, yet she keeps her daughters in school. She labors not for wealth, but for survival, for dignity.

1 Rittel, Horst W. J., and Melvin M. Webber. "Dilemmas in a General Theory of Planning." *Policy Sciences* 4, no. 2 (June 1973): 155–69. https://doi.org/10.1007/BF01405730.

> *"You don't sit—you keep moving. You sit only in new places,"* she says. *"We just need the government to come in and help us. Everything in Kilembe is in the hands of the government."*

She speaks of betrayal—how the white men once respected the river, kept floods at bay. Now, the floods swallow homes. She wonders why the government can't seem to care.

John, once a child of Kilembe's golden age, now supervises a skeletal crew in the mines his father once worked. He walks the same paths, sees the same mountains—but through a different lens. Where his father found pride and prosperity, John finds rot and ruin. Yet, he dares to hope. He speaks of mining regulations, of environmental protections, of a future where Kilembe's wealth enriches rather than exploits.

> *"We cannot just leave,"* he says. *"For to do so, we forfeit our hard-earned salaries."*

Lillian and Abraham, too, speak with a mix of nostalgia and urgency. Their reflections, like those of Judith and John, are not merely recollections—they are pleas. Pleas for justice, for restoration, for a chance to thrive again.

Art and Analysis: Bearing Witness
While residents suffer, scholars and artists like Dr. Abraham Mwesigye and Dr. Lillian Nabulime bear witness.

Abraham's research into the health impacts of mining lays bare the truth: poisoned water, cancer, respiratory disease, and dying soil. He studies the contaminants embedded in locals' toenails, metaphors of how deeply Kilembe's toxins have entered human life.

> *"Artisanal mining with mercury is a silent killer,"* he says. *"What is needed is awareness, and policy. We must protect both people and the land."*

Lillian sees the land as wounded, defiled by greed. As a sculptor, she molds these truths into clay and wood. Her sculptures—rooted in the stories of women, illness, survival—stand like sentinels against silence.

> *"I develop sculptures that give women confidence,"* she says. *"We must tell our stories through materials that withstand time and weather."*

Her art gives form to stories long buried beneath the rubble.

From Prosperity to Poison
What was once Uganda's industrial crown jewel now poisons its soil, air, and water. Abraham warns of cancer clusters, children suffering respiratory illnesses, and biodiversity stripped bare. Waterways once used for irrigation now carry toxic runoff. The company houses, once symbols of upward mobility, are now crumbling shelters of abandonment.

> *"Everything is being stolen—water pipes, door frames—because we are desperate," says one resident.*

The environmental toll is incalculable. Polluted rivers feed into communities' drinking sources. Fields lie fallow. Crops fail. And as the land dies, so does the hope it once carried.

A Path Forward?
Yet, even in despair, Kilembe's people hold a flame of hope. They speak of reopening the mines—but this time, responsibly. They dream of regulations that protect workers, of compensation for unpaid wages, of clean water and functioning schools.

Abraham and Lillian Nabulime call for public dialogue—a table where community members, government officials, environmentalists, and investors meet not to exploit, but to restore. They envision coalitions that can advocate for land rights, human rights, and ecological restoration.

> *"That's what has been done in many developed countries,"* Abraham reminds us.
> *"Mining and sustainability are not enemies—they are uneasy partners that must be forced to coexist."*

The Echoes of Kilembe
The story of Kilembe is more than copper. It is the story of a nation's unfulfilled promises, of landscapes wounded and voices rising from the rubble. Judith, John, Abraham, and Lillian do not just remember the dances of the past—they endure the silence of the present and dare to imagine a different tune for the future.

Their voices are not the only ones—but they are enough to start a reckoning. Enough to call us to remember, to listen, and to act. Kilembe was once a place of dancing and drinking. Today, it waits. It wants. And through its people, it speaks.

Design in Context

So, where and how does design come into this project space of competing interests and concrete consequences? First, let's take a step back and orient ourselves to what design is. This is not easy because design is many things for many people at many different points in time.

Most people would define design as revolving around problem-solving, a discipline that makes physical things, of form and material, to solve problems of need or desire: buildings, objects, perhaps even landscapes like parks. But design also works on systems and processes: temporal, contingent constructions like software, safety protocols, or even how goods are moved around the world. True, but beyond space, material, and time, design also works on the imaginary. It can craft visions of new potential realities in challenge to existing ones, where prison is abolished, economies perform within planetary boundaries, or national parks are returned to and managed by Indigenous groups. Design is also the way you arrive at these above constructions or interventions: the methods one employs to articulate and navigate competing interests and value systems, even the ways of listening and documenting such systems and worldviews.

So, one might argue that design is the intentional process of imagining, structuring, and realizing solutions that shape experiences, systems, and environments. That it integrates creativity, functionality, and context across scales and disciplines, responding to cultural, ecological, technological, and human needs.

But I would take issue with a few elements here. One is the problematic framing of design as solutionist. When dealing with problems at the landscape scale, at least, solutions are often the paradoxical crucible for new problems. Further, solutionist thinking can impart a god-like complex in the designer's mind. This fosters mentalities like domination instead of co-creation, colonialism instead of autonomy and reciprocity, or saviorism instead of empowerment, further perpetuating a way of being in the world with pyramidal hierarchies of exploitation, rather than more horizontal relationships and interdependencies of care and regeneration.

Secondly, hinted at in the first, is the necessity to consider and infuse the design process with political awareness. Design is never neutral, but rather entangled within multiple histories of power and empire: "Modern design has largely been complicit with capitalism, development, and modernity's violent imposition of a single world."[2] In other words, design "helps bring particular worlds into being while foreclosing others."[3] Any definition of design should acknowledge its role in buttressing or subverting dominant, extractive, and homogenizing power structures. Design is always political because it shapes what is possible and what is excluded from this space of possibility.

Lastly, I see design as not merely a technical or creative act but a way of shaping reality itself. What's most inspiring, exciting, and powerful about design is that it can be a practice more than mere process. Design is always engaged in constructing particular ways of being, knowing, and relating to the world. Moreover, this means there is not a universal approach to design. Beyond a singular view, it should support diverse cultural, ecological, and ontological perspectives in a relational approach. Design can build worlds.

This is particularly true in the context of environmental justice. As demonstrated in the prior chapter, the green energy transition is comprised of "the potential contribution of markets (the green economy), the state (green technocracy) and technology (the eco-modernist utopia/dystopia, depending on your perspective)."[4] This is a multifaceted system of relationships that requires a multifaceted strategy

---

2 Escobar, Arturo. *Designs for the Pluriverse: Radical Interdependence, Autonomy, and the Making of Worlds.* New Ecologies for the Twenty-First Century. Durham: Duke University Press, 2018.
3 Escobar, *Designs for the Pluriverse.*
4 Rodríguez, Iokiñe, Mariana Walter, and Leah Temper, eds. *Just Transformations: Grassroots Struggles for Alternative Futures.* London ; Las Vegas, NV: Pluto Press, 2024.

of engagement. As social geographers Karen O'Brien and Linda Sygna contend, any response to climate change must include:

> *"a shift in society's value-normative system and shifting relations across the personal (i.e. beliefs, values, worldviews), political (i.e. systems and structures) and practical (i.e. behaviours and technical responses) levels simultaneously."*[5]

Not only are there asymmetries in power dynamics across these levels, but such hegemonic power can take many forms. Tapping into the work of sociologists and historians of the 70s and 80s, we understand three discrete categories of power—the visible, the hidden, and the invisible.[6] This includes visible formal institutions like legislative assemblies, hidden people and power networks like coalitions between nation-states and private corporations, and invisible discursive practices like cultural violence against a particular groups' beliefs and values. Thus, the strategies of resistance and transformative alternatives must equally shape-shift to confront such "historical, social and political specificities to build emancipatory sustainabilities."[7]

Design possesses meaningful contributions to assist such multi-scalar, multimedia efforts. Design is a way of seeing, shaping, and weaving the world—an ongoing dialogue between intention and possibility. Design is a relational and worldmaking practice that shapes how we live, connect, and imagine possibilities. It is never neutral but always political, embedded in histories and systems of power. To design is to engage with material, social, and ecological realities, ideally in ways that sustain plural ways of being and knowing. Design moves through materials and ideas, through systems and stories, tracing the contours of what is and what could be.

Counternarratives

With that important definitional orientation, let us return to the topics of this volume. In large part, the preceding chapters have been explorations in defining and describing the problem. They render how the promises of the green energy transition fall short, or worse, perpetuate much of the same violence–ecologically, socioculturally, economically–that has been enacted over the last 200 years or more under fossil fuel capitalism's ruthless reign. Or if we consider this more fundamentally, the green energy transition could be said to be perpetuating the last 500 years of colonialist extractivism.

It doesn't have to be this way. So how to short circuit what might seem ingrained in our understanding of humanity? This volume invites you to consider four

---

5 O'Brien, Karen and Linda Sygna. "Responding to climate change: The three spheres of transformation." *Proceedings of Transformation in a Changing Climate*, 19-21 June 2013, Oslo, Norway.

6 Rodríguez, Walter, and Temper build on the following in *Just Transformations*, respectively: Weber, Max. "The Three Types of Legitimate Rule." In Etzioni, A. (ed.) *Complex Organizations: A Sociological Reader*, 1971. Giddens, Anthony. *The Constitution of Society: Outline of the Theory of Structuration*. Polity Press, 1984. Foucault, Michel. "The Order of Discourse." In Young, R. (ed.) *Untying the Text: A Poststructuralist Reader*. Routledge, 1971.

7 Rodríguez, Walter, and Temper, "*Just Transformations.*"

relational counternarratives in line with the book's four sections, or four instructional diagrams that hint at alternatives to the dominant narratives and practices of today. Some are technological or logistical, imploring us to rework current systems, but others are importantly social and cultural, imploring us to fundamentally redesign current relations. They aspire to infiltrate the visible, hidden, and invisible power structures framing our day-to-day. All ask us to expand what's possible. And all are prototypical, so as to be fed by local soil, with local and distant communities overlapping in solidarity. As Cameroonian historian and political theorist Achille Mbembe writes, let us "walk along together with other bodies and, doing so, . . . re-create the world."[8]

It's notable that the counternarratives reuse a series of preexisting AI-generated images, some of which were originally produced in early 2023 by students and faculty through an interactive exhibition at the University of Virginia on visioning alternative futures around the green energy transition. The images are, by nature, weird, not quite representing reality as we know it; at times, they even blur facial details and mutate limbs for an uncanny effect. In a way, these mutations reflect the landscape-scale aberrations related to green energy technologies: the shutter effect, discussed in Chapter 2, is one such impact of large-scale wind projects that disrupts circadian rhythms and imposes new abnormal cycles on affected communities. Chronoception, or time-consciousness, is collapsed through the murky ethics and algorithms that amalgamate preexisting art and imagery into new authorless simulacra. Viewer and author—each with their own histories and hopes—thus meet in collaboration.

Most importantly, these counternarratives attempt to embody core principles as explored above and apply them to the subjects and dynamics of the preceding nonfictional chapters: 1) Pluriversality, or the honoring of multiple ways of knowing and designing, that there is no one world but rather "a world where many worlds fit,"[9] 2) Regeneration, or the shifting from extractive, linear models to living, evolving, circular systems, 3) Reciprocity, or the design with interdependencies and relations prioritized, and 4) Temporal Depth, or the recognition of deep time, long histories, and future trajectories, privileging repair and adaptation. As such they are not to be read as manuals nor prescriptions, but rather visualized practices and processes to inspire local, site-specific, bottom-up change. For, as writer and activist adrienne maree brown contends, "the strongest solutions and innovations come from collective genius. We are not meant to do this work alone."[10]

Underpinning this theory of change is the understanding of social metabolism, or the flow of material and energy between and within society and the biophysical world—the inputs, throughputs, and outputs of human activities within one or

---

8 Mbembe, Achille. *Necropolitics*. Duke University Press, 2020. https://doi.org/10.1515/9781478007227.
9 This phrase was first proffered in the *The Fourth Declaration of the Lacandon Jungle* by the rebels of the Zapatista Army of National Liberation (E.Z.L.N) in Mexico's Chiapas region. The Eraña, Ángeles, and Axel Barceló, 'A World Where Many Worlds Fit', in Ernie Lepore, and Luvell Anderson (eds), *The Oxford Handbook of Applied Philosophy of Language*, Oxford Handbooks, 2024. Accessed May 26, 2025: https://doi.org/10.1093/oxfordhb/9780192844118.013.2.
10 brown, adrienne m. *Emergent Strategy: Shaping Change, Changing Worlds*. Chico: AK Press, 2017.

more environments.[11] In natural systems this is largely circular, outputs getting utilized as inputs (waste as food, for instance), but in human systems they are largely linear. Much of the material and energy metabolized linearly results in great waste, inefficiencies, and, as a result, conflicts. Mines are dug, matter is processed, and wasted is eventually disposed, with environmental conflict at every step.[12] If there was one solution to our climate and social crisis, it might be summed up as "reduce energy demand and build community." Terms such as *metabolic rift*,[13] *circularity gap*,[14] and *entropy hole*[15]—as leading scholars

11 González De Molina, Manuel, and Víctor M. Toledo. *The Social Metabolism: A Socio-Ecological Theory of Historical Change*. Vol. 3. Environmental History. Cham: Springer International Publishing, 2014. https://doi.org/10.1007/978-3-319-06358-4. Martinez-Alier, Juan, and Juan Martinez-Alier. *Ecological Economics: Energy, Environment and Society*. Edited by Klaus Schlüpmann. 2.] Repr., [Paperback ed. Oxford: Basil Blackwell, 1994.
12 Martínez-Alier, Joan. *Land, Water, Air and Freedom*, (Cheltenham, UK: Edward Elgar Publishing, 2023) accessed May 29, 2025, https://doi.org/10.4337/9781035312771
13 Marx, Karl. *Capital: A Critique of Political Economy*. Penguin Classics. London: Penguin books, 1992. Foster, John Bellamy. "Marx's Theory of Metabolic Rift: Classical Foundations for Environmental Sociology." *American Journal of Sociology* 105, no. 2 (September 1999): 366–405. https://doi.org/10.1086/210315.
14 Haas, Willi, Fridolin Krausmann, Dominik Wiedenhofer, and Markus Heinz. "How Circular Is the Global Economy?: An Assessment of Material Flows, Waste Production, and Recycling in the European Union and the World in 2005." *Journal of Industrial Ecology* 19, no. 5 (October 2015): 765–77. https://doi.org/10.1111/jiec.12244. "The Circularity Gap Report 2025." Accessed May 28, 2025. https://www.circularity-gap.world/2025.
15 Georgescu-Roegen, Nicholas. *The Entropy Law and the Economic Process*. Cambridge, Mass: Harvard University Press, 1971.

conceptualize our broken relationships to the biophysical limits of our natural world—point to augmenting our global, societal activities towards convivial communities living within planetary boundaries.

As such, collectively, landscape is understood not as a site not of expansion, but of generative contraction. Landscape is not just green space, but rather a multispecies commons for living together differently, ecologically and socially. Landscape is pedagogy, where every site teaches what it participates in. It's a tool for autonomy, not just amenity. And, finally, landscape is a time machine—where people practice new ways of being and remember what came before, where histories are held and futures rehearsed.

"One can inhabit a place . . . only by allowing oneself to be inhabited by it." But as Achille Mbembe reminds, let us not forget, "inhabiting a place is not the same thing as belonging to this place."[16]

Let us now venture through back through the supply chain, countering the dominant narratives of today and ending where we began.

16 Mbembe, *Necropolitics*.

BIBLIOGRAPHY

Brown, Adrienne M. *Emergent Strategy: Shaping Change, Changing Worlds*. Chico: AK Press, 2017.

"CGR 2025." Accessed May 28, 2025. https://www.circularity-gap.world/2025.

Eraña, Ángeles, and Axel Barceló. "A World Where Many Worlds Fit." In *The Oxford Handbook of Applied Philosophy of Language*, edited by Ernie Lepore and Luvell Anderson, 1st ed., 59–78. Oxford University Press, 2024. https://doi.org/10.1093/oxfordhb/9780192844118.013.2.

Escobar, Arturo. *Designs for the Pluriverse: Radical Interdependence, Autonomy, and the Making of Worlds*. New Ecologies for the Twenty-First Century. Durham: Duke University Press, 2018.

Foster, John Bellamy. "Marx's Theory of Metabolic Rift: Classical Foundations for Environmental Sociology." *American Journal of Sociology* 105, no. 2 (September 1999): 366–405. https://doi.org/10.1086/210315.

Foucault, Michel. "The Order of Discourse." In Young, R. (ed.) *Untying the Text: A Poststructuralist Reader*. Routledge, 1971.

Georgescu-Roegen, Nicholas. *The Entropy Law and the Economic Process*. Cambridge, Mass: Harvard University Press, 1971.

Giddens, Anthony. *The Constitution of Society: Outline of the Theory of Structuration*. Berkeley: University of California Press, 1984.

González De Molina, Manuel, and Víctor M. Toledo. *The Social Metabolism: A Socio-Ecological Theory of Historical Change*. Vol. 3. Environmental History. Cham: Springer International Publishing, 2014. https://doi.org/10.1007/978-3-319-06358-4.

Haas, Willi, Fridolin Krausmann, Dominik Wiedenhofer, and Markus Heinz. "How Circular Is the Global Economy?: An Assessment of Material Flows, Waste Production, and Recycling in the European Union and the World in 2005." *Journal of Industrial Ecology* 19, no. 5 (October 2015): 765–77. https://doi.org/10.1111/jiec.12244.

Martínez-Alier, Joan. *Land, Water, Air and Freedom*, (Cheltenham, UK: Edward Elgar Publishing, 2023) accessed May 29, 2025, https://doi.org/10.4337/9781035312771.

Martínez Alier, Juan, and Klaus Schlüpmann. *Ecological Economics: Energy, Environment, and Society*. Oxford, UK ; Cambridge, Mass., USA: B. Blackwell, 1990.

Marx, Karl. *Capital: A Critique of Political Economy*. Penguin Classics. London: Penguin books, 1992.

Mbembe, Achille. *Necropolitics*. Duke University Press, 2020. https://doi.org/10.1515/9781478007227.

O'Brien, Karen and Linda Sygna. "Responding to climate change: The three spheres of transformation." *Proceedings of Transformation in a Changing Climate*, 19-21 June 2013, Oslo, Norway.

Rittel, Horst W. J., and Melvin M. Webber. "Dilemmas in a General Theory of Planning." Policy Sciences 4, no. 2 (June 1973): 155–69. https://doi.org/10.1007/BF01405730.

Rodríguez, Iokiñe, Mariana Walter, and Leah Temper, eds. *Just Transformations: Grassroots Struggles for Alternative Futures*. London Las Vegas, NV: Pluto Press, 2024.

Weber, Max. "The Three Types of Legitimate Rule." In Etzioni, A. (ed.) *Complex Organizations: A Sociological Reader*, 1971.

IMAGE CITATIONS + CREDITS

Listed by Page Number

266 Photo: Matthew Seibert.

Map sources: Esri, Maxar, USGS, Earthstar Geographics, TomTom, Garmin, FAO, NOAA, (c) OpenStreetMap contributors, and the GIS User Community.

268 Photo: Matthew Seibert.

270 Photo: tranSci Lab team.

271 Photo: Matthew Seibert.

278 Photo: Matthew Seibert.

Counternarratives:
Holding Histories and Rehearsing Futures

CENTER THE PERIPH

by Matthew Seibert + Julia MacNelly

ERY

or, How to Invert a Mine

1. Organize

Change begins in the search for each other. We gather and talk. We air frustrations, trade resources. We call it organizing, but it often begins in small and ordinary ways. It can be slow and uncertain as much as urgent and direct. When we talk about sustainability, we often think about its objects and measurables: emissions, waste streams, batteries. But solidarity—voluntary, prosocial behavior for the benefit of others—is inherent to sustainability.[1] *It asks something from us: our time, our attention, our tools, our care. It is the awkward work of staying in the room when the story does not resolve neatly.*[2]

2. Strategize, Act, and Build

Strategy gives shape to the mess of idealism. Without structure, movements fade.[3] From childcare swaps to meals left on porches, what carries the work forward might not always be the most visible part, but the spaces, routines, and forms of care that persist beyond the crisis are themselves all kinds of power.[4]

3. Cultivate Public Dialogue

Dialogue in activism can emerge through disruption, irony, visual performance, emotional appeal, or internal struggle.[5] Protest, humor, and spectacle all unsettle what counts as legitimate speech. Such moments open dialogic space not by persuading the public, but by troubling its terms. Dialogue is not orderly; Mikhail Bakhtin has called it "polyphonic."[6] Even violent disruptions might become conduits for discourse. Movements fracture and reconverge through struggle, as radical flanks push moderate demands into view or factions argue over core tenets. We lean closer to each other, not to come to consensus, but to come into contact.[7]

5. Let New Lexicons Emerge

By building a shared lexicon, we name what we value, confront dominant narratives, and clarify what we are striving toward (or resisting).[8] Through dialogue and storytelling, we claim our right to define wellbeing on our own terms, affirming our worth and insisting on participation in decisions that shape our lives.[9] By naming what matters, we become linked to one another, to our places, and to the forces and systems that structure our lives. Without space for dialogue, imposed definitions of development erase local meanings of dignity and belonging, reproducing exclusion and dependency. Shared stories, therefore, become ways of resisting erasure.

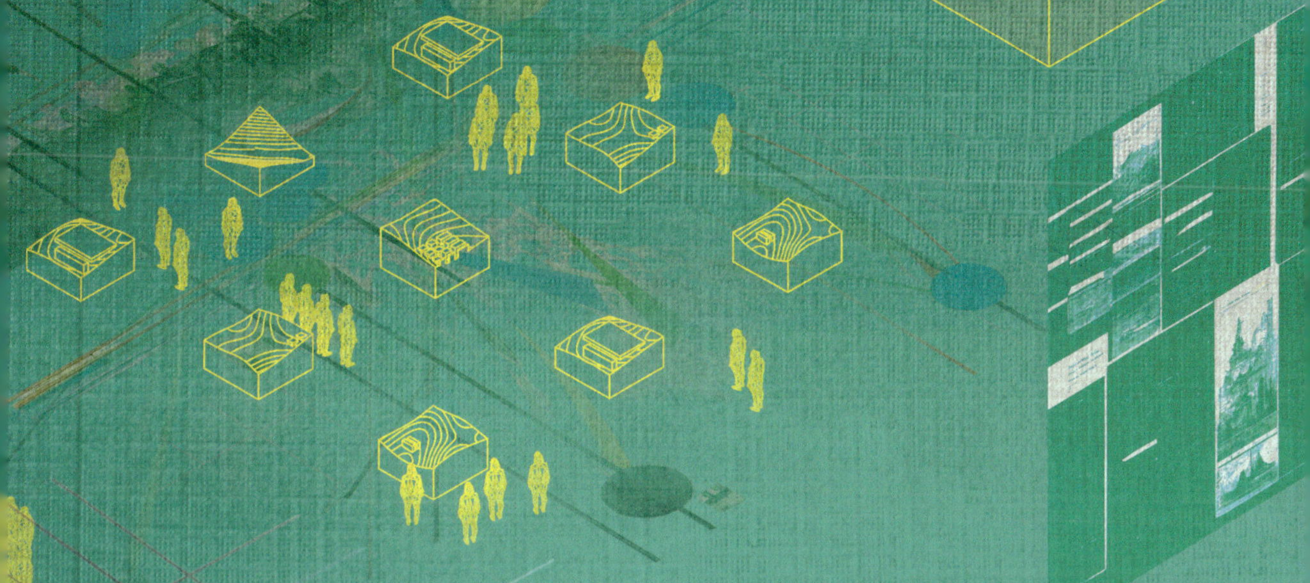

4. Define Values

The energy transition must be shaped by values grounded in community, care, and accountability. These values are not abstract, but rooted in place, history, and survival. Ecological integrity demands restraint and reciprocity, rather than treating ecosystems as mere resources. It requires community self-determination, letting people shape decisions about their land and water, considering the needs of future generations as well as their own. Transparency and accountability reveal who holds the power and benefits from energy projects, while economic equity demands that deep inequalities must be addressed, rather than merely opening new markets. Carbon is not the only metric of sustainability; a project can be "green" and still destroy a watershed or a community. Metrics of harm, collective care, and continuity are equally important to measuring sustainability.

7. Choreograph Inclusive Relationships with Land

Moving beyond "consultation" toward shared governance or even community-led resource management means expanding the scope of a mining project to acknowledge more complex relationships to land.[13] *Mine closures, for example, are not just financial transactions, but social and environmental reckonings. Mines left unrehabilitated threaten the health of ecosystems and communities, and transferring closure responsibilities to smaller, localized, or Indigenous operators complicates accountability.*[14] *Long-term land reclamation plans should be ecologically regenerative and culturally reparative, not just technically compliant.*

6. Reform Mining Governance

Sovereignty and land rights are ongoing struggles. They are legal, material, and ancestral. They are often violated. If new energy development ignores land theft, it is not transition, but repetition. While state-led prior consultations rarely prevent harm in a meaningful way, indigenous community mechanisms like local assemblies rooted in customary law have real power to protect lands and resist extraction.[10] *Recognizing the free, prior, and informed consent of Indigenous peoples or affected communities includes honoring their right to say no or to negotiate equitable terms.*[11] *In practice, this may include adopting resource nationalism policies like higher royalties, public ownership stakes, and domestic value-added processing.*[12]

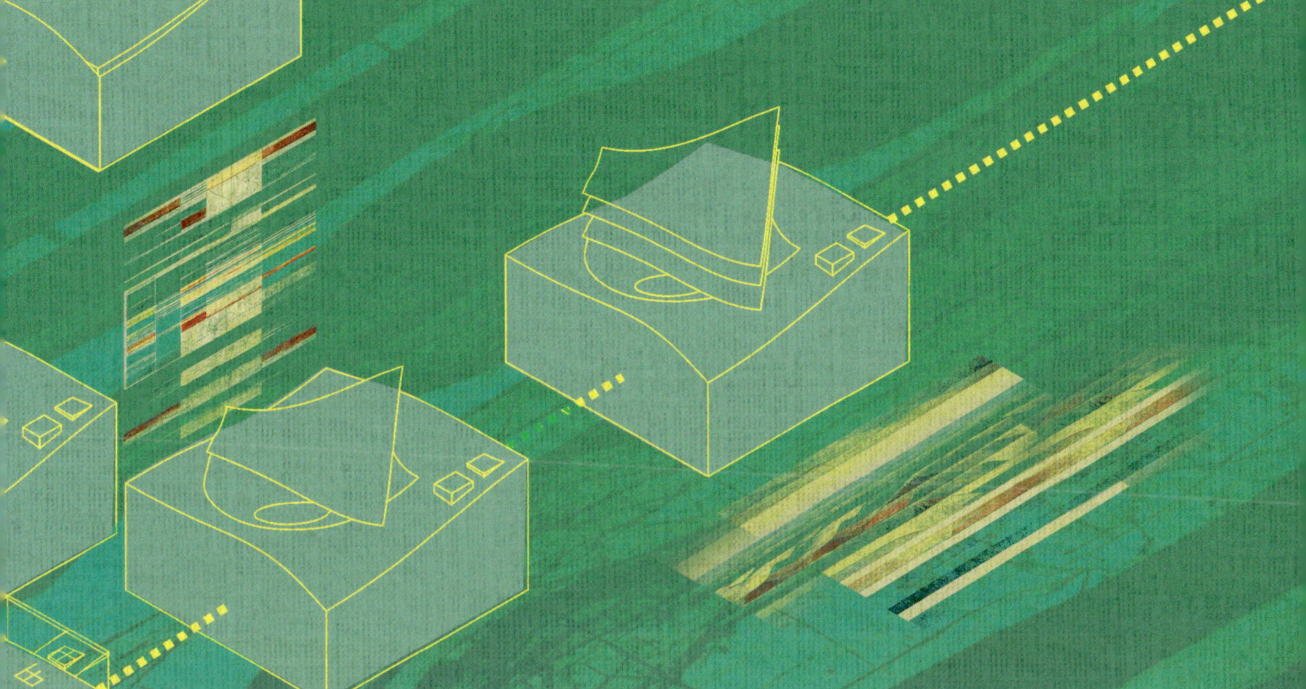

8. Reframe Land Stewardship as Dynamic

Land stewardship involves acts of care over time. Remediation isn't a one-time fix but an ongoing, adaptive process.[15] *Sustainable environmental management depends on well-defined principles: institutional support, clear ecological goals, consistent monitoring, legal prioritization of vulnerable species, ongoing feedback loops to guide corrective action, and the inclusion of Tribal Nations and community stakeholders at every phase.*[16] *Maintaining ecological health is proactive, iterative, and interdisciplinary.*

9. Remediate through Reorientation

Through community collaboration and consultation, decommissioned mine sites could take on new public uses, such as memorials, ecological sanctuaries, renewable-energy farms, or new typologies like community energy gardens.[17]

ENDNOTES

1 Kastner, Karen, and Ellen Matthies. "On the Importance of Solidarity for Transforming Social Systems towards Sustainability." *Journal of Environmental Psychology* 90 (September 1, 2023): 102067. https://doi.org/10.1016/j.jenvp.2023.102067.

2 Bazzani, Giacomo. "Climate Solidarity: A Framework and Research Agenda for Low-Carbon Behavior." *Sociological Forum* 38, no. 2 (2023): 352–374. https://doi.org/10.1111/socf.12885.

3 Fath, Brian D., Carly A. Dean, and Harald Katzmair. "Navigating the Adaptive Cycle: An Approach to Managing the Resilience of Social Systems." *Ecology and Society* 20, no. 2 (2015). https://www.jstor.org/stable/26270208.

4 Pahl-Wostl, Claudia, et al. "The Role of Crises in Transformative Change towards Sustainability." *Ecosystems and People* 19, no. 1 (December 31, 2023): 2188087. https://doi.org/10.1080/26395916.2023.2188087; Pahl-Wostl, Claudia. "A Conceptual Framework for Analysing Adaptive Capacity and Multi-Level Learning Processes in Resource Governance Regimes." *Global Environmental Change* 19, no. 3 (August 2009): 354–365. https://doi.org/10.1016/j.gloenvcha.2009.06.001.

5 Ganesh, Shiv, and Heather M. Zoller. "Dialogue, Activism, and Democratic Social Change." *Communication Theory* 22, no. 1 (February 1, 2012): 79-80. https://doi.org/10.1111/j.1468-2885.2011.01396.x.

6 Bakhtin, Mikhail. *Problems Of Dostoevsky's Poetics*. Edited and translated by Caryl Emerson. Theory and History of Literature, Volume 8. (University of Minnesota Press, 1984): 7–8.

7 Ganesh and Zoller. "Dialogue, Activism, and Democratic Social Change." 84–86.

8 Chabay, Ilan, et al. "Influence of Narratives of Vision and Identity on Collective Behavior Change." *Sustainability* 11, no. 20 (January 2019): 5680. https://doi.org/10.3390/su11205680.

9 Coetzee, Jan K. "The Power of Words: Defining, Narrating, Policy Making." *Africanus* 36, no. 1 (January 2006): 23–34. https://doi.org/10.10520/EJC22616.

10 Torres-Wong, Marcela, and Adrian Jimenez-Sandoval. "Indigenous Resource Governance as an Alternative to Mining: Redefining the Boundaries of Indigenous Participation." *The Extractive Industries and Society* 9 (March 1, 2022). https://doi.org/10.1016/j.exis.2021.101001.

11 Springer, Jenny, and Fernanda Almeida. "Protected Areas and the Land Rights of Indigenous Peoples and Local Communities: Current Issues and Future Agenda." Rights and Resources Initiative, May 2015; Ruwhiu, Diane, and Lynette Carter. "Negotiating 'Meaningful Participation' for Indigenous Peoples in the Context of Mining." *Corporate Governance* 16, no. 4 (August 1, 2016): 641–654. https://doi.org/10.1108/CG-10-2015-0138; Trebeck, Katherine Anne. "Tools for the Disempowered? Indigenous Leverage Over Mining Companies." *Australian Journal of Political Science* 42, no. 4 (December 1, 2007): 541–562. https://doi.org/10.1080/10361140701513604.

12 Koch, Natalie, and Tom Perreault. "Resource Nationalism." *Progress in Human Geography* 43, no. 4 (August 1, 2019): 611–631. https://doi.org/10.1177/0309132518781497; Pryke, Sam. "Explaining Resource Nationalism." *Global Policy* 8, no. 4 (2017): 474–482. https://doi.org/10.1111/1758-5899.12503.

13 Horowitz, Leah S., et al. "Indigenous Peoples' Relationships to Large-Scale Mining in Post/Colonial Contexts: Toward Multidisciplinary Comparative Perspectives." In *Mining and Indigenous Livelihoods*. (London: Routledge, 2024).

14 Kung, Anthony, et al. "Indigenous Co-Ownership of Mining Projects: A Preliminary Framework for the Critical Examination of Equity Participation." *Journal of Energy & Natural Resources Law* 40, no. 4 (October 2, 2022): 433-435. https://doi.org/10.1080/02646811.2022.2029184.

15 Young, Renee E. et al. "International Principles and Standards for the Ecological Restoration and Recovery of Mine Sites." *Restoration Ecology* 30, no. S2 (2022): e13771. https://doi.org/10.1111/rec.13771.

16 Burger, Joanna. "Environmental Management: Integrating Ecological Evaluation, Remediation, Restoration, Natural Resource Damage Assessment and Long-Term Stewardship on Contaminated Lands." *Science of The Total Environment* 400, no. 1 (August 1, 2008): 15. https://doi.org/10.1016/j.scitotenv.2008.06.041.

17 Chu, Edward H. "The Land Revitalization Initiative: Landscape Design and Reuse Planning in Mine Reclamation." In Alan Berger (ed.), *Designing the Reclaimed Landscape*. (London: Routledge, 2007): 141–153.

BIBLIOGRAPHY

Aldrich, Daniel P. "The Right Way to Build Resilience to Climate Change." *Current History* 117, no. 795 (2018): 16–21.

The Nature Conservancy. "TNC Announces 17 New Clean Energy Projects on Former Appalachian Coal Mines with ENGIE and Sun Tribe Development." Accessed May 20, 2025. https://www.nature.org/en-us/newsroom/cumberland-clean-energy-former-mines/.

Bakhtin, Mikhail. *Problems Of Dostoevskys Poetics*. Edited and translated by Caryl Emerson. Theory and History of Literature, Volume 8. University of Minnesota Press, 1984. http://archive.org/details/bakhtin-mikhail-problems-of-dostoevskys-poetics-1984.

Battaïa, Olga, Alexandre Dolgui, Sunderesh S. Heragu, Semyon M. Meerhov, and Manoj Kumar Tiwari. "Design for Manufacturing and Assembly/Disassembly: Joint Design of Products and Production Systems." *International Journal of Production Research* 56, no. 24 (December 17, 2018): 7181–7189. https://doi.org/10.1080/00207543.2018.1549795.

Bazzani, Giacomo. "Climate Solidarity: A Framework and Research Agenda for Low-Carbon Behavior." *Sociological Forum* 38, no. 2 (2023): 352–374. https://doi.org/10.1111/socf.12885.

Burger, Joanna. "Environmental Management: Integrating Ecological Evaluation, Remediation, Restoration, Natural Resource Damage Assessment and Long-Term Stewardship on Contaminated Lands." *Science of The Total Environment* 400, no. 1 (August 1, 2008): 6–19. https://doi.org/10.1016/j.scitotenv.2008.06.041.

Chabay, Ilan, Larissa Koch, Grit Martinez, and Geeske Scholz. "Influence of Narratives of Vision and Identity on Collective Behavior Change." *Sustainability* 11, no. 20 (January 2019): 5680. https://doi.org/10.3390/su11205680.

Chu, Edward H. "The Land Revitalization Initiative: Landscape Design and Reuse Planning in Mine Reclamation." In Alan Berger (ed.), *Designing the Reclaimed Landscape*. London: Routledge, 2007.

Coetzee, Jan K. "The Power of Words: Defining, Narrating, Policy Making." *Africanus* 36, no. 1 (January 2006): 23–34. https://doi.org/10.10520/EJC22616.

Fath, Brian D., Carly A. Dean, and Harald Katzmair. "Navigating the Adaptive Cycle: An Approach to Managing the Resilience of Social Systems." *Ecology and Society* 20, no. 2 (2015). https://www.jstor.org/stable/26270208.

Ganesh, Shiv, and Heather M. Zoller. "Dialogue, Activism, and Democratic Social Change." *Communication Theory* 22, no. 1 (February 1, 2012): 66–91. https://doi.org/10.1111/j.1468-2885.2011.01396.x.

Horowitz, Leah S., Arn Keeling, Francis Lévesque, Thierry Rodon, Stephan Schott, and Sophie Thériault. "Indigenous Peoples' Relationships to Large-Scale Mining in Post/Colonial Contexts: Toward Multidisciplinary Comparative Perspectives." In *Mining and Indigenous Livelihoods*. Routledge, 2024.

Igogo, Tsisilile, Kwame Awuah-Offei, Alexandra Newman, Travis Lowder, and Jill Engel-Cox. "Integrating Renewable Energy into Mining Operations: Opportunities, Challenges, and Enabling Approaches." *Applied Energy* 300 (October 2021): 117375. https://doi.org/10.1016/j.apenergy.2021.117375.

Kastner, Karen, and Ellen Matthies. "On the Importance of Solidarity for Transforming Social Systems towards Sustainability." *Journal of Environmental Psychology* 90 (September 1, 2023): 102067. https://doi.org/10.1016/j.jenvp.2023.102067.

Klinger, Julie Michelle. "Groundbreaking, without Breaking New Ground: Redesigning e-Waste Recycling Could Clean up the Supply Chain While Building the Foundations for a Hemispheric Green Future." *NACLA Report on the Americas* 52, no. 2 (April 2, 2020): 158–164. https://doi.org/10.1080/10714839.2020.1768734.

Koch, Natalie, and Tom Perreault. "Resource Nationalism." *Progress in Human Geography* 43, no. 4 (August 1, 2019): 611–631. https://doi.org/10.1177/0309132518781497.

Kung, Anthony, Sarah Holcombe, Joel Hamago, and Deanna Kemp. "Indigenous Co-Ownership of Mining Projects: A Preliminary Framework for the Critical Examination of Equity Participation." *Journal of Energy & Natural Resources Law* 40, no. 4 (October 2, 2022): 413–435. https://doi.org/10.1080/02646811.2022.2029184.

Onifade, Moshood, Tawanda Zvarivadza, John A. Adebisi, Khadija Omar Said, Oluwatobi Dayo-Olupona, Abiodun Ismail Lawal, and Manoj Khandelwal. "Advancing toward Sustainability: The Emergence of Green Mining Technologies and Practices." *Green and Smart Mining Engineering* 1, no. 2 (June 2024): 157–174. https://doi.org/10.1016/j.gsme.2024.05.005.

Pahl-Wostl, Claudia. "A Conceptual Framework for Analysing Adaptive Capacity and Multi-Level Learning Processes in Resource Governance Regimes." *Global Environmental Change* 19, no. 3 (August 2009): 354–365. https://doi.org/10.1016/j.gloenvcha.2009.06.001.

Pahl-Wostl, Claudia, Oghenekaro Nelson Odume, Geeske Scholz, Ancois De Villers and Ebenezer Forkuo Amankwaa. "The Role of Crises in Transformative Change towards Sustainability." *Ecosystems and People* 19, no. 1 (December 31, 2023): 2188087. https://doi.org/10.1080/26395916.2023.2188087.

Pryke, Sam. "Explaining Resource Nationalism." *Global Policy* 8, no. 4 (2017): 474–482. https://doi.org/10.1111/1758-5899.12503.

Ruwhiu, Diane, and Lynette Carter. "Negotiating 'Meaningful Participation' for Indigenous Peoples in the Context of Mining." *Corporate Governance* 16, no. 4 (August 1, 2016): 641–654. https://doi.org/10.1108/CG-10-2015-0138.

Springer, Jenny, and Fernanda Almeida. "Protected Areas and the Land Rights of Indigenous Peoples and Local Communities: Current Issues and Future Agenda." Rights and Resources Initiative. May 2015.

Stewart, Philip D., and Nissa Shamsi. "Transformative Experience, Conflict Resolution and Sustained Dialogue." *International Journal of Conflict Engagement and Resolution* 3, no. 2 (2015): 158–179.

Toledano, Perrine, Sophie Thomashausen, Nicolas Maennling, and Alpa Shah. "A Framework to Approach Shared Use of Mining-Related Infrastructure." Columbia Center on Sustainable Investment. March 2014.

Tong, Shilu, Jonathan M. Samet, Will Steffen, Patrick L. Kinney, and Howard Frumkin. "Solidarity for the Anthropocene." *Environmental Research* 235 (October 15, 2023): 116716. https://doi.org/10.1016/j.envres.2023.116716.

Torres-Wong, Marcela, and Adrian Jimenez-Sandoval. "Indigenous Resource Governance as an Alternative to Mining: Redefining the Boundaries of Indigenous Participation." *The Extractive Industries and Society* 9 (March 1, 2022). https://doi.org/10.1016/j.exis.2021.101001.

Trebeck, Katherine Anne. "Tools for the Disempowered? Indigenous Leverage Over Mining Companies." *Australian Journal of Political Science* 42, no. 4 (December 1, 2007): 541–562. https://doi.org/10.1080/10361140701513604.

Young, Renee E., George D. Gann, Bethanie Walder, Junguo Liu, Wenhui Cui, Vern Newton, Cara R. Nelson, et al. "International Principles and Standards for the Ecological Restoration and Recovery of Mine Sites." *Restoration Ecology* 30, no. S2 (2022): e13771. https://doi.org/10.1111/rec.13771.

IMAGE CREDITS

All images crafted by Matthew Seibert.

CIRCULARIZE

Designing for Disassembly

the ECONOMY

by Matthew Seibert + Julia MacNelly

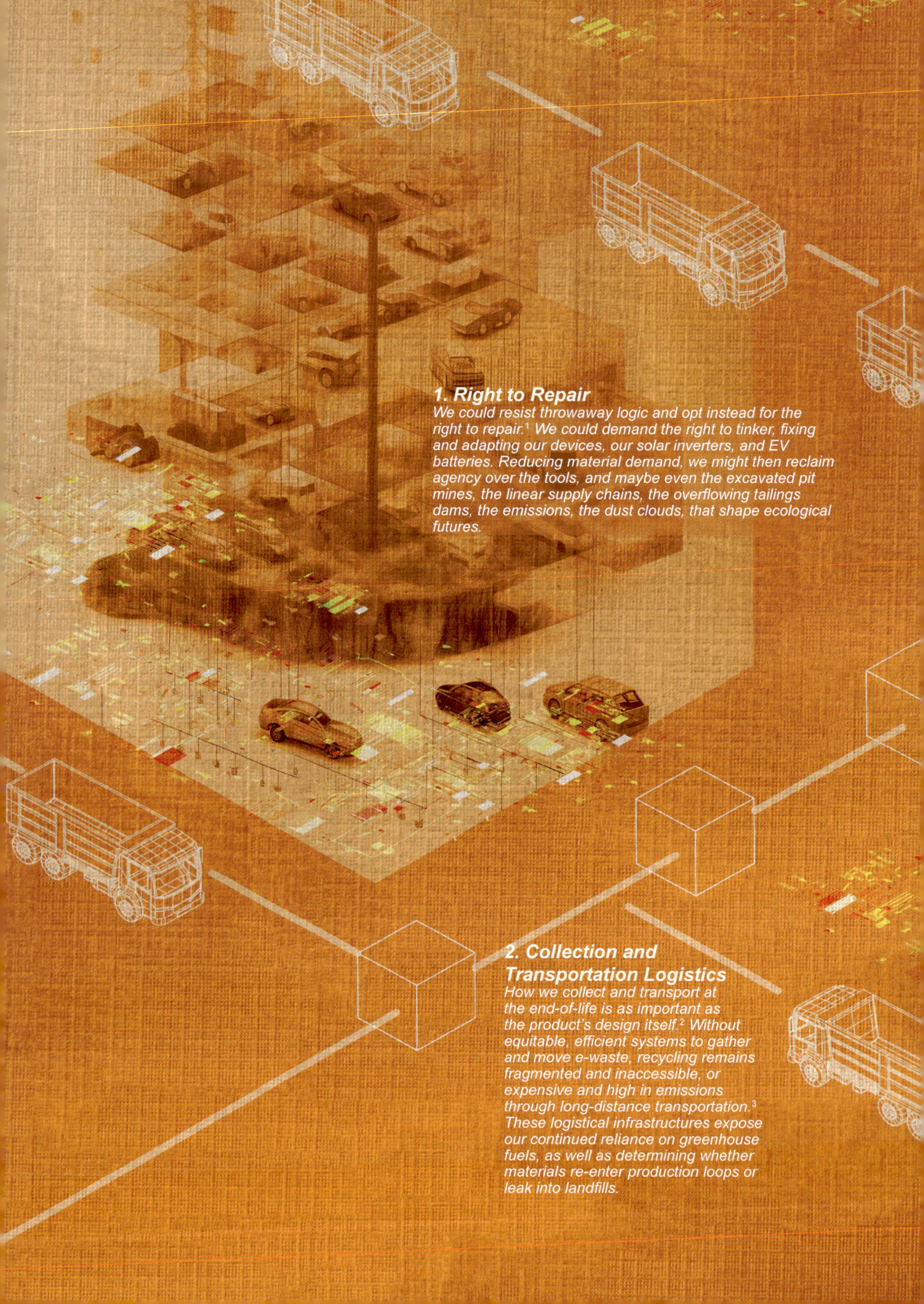

1. Right to Repair

We could resist throwaway logic and opt instead for the right to repair.[1] *We could demand the right to tinker, fixing and adapting our devices, our solar inverters, and EV batteries. Reducing material demand, we might then reclaim agency over the tools, and maybe even the excavated pit mines, the linear supply chains, the overflowing tailings dams, the emissions, the dust clouds, that shape ecological futures.*

2. Collection and Transportation Logistics

How we collect and transport at the end-of-life is as important as the product's design itself.[2] *Without equitable, efficient systems to gather and move e-waste, recycling remains fragmented and inaccessible, or expensive and high in emissions through long-distance transportation.*[3] *These logistical infrastructures expose our continued reliance on greenhouse fuels, as well as determining whether materials re-enter production loops or leak into landfills.*

3. Lifecycle Design

Design can embed circularity from the start, ensuring products are made to be repaired, reused, and ultimately disassembled for recovery.[4] Material durability, modularity, and non-toxic components all disrupt the linear extraction-to-disposal model. In a truly circular system, sustainability isn't an afterthought, but rather the anticipation of a product's afterlife is built into every stage of a product's existence.[5]

4. Modular Design for Easy Disassembly

When components can be efficiently separated without damage (and often without specialized tools), products gain longevity and recyclability.[6] This approach not only reduces waste, but also democratizes maintenance, shifting control from manufacturers to users and communities committed to sustainable material stewardship. This enables efficient repair, recycling, and material recovery: crucial steps in a circular lifecycle that reduce waste, preserve resources, and simplify end-of-life processing.[7]

5. Co-Locating Mining and Recycling Facilities

By largely already possessing the equipment needed for recycling and recovering critical minerals (as well as environmental licenses), mining facilities performing double duty as recycling facilities shortens material loops, reduces transportation emissions, and fosters industrial symbiosis. In the green energy transition, this spatial strategy bridges extraction and regeneration, transforming waste into resource and making critical material flows more transparent, accountable, and sustainable.

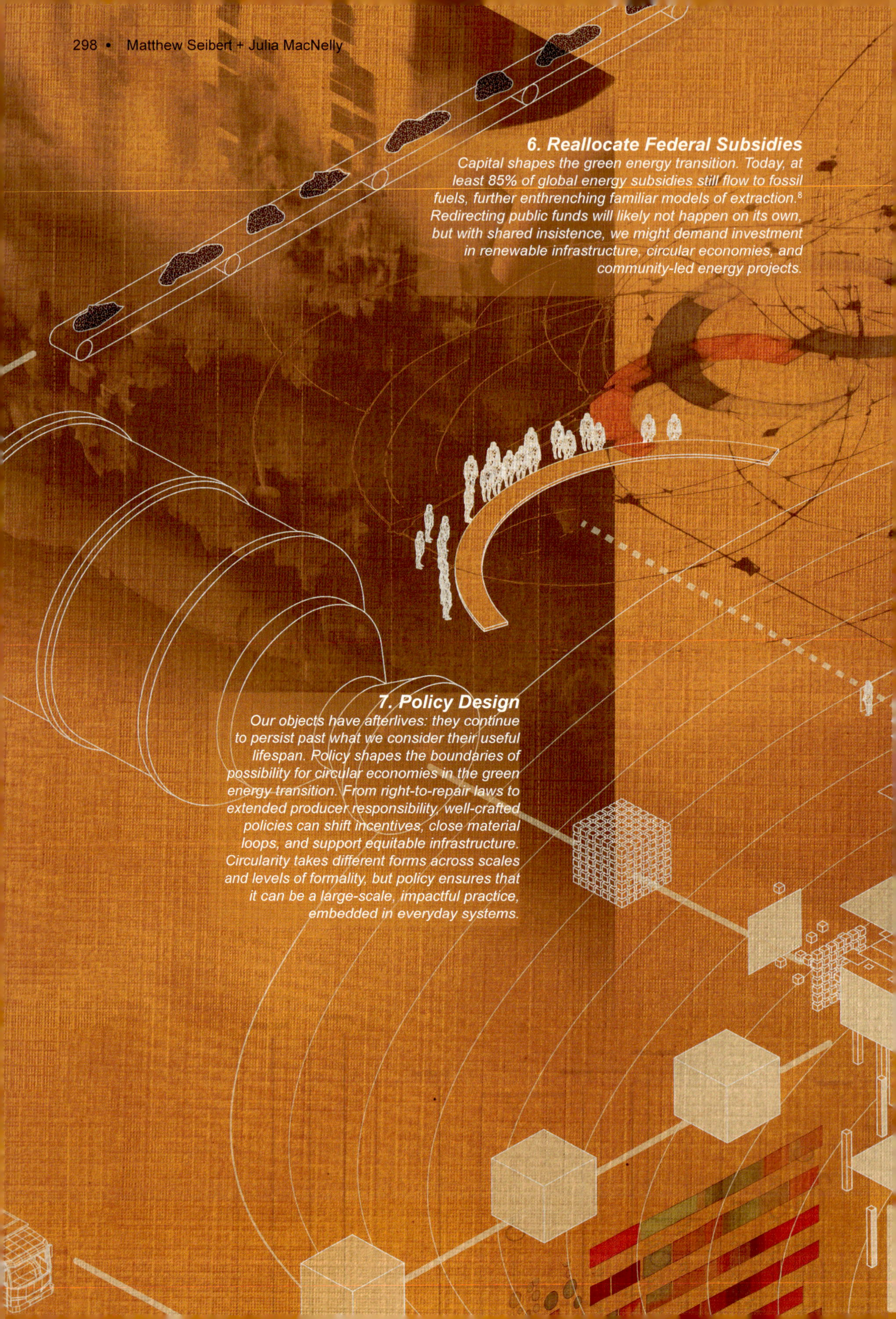

6. Reallocate Federal Subsidies

Capital shapes the green energy transition. Today, at least 85% of global energy subsidies still flow to fossil fuels, further enthrenching familiar models of extraction.[8] Redirecting public funds will likely not happen on its own, but with shared insistence, we might demand investment in renewable infrastructure, circular economies, and community-led energy projects.

7. Policy Design

Our objects have afterlives: they continue to persist past what we consider their useful lifespan. Policy shapes the boundaries of possibility for circular economies in the green energy transition. From right-to-repair laws to extended producer responsibility, well-crafted policies can shift incentives, close material loops, and support equitable infrastructure. Circularity takes different forms across scales and levels of formality, but policy ensures that it can be a large-scale, impactful practice, embedded in everyday systems.

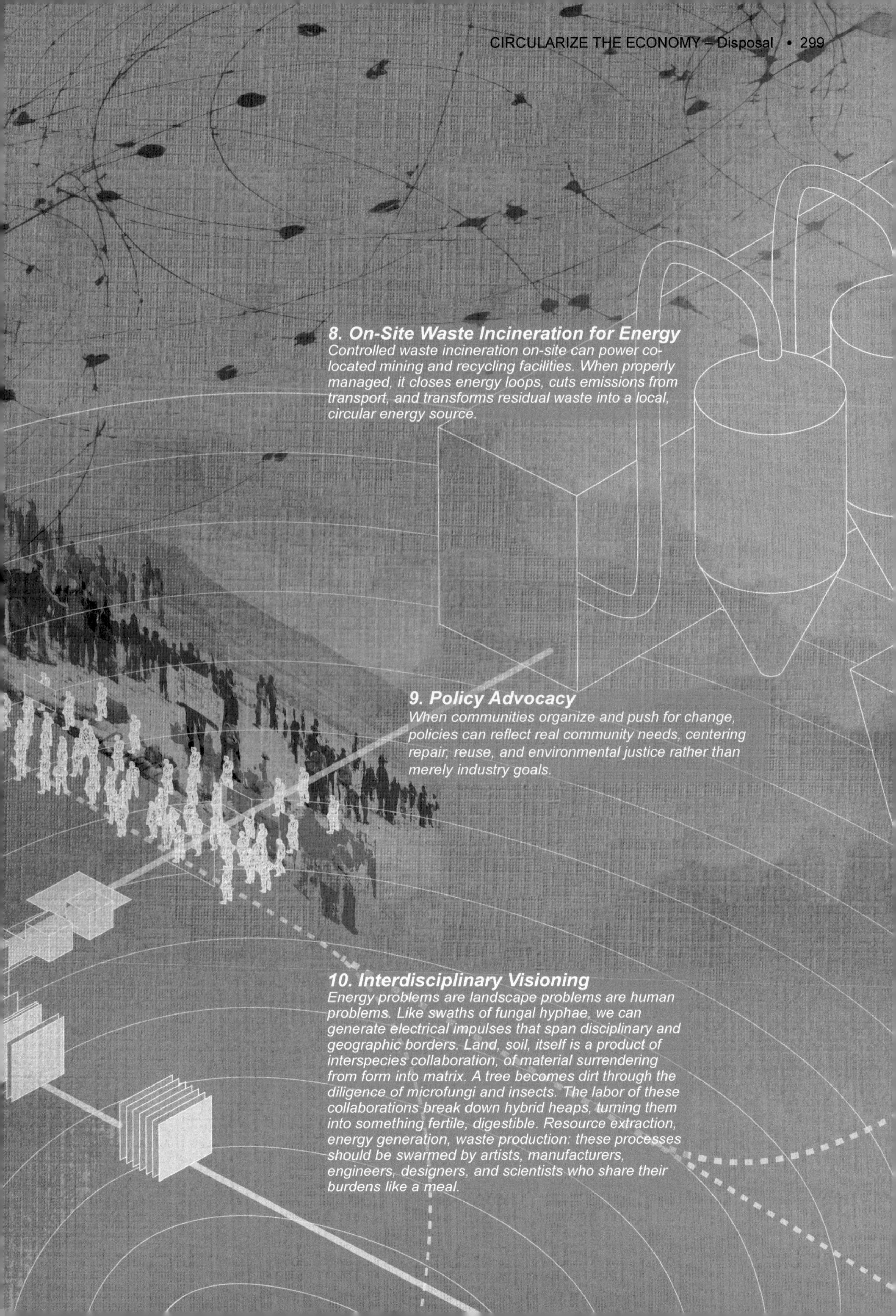

8. On-Site Waste Incineration for Energy

Controlled waste incineration on-site can power co-located mining and recycling facilities. When properly managed, it closes energy loops, cuts emissions from transport, and transforms residual waste into a local, circular energy source.

9. Policy Advocacy

When communities organize and push for change, policies can reflect real community needs, centering repair, reuse, and environmental justice rather than merely industry goals.

10. Interdisciplinary Visioning

Energy problems are landscape problems are human problems. Like swaths of fungal hyphae, we can generate electrical impulses that span disciplinary and geographic borders. Land, soil, itself is a product of interspecies collaboration, of material surrendering from form into matrix. A tree becomes dirt through the diligence of microfungi and insects. The labor of these collaborations break down hybrid heaps, turning them into something fertile, digestible. Resource extraction, energy generation, waste production: these processes should be swarmed by artists, manufacturers, engineers, designers, and scientists who share their burdens like a meal.

ENDNOTES

1 Hernandez, Ricardo J., Constanza Miranda, and Julian Goñi. "Empowering Sustainable Consumption by Giving Back to Consumers the 'Right to Repair.'" *Sustainability* 12, no. 3 (January 2020): 1–15. https://doi.org/10.3390/su12030850; Lloveras, Javier, Mario Pansera, and Adrian Smith. "On 'the Politics of Repair Beyond Repair': Radical Democracy and the Right to Repair Movement." *Journal of Business Ethics* 196, no. 2 (January 1, 2025): 325–344. https://doi.org/10.1007/s10551-024-05705-z; Mirr, Nicholas A. "Defending the Right to Repair: An Argument for Federal Legislation Guaranteeing the Right to Repair Notes." *Iowa Law Review* 105, no. 5 (2020): 2393–2428; Nigam, Swayam. "Right to Repair: A Sustainable Solution for E-Waste Reduction." *Jus Corpus Law Journal* 4, no. 1 (2024): 627-637; Manwaring, Kayleen, et al. "What Does a Right to Repair Tell Us about Our Relationship with Technology?" *Alternative Law Journal* 47, no. 3 (September 1, 2022): 179–186. https://doi.org/10.1177/1037969X221108557.

2 Slattery, Margaret, Jessica Dunn, and Alissa Kendall. "Transportation of Electric Vehicle Lithium-Ion Batteries at End-of-Life: A Literature Review." *Resources, Conservation and Recycling* 174 (November 1, 2021): 105755. https://doi.org/10.1016/j.resconrec.2021.105755.

3 Celik, Ilke, et al. "Sustainable End of Life Management of Crystalline Silicon and Thin Film Solar Photovoltaic Waste: The Impact of Transportation." *Applied Sciences* 10, no. 16 (January 2020): 5465. https://doi.org/10.3390/app10165465.

4 Ziegler, Rafael, et al. "Embedding Circularity: Theorizing the Social Economy, Its Potential, and Its Challenges." *Ecological Economics* 214 (December 1, 2023): 1–12. https://doi.org/10.1016/j.ecolecon.2023.107970.

5 Radavičius, Tadas, et al. "Circular Solar Industry Supply Chain through Product Technological Design Changes." *Insights into Regional Development* 3, no. 3 (September 2021): 10–30. https://doi.org/10.9770/ird.2021.3.3.

6 Desai, Anoop, and Anil Mital. "Evaluation of Disassemblability to Enable Design for Disassembly in Mass Production." *International Journal of Industrial Ergonomics* 32, no. 4 (October 2003): 265–281. https://doi.org/10.1016/s0169-8141(03)00067-2.

7 Gershenson, J. K., G. J. Prasad, and S. Allamneni. "Modular Product Design: A Life-Cycle View." *Journal of Integrated Design and Process Science* 3, no. 4 (November 1, 1999): 13–26. https://doi.org/10.3233/JID-1999-3402.

8 Coady, David, et al. "Global Fossil Fuel Subsidies Remain Large: An Update Based on Country-Level Estimates." *IMF Working Papers* 2019, no. 089 (May 2019): 2. https://doi.org/10.5089/9781484393178.001.

BIBLIOGRPAHY

Celik, Ilke, Marina Lunardi, Austen Frederickson, and Richard Corkish. "Sustainable End of Life Management of Crystalline Silicon and Thin Film Solar Photovoltaic Waste: The Impact of Transportation." *Applied Sciences* 10, no. 16 (January 2020): 5465. https://doi.org/10.3390/app10165465.

Coady, David, Baoping Shang, Nghia-Piotr Le, and Ian Parry. "Global Fossil Fuel Subsidies Remain Large: An Update Based on Country-Level Estimates." *IMF Working Papers* 2019, no. 089 (May 2019): 1. https://doi.org/10.5089/9781484393178.001.

Desai, Anoop, and Anil Mital. "Evaluation of Disassemblability to Enable Design for Disassembly in Mass Production." *International Journal of Industrial Ergonomics* 32, no. 4 (October 2003): 265–281. https://doi.org/10.1016/s0169-8141(03)00067-2.

Duflou, J. R., G. Seliger, S. Kara, Y. Umeda, A. Ometto, and B. Willems. "Efficiency and Feasibility of Product Disassembly: A Case-Based Study." *CIRP Annals* 57, no. 2 (2008): 583–600. https://doi.org/10.1016/j.cirp.2008.09.009.

Gershenson, J. K., G. J. Prasad, and S. Allamneni. "Modular Product Design: A Life-Cycle View." *Journal of Integrated Design and Process Science* 3, no. 4 (November 1, 1999): 13–26. https://doi.org/10.3233/JID-1999-3402.

Hernandez, Ricardo J., Constanza Miranda, and Julian Goñi. "Empowering Sustainable Consumption by Giving Back to Consumers the 'Right to Repair.'" *Sustainability* 12, no. 3 (January 2020): 850. https://doi.org/10.3390/su12030850.

Klinger, Julie Michelle. 2020. "Groundbreaking, without Breaking New Ground: Redesigning e-Waste Recycling Could Clean up the Supply Chain While Building the Foundations for a Hemispheric Green Future." *NACLA Report on the Americas* 52 (2): 158–164. doi:10.1080/10714839.2020.1768734.

Lloveras, Javier, Mario Pansera, and Adrian Smith. "On 'the Politics of Repair Beyond Repair': Radical Democracy and the Right to Repair Movement." *Journal of Business Ethics* 196, no. 2 (January 1, 2025): 325–344. https://doi.org/10.1007/s10551-024-05705-z.

Manwaring, Kayleen, Matthew Kearnes, Bronwen Morgan, Paul Munro, Roberta Pala, and Shanil Samarakoon. "What Does a Right to Repair Tell Us about Our Relationship with Technology?" *Alternative Law Journal* 47, no. 3 (September 1, 2022): 179–186. https://doi.org/10.1177/1037969X221108557.

Mirr, Nicholas A. "Defending the Right to Repair: An Argument for Federal Legislation Guaranteeing the Right to Repair Notes." *Iowa Law Review* 105, no. 5 (2020): 2393–2428.

Nigam, Swayam. "Right to Repair: A Sustainable Solution for E-Waste Reduction." *Jus Corpus Law Journal 4*, no. 1 (2024): 627–637.

Radavičius, Tadas, Arvid van Der Heide, Wolfram Palitzsch, Tom Rommens, Julius Denafas, and Manuela Tvaronavičienė. "Circular Solar Industry Supply Chain through Product Technological Design Changes." *Insights into Regional Development* 3, no. 3 (September 2021): 10–30. https://doi.org/10.9770/ird.2021.3.3.

Slattery, Margaret, Jessica Dunn, and Alissa Kendall. "Transportation of Electric Vehicle Lithium-Ion Batteries at End-of-Life: A Literature Review." *Resources, Conservation and Recycling* 174 (November 1, 2021): 105755. https://doi.org/10.1016/j.resconrec.2021.105755.

Trollman, Hana, James Colwill, and Sandeep Jagtap. "A Circularity Indicator Tool for Measuring the Ecological Embeddedness of Manufacturing." *Sustainability* 13, no. 16 (January 2021): 8773. https://doi.org/10.3390/su13168773.

Ziegler, Rafael, Thomas Bauwens, Michael J. Roy, Simon Teasdale, Ambre Fourrier, and Emmanuel Raufflet. "Embedding Circularity: Theorizing the Social Economy, Its Potential, and Its Challenges." *Ecological Economics* 214 (December 1, 2023): 107970. https://doi.org/10.1016/j.ecolecon.2023.107970.

IMAGE CREDITS

All images crafted by Matthew Seibert.

Searching for Symbiosis

OVERLAP SYSTEMS

by Matthew Seibert + Julia MacNelly

1. Reimagine Industrial Zones as Ecological Landscapes

Demand that manufacturing zones not be developed as wastelands, but rather active, designed systems with ecological, spatial, and cultural value. Employ landscape processes to inform planning in framing the layout, legibility, and regenerative potential of factory sites and logistics corridors. Design for air filtration, water remediation, heat island reduction, and biodiversity.[1]

2. Apply Ecosystem Logic to Industrial Sites

The waste of one industry or process can be the food for another. "Industrial Ecology" logic suggests that industries can be designed and sited to mimic the symbiotic relationships of an ecological network, prioritizing "connectance and cyclicity."[2] Co-locate refining and manufacturing with renewable energy and recycling. Site factories near renewable energy sources (solar/wind/hydro) to reduce emissions, or better yet, identify sites of potential adaptive reuse. Recycling and reprocessing plants used to recover critical minerals from on-site refining facilities play the role of decomposers in the food web, reducing virgin extraction and strengthening interdependence between industrial processes. Spatial proximity translates to efficiency and co-benefits. It is more than just switching energy sources, it is about restructuring systems, behaviors, and spatial configurations to use less energy in the first place.

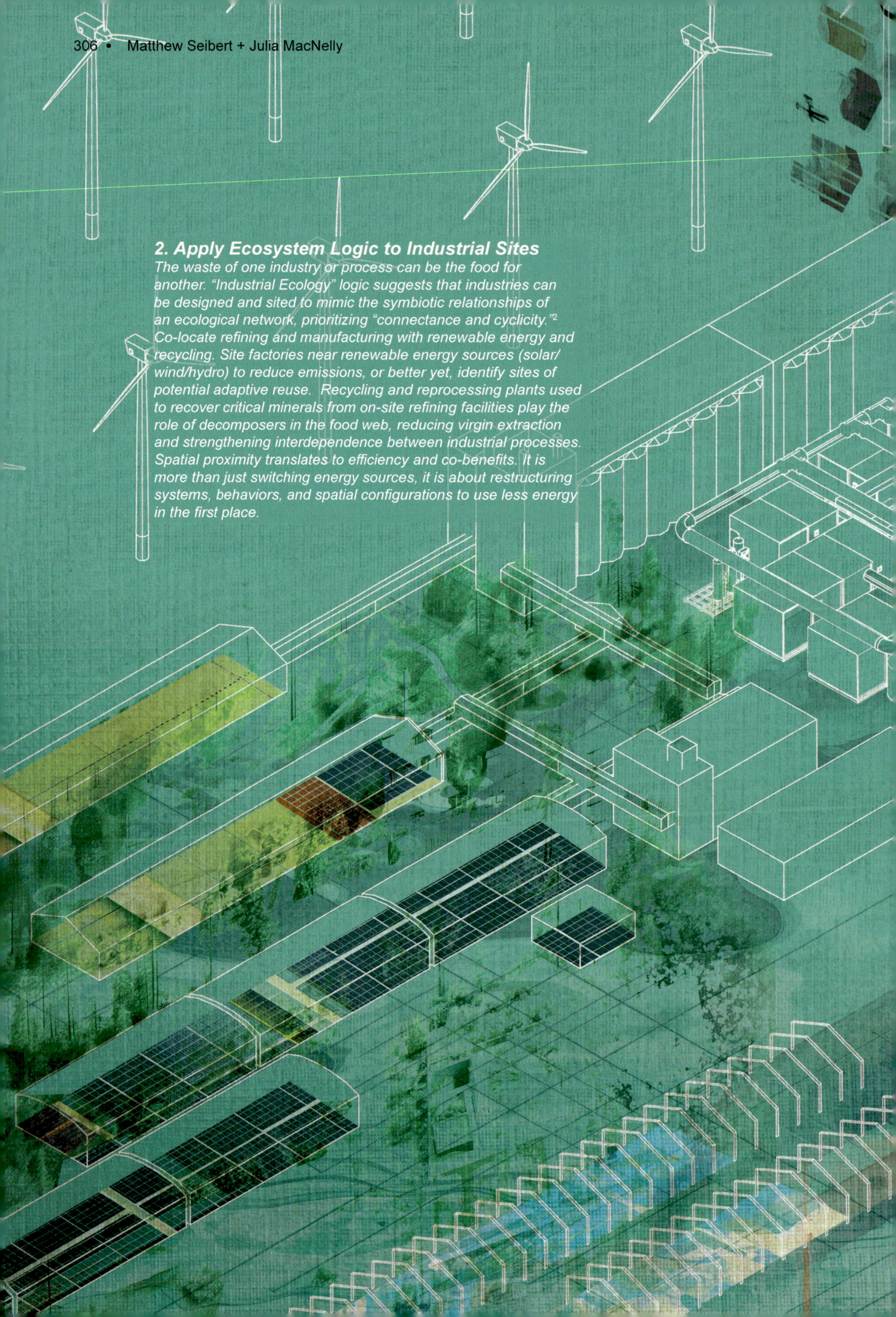

3. Participatory Governance

In extraction zones, "participatory inclusion" is sometimes named but rarely embodied. Who governs the mine? Who governs the ground? Companies do not operate with systemic incentives to center community well-being. The State, too, lacks the will or capacity to mediate. Communities near refining and manufacturing zones are often excluded from planning decisions or lack access to data.[3] For the energy transition to be truly just, it requires a redistribution of material benefits as well as of voice, agency, and world-making rights. Transparency, representation within decision-making bodies, and the power to shape development agendas to local rural needs are core principles of shared governance.[4] Fair and practical systems are needed to involve communities in decision-making.[5] Participation is not just an operational challenge but an ethical terrain.

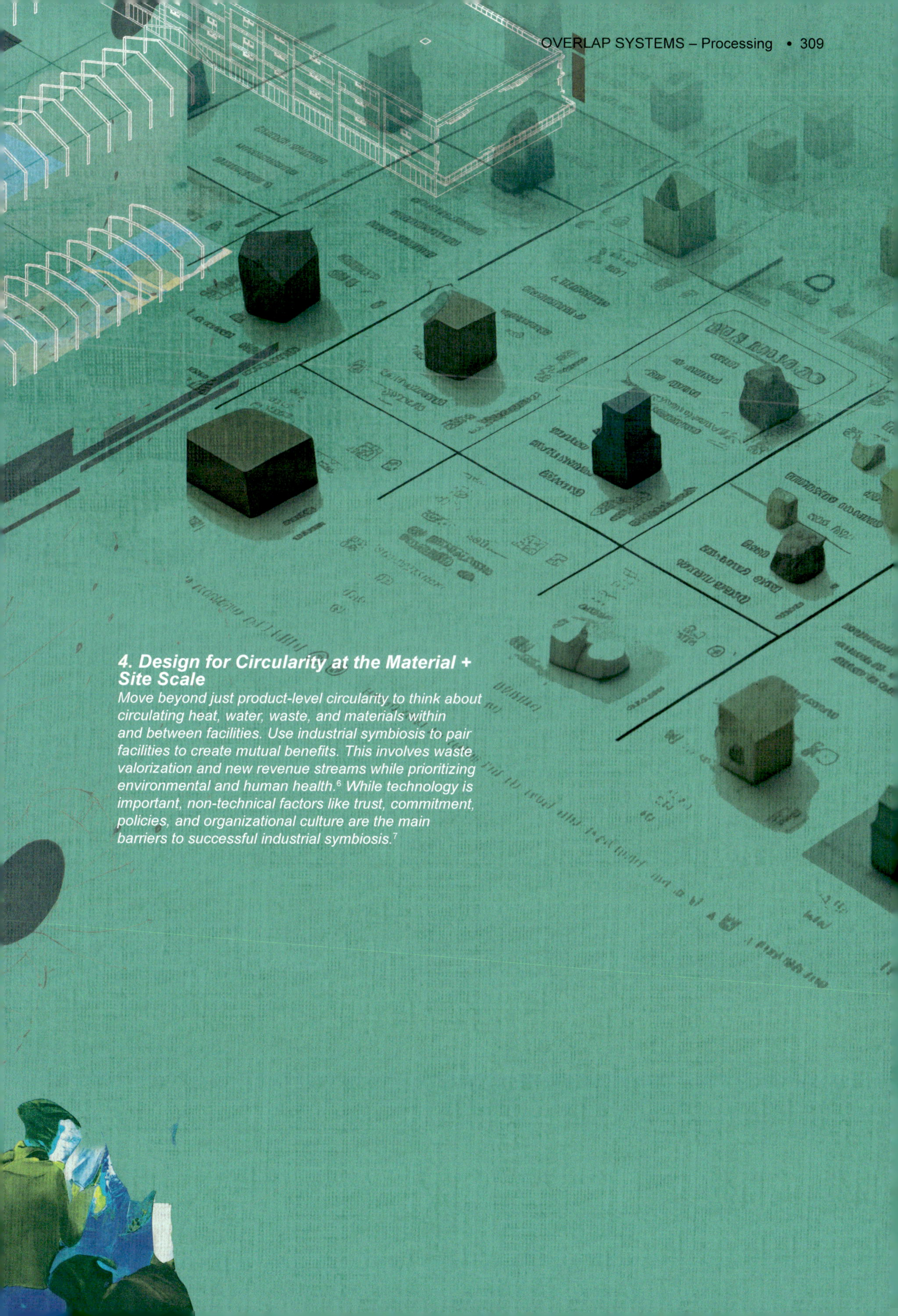

4. Design for Circularity at the Material + Site Scale

Move beyond just product-level circularity to think about circulating heat, water, waste, and materials within and between facilities. Use industrial symbiosis to pair facilities to create mutual benefits. This involves waste valorization and new revenue streams while prioritizing environmental and human health.[6] While technology is important, non-technical factors like trust, commitment, policies, and organizational culture are the main barriers to successful industrial symbiosis.[7]

ENDNOTES

1 Gibbs, David, Pauline Deutz, and Amy Proctor. “Industrial Ecology and Eco-industrial Development: A Potential Paradigm for Local and Regional Development?” *Regional Studies* 39, no. 2 (April 1, 2005): 171–183. https://doi.org/10.1080/003434005200059959.

2 Genc, Olcay, et al. “A Socio-Ecological Approach to Improve Industrial Zones towards Eco-Industrial Parks.” *Journal of Environmental Management* 250 (November 15, 2019): 11-13. https://doi.org/10.1016/j.jenvman.2019.109507; Khaled, Omar, and Mohie Edeen Shalaby. “Intellectual Framework for Transforming Conventional Industrial Zones into Eco- Industrial Parks.” *Journal of Urban Research*, October 1, 2021. https://doi.org/10.21608/jur.2021.73266.1061.

3 Leonard, Llewellyn. “State Governance, Participation and Mining Development: Lessons Learned from Dullstroom, Mpumalanga.” *Politikon* 44, no. 2 (May 4, 2017): 327–345. https://doi.org/10.1080/02589346.2016.1245526.

4 Kasimba, Sam A, and Päivi Lujala. “Community Based Participatory Governance Platforms and Sharing of Mining Benefits: Evidence from Ghana.” *Community Development Journal* 57, no. 4 (October 1, 2022): 635–654. https://doi.org/10.1093/cdj/bsab021; Erwin, Anna, et al. “Centering Community Voices in Mining Governance.” *Society & Natural Resources* 35, no. 10 (October 3, 2022): 1043–1062. https://doi.org/10.1080/08941920.2022.2053018.

5 Spiller, Beia, Tinzar Htun, and Kriti Shah. “Community Engagement and Participatory Inclusion in Mining: Challenges, Barriers, and Opportunities.” Washington DC: Resources for the Future. March, 2025: 1–43.

6 Castellet-Viciano, Lledó, Vicent Hernández-Chover, Águeda Bellver-Domingo, and Francesc Hernández-Sancho. “Industrial Symbiosis: A Mechanism to Guarantee the Implementation of Circular Economy Practices.” *Sustainability* 14, no. 23 (January 2022): 15872. https://doi.org/10.3390/su142315872.

7 Van Eetvelde, Greet. “Industrial Symbiosis.” In Krämer, Stephan, and Sebastian Engell (eds.), *Resource Efficiency of Processing Plants Part III*. (Wiley-VCH Verlag GmbH & Co., 2018): 441–469. https://onlinelibrary.wiley.com/doi/chapter-epub/10.1002/9783527804153.ch17.

BIBLIOGRAPHY

Castellet-Viciano, Lledó, Vicent Hernández-Chover, Águeda Bellver-Domingo, and Francesc Hernández-Sancho. “Industrial Symbiosis: A Mechanism to Guarantee the Implementation of Circular Economy Practices.” *Sustainability* 14, no. 23 (January 2022): 15872. https://doi.org/10.3390/su142315872.

Erwin, Anna, Zhao Ma, Emma Patricia Salas O’Brien, Jonathan Bauchet, Nelly Ramírez Calderón, Eliseo Zeballos Zeballos, Glenn Roberto Arce Larrea, Ruxandra Popovici, Laura Zanotti, and Chelsea A. Silva. “Centering Community Voices in Mining Governance.” *Society & Natural Resources* 35, no. 10 (October 3, 2022): 1043–1062. https://doi.org/10.1080/08941920.2022.2053018.

Genc, Olcay, Guido van Capelleveen, Ercan Erdis, Onur Yildiz, and Devrim Murat Yazan. “A Socio-Ecological Approach to Improve Industrial Zones towards Eco-Industrial Parks.” *Journal of Environmental Management* 250 (November 15, 2019): 109507. https://doi.org/10.1016/j.jenvman.2019.109507.

Gibbs, David, Pauline Deutz, and Amy Proctor “Industrial Ecology and Eco-industrial Development: A Potential Paradigm for Local and Regional Development?” *Regional Studies* 39, no. 2 (April 1, 2005): 171–183. https://doi.org/10.1080/003434005200059959.

Kasimba, Sam A, and Päivi Lujala. “Community Based Participatory Governance Platforms and Sharing of Mining Benefits: Evidence from Ghana.” *Community Development Journal* 57, no. 4 (October 1, 2022): 635–654. https://doi.org/10.1093/cdj/bsab021.

Khaled, Omar, and Mohie Edeen Shalaby. “Intellectual Framework for Transforming Conventional Industrial Zones into Eco- Industrial Parks.” *Journal of Urban Research*, October 1, 2021. https://doi.org/10.21608/jur.2021.73266.1061.

Leonard, Llewellyn. “State Governance, Participation and Mining Development: Lessons Learned from Dullstroom, Mpumalanga.” *Politikon* 44, no. 2 (May 4, 2017): 327–345. https://doi.org/10.1080/02589346.2016.1245526.

Mulrow, John S., Sybil Derrible, Weslynne S. Ashton, and Shauhrat S. Chopra. “Industrial Symbiosis at the Facility Scale.” *Journal of Industrial Ecology* 21, no. 3 (2017): 559–571. https://doi.org/10.1111/jiec.12592.

Spiller, Beia, Tinzar Htun, and Kriti Shah. “Community Engagement and Participatory Inclusion in Mining: Challenges, Barriers, and Opportunities.” Washington DC: Resources for the Future. March 2025.

Van Eetvelde, Greet. “Industrial Symbiosis.” In Krämer, Stephan and Sebastian Engell (eds.), *Resource Efficiency of Processing Plants Part III*. Wiley-VCH Verlag GmbH & Co., 2018: 441–469. https://onlinelibrary.wiley.com/doi/chapter-epub/10.1002/9783527804153.ch17.

IMAGE CREDITS

All images crafted by Matthew Seibert.

REBUILD COMMU

the Cultural Project of a True Transition

UCE ENERGY

by Matthew Seibert + Julia MacNelly

NITY

1. Site Solar at High-Benefit, Low-Harm Land

Solar farms are too often built atop previously undisturbed landscapes, from pristine desert habitat to highly arable land.[1] There are, however, 226 million acres of high-benefit, low-harm land in the U.S.[2] This means not only avoiding ecological, cultural, and/or economic loss but achieving co-benefits to energy production like water conservation, opportunities to repair / revitalize land and community, etc.[3] Such sites include:[4]

- *abandoned agricultural land*
- *highway and transmission lines rights-of-way*
- *parking lots*
- *brownfields and Superfund sites*
- *oil and gas fields*
- *abandoned mines*
- *rooftops*
- *landfills*
- *irrigation canals.*

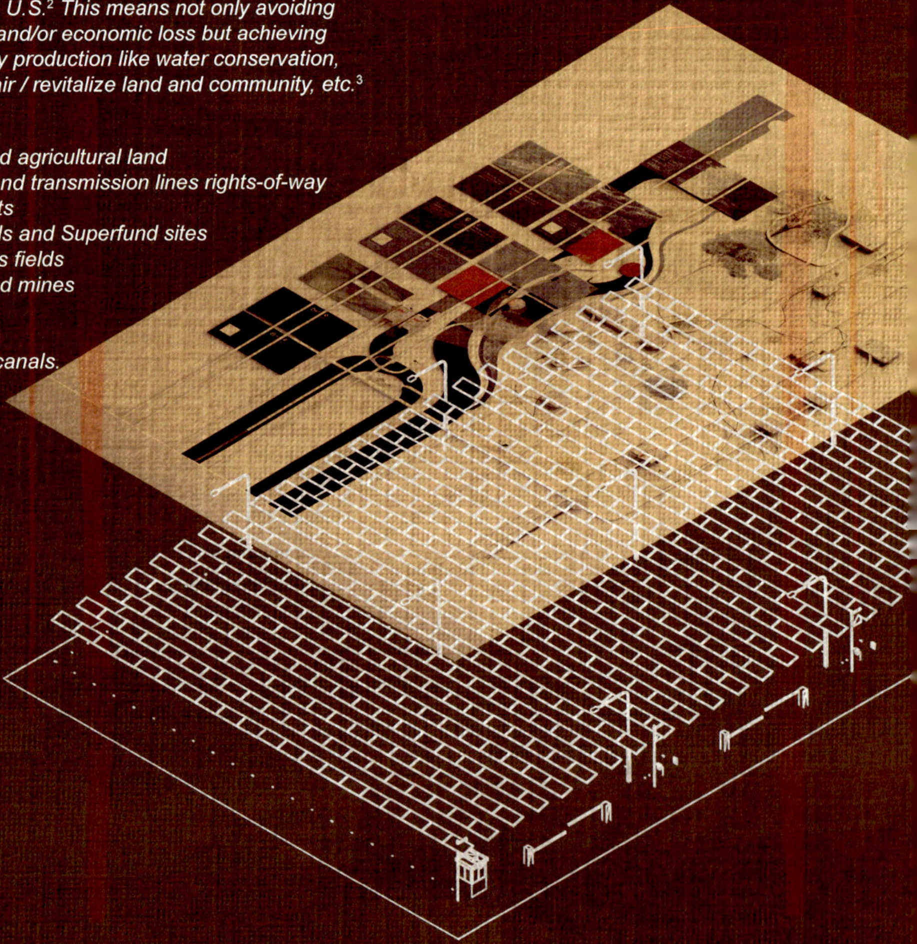

2. Foster Relational Infrastructures

Our neighborhood and residential infrastructure could become scaffolding to support people and ecologies in relation: shared heating systems, civic composting, and multi-generational housing all represent flexible, useful spaces that could come to reflect evolving social ties and collective histories. Moving from centralized, resource-intensive infrastructure to distributed, care-based networks means entering into ongoing negotiations and reimaginings of how people and ecologies might coexist.[5]

3. Encourage Slowness[6]

We need places that resist speed, extraction, and productivity. Spaces hold the possibility for deep time experiences; spaces that become refuges for grief, ritual, rest, celebration, and storytelling.[7]

4. Support Communal Living and Shared Resources

Celebrate "enough." It is a limit as well as a practice. "Enough" is sensual: it requires presence, sitting with the textures of a place. It is low-energy systems, materials with memory, modest footprints.[8] In the space of this shift, aesthetics might evolve beyond "green" high-tech solutions toward rituals of reuse, simplicity, and time-richness.[9]

5. Low-Carbon Leisure as Climate Infrastructure

The neoliberal project has resulted in a population stressed, alone, and despairing.[10] Pools, parks, libraries, plazas, public beaches, and playgrounds are all terrains of encounter, places that invite connection while reducing carbon emissions.[11] Climate action can take the form of spaces of leisure.

6. Decentralize Energy Systems and Tie to Place

Community solar and energy cooperatives reduce transmission losses and increase local resilience, recasting energy as a shared responsibility, rather than a utility.[12] *In a decentralized energy system, we start thinking of a neighborhood microgrid as an infrastructure and a commons. Community meetings become governance spaces as well as sites of mutual care, where we cooperatively manage excess and demand.*

We stop referring to it as "the grid."

7. Identify Synergies and Co-Benefits

We can layer renewable energy production with care for both land and labor. On an agrisolar site in a formerly abandoned monoculture field, rows of crops thrive under the microclimate of the elevated solar panels, while farmers move comfortably in the shade.[13] Hammocks are strung up between panel support poles. In a former oil field, rows of solar panels now replace pumpjacks, and wind turbines whir beside an old access road. Native grasses reclaim the margins.

ENDNOTES

1 Davis, Noah. "BLM's Western Solar Plan Puts Western Big Game Herds at Risk." *Theodore Roosevelt Conservation Partnership* (blog), April 11, 2024. https://www.trcp.org/2024/04/11/blms-western-solar-plan-puts-western-big-game-herds-at-risk/.

2 Bozuwa, Johanna, et al. "Planning to Build Faster: A Solar Energy Case Study Toward a Democratically Rooted, Nationwide Strategy to Accelerate Solar Deployment in the United States." Climate & Community Institute: The Roosevelt Institute, 2024.

3 Bozuwa, Johanna, et al. "Building Public Renewables in the United States." Climate + Community Project. 2023. 7-8. https://climateandcommunity.org/research/public-renewables-in-the-us/.

4 Bozuwa et al. "Planning to Build Faster."

5 Tabb, Phillip James. *Biophilic Urbanism | Designing Resilient Communities for the Future*. New York: Routledge, Taylor & Francis Group, 2021: 144-145; Holtzman, Gilo. "Community By Design, By The People: Social Approach To Designing and Planning Cohousing and Ecovillage Communities." *Journal of Green Building* 9, no. 3 (October 1, 2014): 77. https://doi.org/10.3992/1943-4618-9.3.60; Chopra, Simran, Rachel E. Clarke, Adrian K. Clear, Sara Heitlinger, Ozge Dilaver, and Christina Vasiliou. "Negotiating Sustainable Futures in Communities through Participatory Speculative Design and Experiments in Living." In *Proceedings of the 2022 CHI Conference on Human Factors in Computing Systems*, 1–17. CHI '22. New York, NY, USA: Association for Computing Machinery, 2022. https://doi.org/10.1145/3491102.3501929.

6 Shams, Shahrzad, Deepak Bhargava, and Harry W. Hanbury. "The Cultural Contradictions of Neoliberalism:" Roosevelt Institute, 2024: 1-49.

7 Gamba, Fiorenza, and Sandro Cattacin. "Urbans Rituals as Spaces of Memory and Belonging: A Geneva Case Study." *City, Culture and Society* 24 (March 1, 2021): 100385. https://doi.org/10.1016/j.ccs.2021.100385.

8 Nelson, Anitra. *Small Is Necessary: Shared Living on a Shared Planet.* Pluto Press, 2018. https://library.oapen.org/handle/20.500.12657/30716; Gausset, Quentin, and Pia Duus Jensen. "Living Sustainably in a Danish Eco-Community: How Social and Physical Infrastructures Affect Carbon Footprints." *Npj Climate Action* 3, no. 1 (May 8, 2024): 1–7. https://doi.org/10.1038/s44168-024-00113-5.

9 Geiger, Sonja M, Jan-Philipp Freudenstein, Gerrit von Jorck, Stefanie Gerold, and Ulf Schrader. "Time Wealth: Measurement, Drivers and Consequences." *Current Research in Ecological and Social Psychology* 2 (January 1, 2021): 100015. https://doi.org/10.1016/j.cresp.2021.100015.

10 Shams, et al. "The Cultural Contradictions of Neoliberalism."

11 Bozuwa, Johanna. "Why We Should Care About Low-Carbon Leisure." *The American Prospect*, June 13, 2024. https://prospect.org/api/content/86288208-29cf-11ef-bf6a-12163087a831.

12 Jiménez-Estévez, Guillermo, Alejandro Navarro-Espinosa, Rodrigo Palma-Behnke, Luigi Lanuzza, and Nicolás Velázquez. "Achieving Resilience at Distribution Level: Learning from Isolated Community Microgrids." *IEEE Power and Energy Magazine* 15, no. 3 (May 2017): 64–73. https://doi.org/10.1109/MPE.2017.2662328; Gajdzik, Bożena, Magdalena Jaciow, Radosław Wolniak, Robert Wolny, and Wieslaw Wes Grebski. "Diagnosis of the Development of Energy Cooperatives in Poland—A Case Study of a Renewable Energy Cooperative in the Upper Silesian Region." *Energies* 17, no. 3 (January 2024): 1–27. https://doi.org/10.3390/en17030647; Adesanya, Adewale A., Roman V. Sidortsov, and Chelsea Schelly. "Act Locally, Transition Globally: Grassroots Resilience, Local Politics, and Five Municipalities in the United States with 100% Renewable Electricity." *Energy Research & Social Science* 67 (September 2020): 1–10. https://doi.org/10.1016/j.erss.2020.101579.

13 The Nature Conservancy. "Power of Place National: Executive Summary." May 2023: 14.

BIBLIOGRAPHY

Adesanya, Adewale A., Roman V. Sidortsov, and Chelsea Schelly. "Act Locally, Transition Globally: Grassroots Resilience, Local Politics, and Five Municipalities in the United States with 100% Renewable Electricity." *Energy Research & Social Science* 67 (September, 2020):101579. https://doi.org/10.1016/j.erss.2020.101579.

Al-Kodmany, Kheir. "Bridging the Gap Between Technical and Local Knowledge: Tools for Promoting Community-Based Planning and Design." *Journal of Architectural and Planning Research* 18, no. 2 (2001): 110–130.

Bozuwa, Johanna. "Why We Should Care About Low-Carbon Leisure." *The American Prospect*. June 13, 2024. https://prospect.org/api/content/86288208-29cf-11ef-bf6a-12163087a831/.

Bozuwa, Johanna, Sarah Knuth, Grayson Flood, Patrick Robbins, and Olúfẹ́mi O. Táíwò. "Building Public Renewables in the United States." Climate + Community Project. 2023. https://climateandcommunity.org/research/public-renewables-in-the-us/.

Bozuwa, Johanna, Dustin Mulvaney, Isabel Estevez, Kristina Karlsson, and Sunny Malhotra. "Planning to Build Faster: A Solar Energy Case Study Toward a Democratically Rooted, Nationwide Strategy to Accelerate Solar Deployment in the United States." Climate & Community Institute: The Roosevelt Institute. 2024.

Bullock-Sieger, Lucy. "Has the US Caught up with European Agrivoltaic Deployment?" *Pv Magazine International* (blog). January 16, 2024. https://www.pv-magazine.com/2024/01/16/has-the-us-caught-up-with-european-agrivoltaic-deployment/.

"Case Studies | Playbook of Solutions." n.d. Accessed May 14, 2025. https://initiatives.weforum.org/playbook-of-solutions/case-studies.

Chopra, Simran, Rachel E. Clarke, Adrian K. Clear, Sara Heitlinger, Ozge Dilaver, and Christina Vasiliou. "Negotiating Sustainable Futures in Communities through Participatory Speculative Design and Experiments in Living." In Proceedings of the 2022 CHI Conference on Human Factors in Computing Systems, 1–17. CHI '22. New York, NY, USA: Association for Computing Machinery. https://doi.org/10.1145/3491102.3501929.

Dashiell, Stephanie, and Mark Buckley. "Green Light Study: Economic and Conservation Benefits of Low-Impact Solar Siting in California." The Nature Conservancy. November, 2019. https://www.scienceforconservation.org/products/green-light-study/.

Davis, Noah. "BLM's Western Solar Plan Puts Western Big Game Herds at Risk." *Theodore Roosevelt Conservation Partnership* (blog). April 11, 2024. https://www.trcp.org/2024/04/11/blms-western-solar-plan-puts-western-big-game-herds-at-risk/.

Duane, Tim, and Siobhan McIntyre. "Water, Work, Wildlife, and Wilderness: The Collaborative Federal Public Lands Planning Framework for Utility-Scale Solar Energy Development in the Desert Southwest." SSRN Scholarly Paper. Rochester, NY: Social Science Research Network. https://papers.ssrn.com/abstract=2004637.

Dutzik, Tony, and Johanna Neumann. "Rooftop Solar on the Rise." Environment America: Research & Policy Center. February 12, 2024. https://environmentamerica.org/center/resources/rooftop-solar-on-the-rise/.

Fields, Margaret. "North Carolina Low-Impact Solar Siting Analysis." ArcGIS StoryMaps. May 18, 2022. https://storymaps.arcgis.com/stories/f165477874084d3cba22ab4dbb40230d.

Gajdzik, Bożena, Magdalena Jaciow, Radosław Wolniak, Robert Wolny, and Wieslaw Wes Grebski. "Diagnosis of the Development of Energy Cooperatives in Poland—A Case Study of a Renewable Energy Cooperative in the Upper Silesian Region." Energies 17, no. 3 (January 30, 2024): 1–27. https://doi.org/10.3390/en17030647.

Gamba, Fiorenza, and Sandro Cattacin. "Urbans Rituals as Spaces of Memory and Belonging: A Geneva Case Study." *City, Culture and Society* 24 (March 30, 2021): 100385. https://doi.org/10.1016/j.ccs.2021.100385.

Gausset, Quentin, and Pia Duus Jensen. "Living Sustainably in a Danish Eco-Community: How Social and Physical Infrastructures Affect Carbon Footprints." *Npj Climate Action* 3, no. 1 (May 8, 2024). https://doi.org/10.1038/s44168-024-00113-5.

Geiger, Sonja M., Jan-Philipp Freudenstein, Gerrit von Jorck, Stefanie Gerold, and Ulf Schrader. "Time Wealth: Measurement, Drivers and Consequences." *Current Research in Ecological and Social Psychology* 2 (July 24, 2021): 100015. https://doi.org/10.1016/j.cresp.2021.100015.

Holtzman, Gilo. "Community By Design, By The People: Social Approach To Designing And Planning Cohousing And Ecovillage Communities." Journal of Green Building 9, no. 3 (October 1, 2014): 60–82. https://doi.org/10.3992/1943-4618-9.3.60.

Jiménez-Estévez, Guillermo, Alejandro Navarro-Espinosa, Rodrigo Palma-Behnke, Luigi Lanuzza, and Nicolás Velázquez. "Achieving Resilience at Distribution Level: Learning from Isolated Community Microgrids." *IEEE Power and Energy Magazine* 15, no. 3 (April 17, 2017): 64–73. https://doi.org/10.1109/MPE.2017.2662328.

The Nature Conservancy. "Power of Place National: Executive Summary." 2023.

Nelson, Anitra. *Small Is Necessary: Shared Living on a Shared Planet.* Pluto Press, 2018. https://library.oapen.org/handle/20.500.12657/30716.

Shams, Shahrzad, Deepak Bhargava, and Harry W. Hanbury. "The Cultural Contradictions of Neoliberalism." Roosevelt Institute. 2024.

Sioshani, Fereidoon P., ed. *Innovation and Disruption at the Grid's Edge: How Distributed Energy Resources Are Disrupting the Utility Business Model.* Academic Press, an imprint of Elsevier, 2017.

Tabb, Phillip James. *Biophilic Urbanism | Designing Resilient Communities for the Future.* New York: Routledge, Taylor & Francis Group, 2021. https://www-taylorfrancis-com.proxy1.library.virginia.edu/books/mono/10.4324/9781003034896/biophilic-urbanism-phillip-james-tabb.

World Economic Forum. "7 Ways We Can Phase out Coal without Harming Economies or Energy Security." February 8, 2023. https://www.weforum.org/stories/2023/02/coal-emissions-energy-security-iea/.

World Economic Forum. "This Is How to Accelerate the Move from Coal to Renewables in Emerging Markets." June 16, 2022. https://www.weforum.org/stories/2022/06/accelerating-coal-to-renewables-in-emerging-markets/.

IMAGE CREDITS

All images crafted by Matthew Seibert.

AUTHOR BIO

Fitsum Areguy is a writer, researcher, and cultural worker living in Southwestern Ontario. He is the co-founder and project director for Textile, a hyper-local arts collective in Kitchener-Waterloo.

Ässia Boukhatmi is a researcher at Bern University of Applied Sciences and the Technical University of Berlin. Her research focuses on investigating digital technologies to foster the circular economy transition in the European photovoltaics industry.

Daniel Carmelo is a scholar, storyteller, illustrator, landscape designer, and creative technologist originally from the piney woodlands of east Texas, currently based in Charlottesville, VA. His work explores cultural technologies for sensory connection and attunement with land. Daniel holds an MLA and MUEP from the University of Virginia, a BS in Communications from LeTourneau University, and is an alumnus of the Wright-Ingram Institute Field Stations Program (Iceland).

Ash Duhrkoop is a PhD candidate in Art and Architectural History at the University of Virginia specializing in twentieth-century and African art, with an interest in ecocriticism.

Billy Fleming is an interdisciplinary scholar of climate justice and the built environment at Temple University's Tyler School of Art and Architecture, and founding co-director of the Climate and Community Institute, a progressive think-tank focused on the political economy of the climate crisis. His research focuses on clean energy supply chains, social and abolitionist movements in rural landscapes, land-based practices of carbon management, and the communities they transform.

Kelly Herbinson is a Mojave Desert biologist and conservationist currently serving as the executive director of the Mojave Desert Land Trust. She spent 12 years working as a field biologist with the endangered desert tortoise, only pausing briefly to pursue master's degrees focused on her two other passions: ant behavior and nonfiction writing.

Hugo Kamya, PhD, is a professor in the School for Social Work at Smith College, Northampton. He is a psychologist, clinical social worker, and couple's and family therapist. He lives and conducts his clinical practice as a narrative therapist using stories and storytelling in Arlington, Massachusetts.

Julia MacNelly holds a Master of Landscape Architecture from the University of Virginia (UVA). Her research focuses on the way in which material management can give broader insights into the planetary-scale issue of climate change. Before attending UVA, she worked with a community land trust and land bank in Richmond, Virginia, engaging with land stewardship and affordable housing development. With a multidisciplinary art background, Julia sees landscape architecture as an opportunity to creatively engage with questions about equitable development, ecological stewardship, and land access.

Vusumuzi Maphosa is Director of Information and Communication Technologies (ICT) at National University of Science and Technology, Zimbabwe, holding a PhD in Information Systems. He leads ICT strategy implementation and postgraduate supervision. An accomplished researcher, he serves on journal editorial boards and holds leadership roles in IEEE, ISOC Zimbabwe and CSZ. His expertise spans AI, data mining, and ICT4D.

Roger Nyffenegger is a PhD researcher on Circular Business Models in the solar industry at Maastricht Sustainability Institute and Bern University of Applied Sciences. He contributes to the EU-Horizon projects CIRCUSOL and RETRIEVE, as well as Swiss PV Circle, and holds a double degree from HSG St. Gallen and HEC Paris.

Javier Arroyo Olea is from the Biobío Region, Chile. He is a professor of History and Geography, holding diplomas in the areas of Memory and Environment and the author of works that address various processes of the Chilean dictatorship and post-dictatorship. He is a member of the editorial team of Resumen.cl and of the Latin American Observatory of Environmental Conflicts (OLCA).

María Paz López Ponce is a Chilean environmental biologist from the University of Chile and holds a diploma in Environmental Education from the Alberto Hurtado University. Since 2017, she has worked as an environmental educator and, since 2020, as a researcher at OLCA.

Douglas Robb is an Assistant Professor of Landscape Architecture at the University of Calgary. His research centers on the production of just and inclusive climate futures, spanning topics of energy landscapes, critical resource studies, and degrowth. Doug's work is grounded in place-based methods of drawing, writing, and designing, with a focus on rural and Indigenous lands across western Canada.

Matthew Seibert is an Associate Professor of Landscape Architecture at the University of Virginia. His research and teaching challenge dominant modes of knowledge production with specific attention to our relations with our host landscapes. This he sees as the first, fundamental step in a theory of change towards a just, more promising future.

Theodore Teichman is scholar, landscape designer, and soundscape composer, originally from the arid shrublands and sprawl of inland California, currently based in Charlottesville, VA. Their research explores a queer ethics of care for novel hybrid ecologies through centering the latent agencies of materials and landscapes. Theodore holds an MLA from the University of Virginia and a transdisciplinary BSA specializing in soundscape studies from Carnegie Mellon University. They are a Fellow of the Leifur Eiriksson Foundation, and an inaugural Fellow of the tranSci Lab at the University of Virginia.

D

E

F

G

R

S

T

U

V

W

Z